Mixed Plantations of Eucalyptus and Leguminous Trees

Elke Jurandy Bran Nogueira Cardoso
José Leonardo de Moraes Gonçalves
Fabiano de Carvalho Balieiro
Avílio Antônio Franco
Editors

Mixed Plantations of Eucalyptus and Leguminous Trees

Soil, Microbiology and Ecosystem Services

Springer

Editors
Elke Jurandy Bran Nogueira Cardoso
Department of Soil Science
University of São Paulo
"Luiz de Queiroz" College of Agriculture
Piracicaba, SP, Brazil

José Leonardo de Moraes Gonçalves
Department of Forest Science
University of São Paulo
"Luiz de Queiroz" College of Agriculture
Piracicaba, SP, Brazil

Fabiano de Carvalho Balieiro
EMBRAPA Soils
Brazilian Agricultural Research Corporation
Rio de Janeiro, RJ, Brazil

Avílio Antônio Franco
EMBRAPA Agrobiology
(Retired Researcher)
Brazilian Agricultural Research Corporation
Seropédica, RJ, Brazil

ISBN 978-3-030-32364-6 ISBN 978-3-030-32365-3 (eBook)
https://doi.org/10.1007/978-3-030-32365-3

This Springer imprint is published by the registered company Springer Nature Switzerland AG
The registered company address is: Gewerbestrasse 11, 6330 Cham, Switzerland

Preface

Mixed Plantations of Eucalyptus and *Acacia mangium*

We indicate this book to foresters, entrepreneurs in forestry, producers of wood, cellulose or other forest sub-products, agricultural engineers, agronomists, biologists, people working in Environmental Conservation and Natural Life Preservation, as well as the graduate and post-graduate students in these areas and socio-economic planners.

Here, we present a compilation as complete as possible on research, themes regarding eucalypt plantations, especially when in consortia with leguminous tropical trees. Such mixed stands generally can be productive and economically viable and are ecologically sustainable and favorable for the social-economic guarantees of rural workers and entrepreneurs. Eucalypt is one of the most cultivated forest trees in Brazil. However, most of the time, it is produced as monoculture and needs continuous fertilizer applications, including nitrogenous compounds, to maintain productivity. These plantations, usually propagated by vegetative means, as clones, receive criticism from various sectors of society for their low genetic diversity. Among forest producers, the main concern is the high susceptibility of homogeneous stands to abiotic and biotic stress. In consequence, during the last decades, several research teams all over the tropical countries have been investigating changes in the eucalypt management system to provide a more attractive and sustainable activity, achieving economic gains and diminishing its environmental impacts. Our proposal is mostly restricted to the Brazilian research experience on this topic, with special consideration of the need to diminish the use of industrial products as fertilizers and pesticides, but valuating soil health, biological gains, and ecosystem services.

Considering those prerequisites, it becomes easy to answer the main question of the first chapter: "Why mixed forest plantations?" Our response is as follows: "Because, when comparing all investigated management systems, this one is socially more advantageous and benefitting the rural entrepreneur economically, besides its high sustainability."

Considering some of these advantages, we want to present the main line of thought of the sequence of the book chapters. Thus, regarding growth and productivity of the trees, the most outstanding combination is *Eucalypt* in consortium with *Acacia mangium,* an exotic species, already known for its expressive productivity. Although, in biology, one always gets some variation, the best responses occur with mixed plantations in deep and sandy soils of low fertility, with a hot and rainy climate, which corresponds to most of the Brazilian regions used for eucalypt production.

The main reason for these superior results is that most legumes present a natural association with soil bacteria, generically known as rhizobia, which nodulate the roots of these trees and have a high potential of fixing atmospheric nitrogen, transforming it into ammonia in the root nodules. The forthcoming fixed nitrogenous compounds cycle throughout the whole tree and generally are exported and divided with other plants in their neighborhood. During litterfall, some of the fixed N reaches the soil causing its enrichement in N. The symbiotically fixed N can supply nitrogen needs of all plants, at the same time retaining nutrient reserves for the next tree rotation. Thus, the enriched soil in C and N turns the nutrient cycling much more dynamic.

Chapters 5–10 are all dedicated to discussing biological aspects of this kind of management, providing us with a constant increase in knowledge on the most adequate strategies to be applied to better soil health and plant growth. The perfect functioning of the soil as producer correlates directly with the size and diversity of the soil bacterial community structure, with each plant recruiting in its rhizosphere the most helpful bacteria, with functions of mineralizing the soil organic matter, acting on enzymes related to plant nutrition and on biological control of pathogens, among others. Right thereafter, we present the processes inherent to biological N fixation (BNF), selection of the most adequate rhizobial strains, and the transfer from soil to plant or from plant to plant mediated by mycorrhizal fungi, having a synergic interaction with BNF. Mycorrhizal fungi transport all kinds of nutrients to plants but are more active regarding the ones with slow mobility in soil, as phosphorus (P). Here, it is important to highlight another trait of *Eucalyptus* and *Acacia*: both host plants form symbioses with both kinds of mycorrhiza, the arbuscular (AM) and the ectomycorrhiza (ECM), which is exceptional among plants, that normally associate only with one kind of mycorrhiza, if at all.

Following the chapters, we now come to the important contribution of insects and other soil invertebrates to soil health, nutrient cycling and organic matter decomposition, showing how they influence the plants and how they become affected, in numbers and diversity, besides the influence of the climate. Right afterwards, there follows a chapter on bio-indicators of soil health, showing many recent results on the extreme relevance of the microbiological phenomena and the innumerable ecosystem services derived from the ecologically correct management, with benefits to soil, plants, and workers, still contributing to economic gains.

Chapter 11 discusses the problems that may derive from the introduction of exotic possible biological species into any ecosystem different from its origin, which resides in its invasiveness, sometimes causing severe ecological and economic problems, a phenomenon that has been occurring regarding specimens of plants and animals of most categories. Although most of our researchers did not perceive such behavior regarding *A. mangium*, we feel ethically compromised to tell what a few other researchers have reported, and it is always worthwhile to use prevention and caution in such situations. Generally, however, our experience says that the danger of this species becoming invasive should be neglected, when being used adequately. We belief that invasiveness may only occur in open lands and not when used in forests.

Chapter 12 presents the use of Brazilian leguminous trees to substitute *A. mangium*. So far, the best choice has always been this species (*A. mangium*), and it is the only one about which we have already compiled a great amount of experimental reports and practical experience, as well in Brazil as in many other countries. The approval of employing this species was almost unanimous by all researchers or farmers who had the opportunity of following its performance in the field, and for many decades.

Nevertheless, nothing impedes to test other species, as, for example, the national leguminous trees, which are available in great numbers. Maybe in the future, we can select some of them, which present the same advantages or are even better in such a consortium, as indicated by the author of this chapter, one of the very few scientists who worked in this area.

Finally, all this information is complemented by the last chapter, which describes the Brazilian legal structure and presents the regulations for the exploration of forests, either Eucalypt plantations in monoculture or in consortia, which we hope, will help the interested people to make the best choices on this activity.

Piracicaba, SP, Brazil — Elke Jurandy Bran Nogueira Cardoso
Piracicaba, SP, Brazil — José Leonardo de Moraes Gonçalves
Rio de Janeiro, RJ, Brazil — Fabiano de Carvalho Balieiro
Seropédica, RJ, Brazil — Avílio Antônio Franco

Contents

Contributors

Nina Attias Department of Animal Biology, Federal University of Mato Grosso do Sul, Campo Grande, MS, Brazil

Dilmar Baretta Santa Catarina State University, Chapecó, SC, Brazil

Maurel Bheling Embrapa Agrosilvopastoral, Brazilian Agricultural Research Corporation, Sinop, MT, Brazil

Daniel Bini State University of the Central West, Guarapuava, PR, Brazil

Bruno Bordon Department of Forest Sciences, University of São Paulo, "Luiz de Queiroz" College of Agriculture, Piracicaba, SP, Brazil

Elke Jurandy Bran Nogueira Cardoso Department of Soil Science, University of São Paulo, "Luiz de Queiroz" College of Agriculture, Piracicaba, SP, Brazil

Carolina Braga Brandani College of Agricultural, Consumer and Environmental Sciences - Animal and Range Sciences, New Mexico State University, Las Cruces, NM, USA

Fernando Vieira Cesário Fluminense Federal University, Niterói, RJ, Brazil

Caio Tavora Coelho da Costa Rachid Institute of Microbiology, Federal University of Rio de Janeiro, Ilha do Fundão, RJ, Brazil

Arthur Prudêncio de Araujo Pereira Department of Soil Science, University of São Paulo, "Luiz de Queiroz" College of Agriculture, Piracicaba, SP, Brazil

Soil Science Department (Pici Campus), Federal University of Ceará, Fortaleza, Ceará, Brazil

Bruna Andréia de Bacco Lopes Department of Soil Science, University of São Paulo, "Luiz de Queiroz" College of Agriculture, Piracicaba, SP, Brazil

Fabiano de Carvalho Balieiro EMBRAPA Soils, Brazilian Agricultural Research Corporation, Rio de Janeiro, RJ, Brazil

Sergio Miana de Faria EMBRAPA Agrobiology, Brazilian Agricultural Research Corporation, Seropédica, RJ, Brazil

Helena de Godoy Bergallo Department of Ecology, Institute of Biology Roberto Alcântara Gomes, Rio de Janeiro State University, Rio de Janeiro, RJ, Brazil

José Leonardo de Moraes Gonçalves Department of Forest Sciences, University of São Paulo, "Luiz de Queiroz" College of Agriculture, Piracicaba, SP, Brazil

Luiz Fernando Duarte de Moraes EMBRAPA Agrobiology, Brazilian Agricultural Research Corporation, Seropédica, RJ, Brazil

Ciro José Ribeiro de Moura Federal University of Rio de Janeiro, Rio de Janeiro, RJ, Brazil

Ivanka Rosado de Oliveira Department of Forest Sciences, University of São Paulo, "Luiz de Queiroz" College of Agriculture, Piracicaba, SP, Brazil

Renata Evangelista de Oliveira Department of Rural Development, Federal University of São Carlos, Araras, SP, Brazil

Alexandre de Vicente Ferraz Institute of Forest Science and Research (IPEF), Piracicaba, SP, Brazil

Avílio Antônio Franco EMBRAPA Agrobiology (Retired Researcher), Brazilian Agricultural Research Corporation, Seropédica, RJ, Brazil

Antonio Carlos Gama-Rodrigues Darcy Ribeiro State University of Norte Fluminense, Campos dos Goytacazes, RJ, Brazil

Cintia Carla Niva Embrapa Cerrados, Brazilian Agricultural Research Corporation, Planaltina, DF, Brazil

Luís Carlos Iuñes Oliveira Filho Santa Catarina State University, Chapecó, SC, Brazil

Ranieri Ribeiro Paula Center for Agricultural Sciences and Engineering, Federal University of Espirito Santo, Alegre, ES, Brazil

Pâmela Niederauer Pompeo Santa Catarina State University, Chapecó, SC, Brazil

Rachel Bardy Prado EMBRAPA Soils, Brazilian Agricultural Research Corporation, Rio de Janeiro, RJ, Brazil

Agnès Robin Department of Soil Science, University of São Paulo, "Luiz de Queiroz" College of Agriculture, Piracicaba, SP, Brazil

Eco & Sols, Univ. Montpellier, CIRAD, INRA, IRD, Montpellier SupAgro, Montpellier, France

José Henrique Tertulino Rocha Agriculture and Forest Engineering College (FAEF), Garça, SP, Brazil

Emanuela Gama Rodrigues North Fluminense State University, Campos dos Goytacazes, RJ, Brazil

Maiele Cintra Santana Department of Soil Science, University of São Paulo, "Luiz de Queiroz" College of Agriculture, Piracicaba, SP, Brazil

Felipe Martini Santos Federal Rural University of Rio de Janeiro, Seropédica, RJ, Brazil

Antonio Marcos Miranda Silva Department of Soil Science, University of São Paulo, "Luiz de Queiroz" College of Agriculture, Piracicaba, SP, Brazil

Eduardo Vinicius Silva Federal Rural University of Rio de Janeiro, Seropédica, RJ, Brazil

Helena Carrascosa Von Glehn São Paulo Environmental System, São Paulo, SP, Brazil

Maurício Rumenos Guidetti Zagatto Department of Soil Science, University of São Paulo, "Luiz de Queiroz" College of Agriculture, Piracicaba, SP, Brazil

Maria Jose Brito Zakia Department of Forest Sciences, São Paulo State University, Botucatu, SP, Brazil

Jerri Edson Zilli EMBRAPA Agrobiology, Brazilian Agricultural Research Corporation, Seropédica, RJ, Brazil

About the Editors

Elke Jurandy Bran Nogueira Cardoso holds a PhD degree in Plant Pathology from the Ohio State University (USA) and a post doc from Göttingen University (Germany). She is currently senior professor of Soil Microbiology and Biotechnology in the Department of Soil Science at the University of São Paulo (Brazil), where she has been working since 1971 and has supervised more than 100 post-graduate and post-doctoral students. Her research interests are focused on ecology and agricultural sustainability, forestry, plant growth-promoting microorganisms, plant microbiomes, organic residues in agriculture and forestry, and soil remediation. She has published more than 130 scientific articles in international journals, about 20 books and book chapters, many in cooperation with colleagues from other countries.

José Leonardo de Moraes Gonçalves is full professor in the Department of Forest Sciences at the University of São Paulo. He has undertaken extensive research on soil fertility, site preparation, and plant nutrition, focusing on the tropical and subtropical eucalyptus and pine plantations, including genotypic adaptation and phenotypic acclimatization of plantations in regions with high water, thermal, and nutritional stresses. Ecosystem management and sustained production of wood and non-wood products from planted forests always were among his main objectives, having been a pioneer in forest research with minimum tilling in the early 1990s. He has edited three books and published more than one hundred scientific articles in journals of wide dissemination and international impact. He supervised 25 masters, 24 doctors, and 5 post-doctorates.

Fabiano de Carvalho Balieiro holds a PhD in Soil Science from the Federal Rural University of Rio de Janeiro (Brazil), where he also was a post doc fellow. He currently works as a researcher at the National Center for Soil Research at Embrapa Solos (Brazil), and his research is focused on nutrient cycling, soil organic matter dynamics and soil ecology in different crop systems (eucalyptus, mixed forest plantations, and sugarcane), and native vegetation (Amazon and Atlantic Forests and Cerrado biomes). Dr. Balieiro has published 27 articles in high impact factor journals over the past 5 years and edited the book *Manual of Fertilization and Liming for the State Rio de Janeiro* (in Portuguese), a publication on the sustainable agricultural development in Rio de Janeiro State (Brazil).

Avílio Antônio Franco is a graduated Agricultural Engineer from the Federal Rural University of Rio de Janeiro (Brazil) (UFRRJ), and he obtained an MSc in Microbiology from the University of New South Wales (Australia), a PhD in Soil Science from the University of California-Davis (USA), and a post doc from the University of Queensland (Australia). He is retired scientist from the Center for Agrobiology at Embrapa (Brazil) after 40 years of activity. He has supervised 33 graduate students at URRJ and Federal University of Rio de Janeiro (UFRJ) and published 92 scientific articles and 25 books and book chapters. His research interests are soil microbiology (Biological Nitrogen Fixation), plant nutrition, land reclamation, and agroforestry. He is an effective member of the Brazilian Academy of Science (ABC) and the Third World Academy of Sciences (TWAS). He has received the Brazilian National Scientific Award.

Chapter 1
Why Mixed Forest Plantation?

Ranieri Ribeiro Paula, Ivanka Rosado de Oliveira, José Leonardo de Moraes Gonçalves, and Alexandre de Vicente Ferraz

1.1 Introduction

Although relevant to Brazilian gross domestic product (GDP), forest plantations currently occupy only a small fraction of Brazilian land with 9.8 million hectares (Bacha 2008; IBGE 2017). Approximately 96% of these lands are occupied by monocultures of species of *Eucalyptus* (75.2%) and *Pinus* (20.6%), and only a few forest species occupy another 400,000 hectares (IBGE 2017). These plantations have expanded over the past 30–40 years on land abandoned by agriculture and livestock, especially in the Atlantic Forest (South, Southeast, Coastal area of Bahia), Cerrado (Southeast, Midwest, and North), and Pampa (South) (Gonçalves et al. 2013; IBGE 2017).

Pastures planted with "African" grass occupy at least 80 million hectares in tropical regions of Brazil, and at least half of those are considered degraded (Boddey et al. 2004). Additionally, current estimates by the Brazilian Government indicate that 4.5 million ha of permanent preservation areas of native vegetation need to be recovered throughout Brazil. In addition, approximately 7.2 million ha of legal reserve should be recovered mainly in the Amazon, 4.8 million ha in the Atlantic

R. R. Paula (✉)
Center for Agricultural Sciences and Engineering, Federal University of Espirito Santo, Alegre, ES, Brazil

I. R. de Oliveira · J. L. M. Gonçalves
Department of Forest Sciences, University of São Paulo, "Luiz de Queiroz" College of Agriculture, Piracicaba, SP, Brazil

A. de Vicente Ferraz
Institute of Forest Science and Research (IPEF), Piracicaba, SP, Brazil

E. J. Bran Nogueira Cardoso et al. (eds.), *Mixed Plantations of Eucalyptus and Leguminous Trees*, https://doi.org/10.1007/978-3-030-32365-3_1

Forest, and 3.7 million ha in the Cerrado—Brazilian savannah (Soares-Filho et al. 2014; Brazil 2017). Public policies are foreseen in the National Plan for the Recovery of Native Vegetation for the recovery of at least 12 million ha until 2030 (Brazil 2017).

The expansion of highly productive monospecific forest plantations faces several challenges. Species are recommended for a given site according to edaphoclimatic adaptation, productivity, quality of wood, and resistance to pests and diseases, among others. In the case of *Eucalypt*, higher wood yields (e.g., annual average increment >40 m^3 ha^{-1} $year^{-1}$) have been achieved by integrating improved genetic material, suitable edaphoclimatic conditions, and good silvicultural practices (Gonçalves et al. 2013). However, breeding programs exist only for a limited number of species, notably *Eucalyptus*, *Corymbia*, *Pinus*, and *Hevea*, and for commercial fruit trees. Several potential native and introduced tree species for forest plantations with economic purposes have little or no level of genetic improvement (Carvalho 2003).

Most of the soils located in the tropical and warm subtropical portions are generally deep and well drained, with high acidity and low fertility, mostly classified as Oxisols and Ultisols (Gonçalves et al. 2013; Guerra et al. 2014). The chemical, physical, and biological characteristics of soils intended for forest plantations will be increasingly limiting as plantations occupy lands degraded by agriculture and livestock. The availability of water is in most cases the main limiting factor for forest production, since other edaphic factors can be overcome through mechanization and fertilization (Stape et al. 2010; Gonçalves et al. 2013). The availability of water may be a limiting factor for the expansion of highly productive forest plantations in the regions with predominance of the tropical and subtropical climates, whose dry season varies between 3 and 7 months. Moreover, the effects of climate change on forest plantations need better understanding. The mortality induced by lack of water in native and exotic trees has been detected in several regions of the world, including Brazil (Laclau et al. 2013; Rowland et al. 2015). To minimize the risks of tree mortality due to water shortage, it is indicated to plant species that are adapted and efficient in the use of resources (Gonçalves et al. 2013). The impacts of highly productive trees and plantations on water resources need to be well understood to promote the sustainable expansion of plantations (Christina et al. 2011; Nouvellon et al. 2011).

Nitrogen is a nutrient commonly limiting the productivity of forest plantations with non-nitrogen-fixing species (Gonçalves et al. 2003; Rennenberg et al. 2009; Laclau et al. 2010; Bouillet et al. 2013; Gonçalves et al. 2013). Although N is naturally available in the soil through the mineralization of organic matter, this is reduced in degraded soils because of the low content and quality of the organic matter (Gonçalves et al. 2003). And availability of N is conditioned by the microbial activity that is regulated by soil moisture content (Rennenberg et al. 2009; Voigtlaender et al. 2019). Nitrogen fertilizer application in forest plantation may be recommended according to organic matter concentrations in soil (Gonçalves 1995). For example, values of 60 kg ha^{-1} and 30 kg ha^{-1} of total nitrogen are indicated for commercial plantations of *Eucalyptus* and *Pinus*, respectively, when the concentration of soil organic matter is less than 1.5%. These amounts decrease to 20 kg ha^{-1} if the concentration of soil organic matter is higher than 4%. For native species of Brazil with a medium and high nitrogen demand 50 kg ha^{-1} of total nitrogen is recommended (Gonçalves 1995). The production and use of N fertilizers involve environmental risks of water pollution and greenhouse gas emissions, as well as being increasingly costly for Brazilian planters due to the fluctuation of the dollar and the influence of the value of oil (Galloway 1998; Dias and Fernandes 2006).

A growing scientific interest in the establishment of more biodiverse forest plantations has been observed around the world and Brazil (Paquette and Messier 2010; Brancalion et al. 2012; Bouillet et al. 2013; Del Río et al. 2016; Dai et al. 2018; Marron and Epron 2019). These plantations may involve only trees and/or trees with agricultural crops and/or livestock pastures. The call for the establishment of mixed forests is mainly associated with the possibility of higher productivity and greater provision of products (e.g., wood for multiple uses, non-timber forest products, fibers, and proteins) and ecosystem services (e.g., soil and water conservation, carbon storage, wildlife feeding). Mixed forest plantations involving N_2-fixing legume species and non-N_2-fixing species such as eucalypt have been proposed to increase productivity and ecosystem services in regions with N-deficient soils (Balieiro et al. 2002; Chaer et al. 2011; Bouillet et al. 2013; Santos et al. 2016; Voigtlaender et al. 2019; Marron and Epron 2019). Tree legumes inoculated with specific bacteria can fix most of the N demanded for growth and transfer the fixed N to the soil and plants in companion, as detailed in Chap. 6. Results of research in Brazil testing intercropped *Eucalyptus* sp. with *Acacia mangium* (Fig. 1.1) increased the nutrient cycling rate, contributing to the soil a large amount of N from the biological nitrogen fixation (FBN) in only one crop rotation (Santos et al. 2016; Voigtlaender et al. 2019), as detailed in Chap. 3.

Fig. 1.1 Mixed forest of *Eucalyptus grandis* and *Acacia mangium* at spacing of 3 m by 3 m (proportion 1:1): (**a**) 5 months and (**b**) 30 months after planting, at the Experimental Station of Forest Sciences of Itatinga, São Paulo, Brazil

In a recent meta-analysis, Marron and Epron (2019) showed that mixed forest plantations involving at least equal proportions of N_2-fixing and non-N_2-fixing species are generally more productive than monospecific plantations when established in sites with low productive capacity. In these sites, poorer in nutrients or water, positive interactions (e.g., facilitation and complementarity) over negative ones (e.g., intraspecific competition) are expected to prevail. Sites with lower growth resource limitations tend to favor fast-growing species more adapted, leading to greater intraspecific competition (Kelty 2006; Forrester et al. 2006; Bouillet et al. 2013). Thus, the zoning of the productive capacity of the forest sites is a necessary measure for the best use of the sites destined to the mixed and monospecific plantations. Even when yield is not increased the benefit of long-term sustainability and other ecosystem services should be considered (see the Chap. 10).

The choice of the right species to minimize the effects of site quality also is important. According to authors such as Gonçalves et al. (2003) and Kelty (2006), the success in terms of production and ecosystem service delivery is obtained more easily by combining species that differ in growth rates, in growth resource requirements, and in the form they obtain resources. The combination of these functional characteristics of the species is necessary to promote a better capture of the resources to maximize nutrient cycling and the recovery of degraded soils in Brazil (Gonçalves et al. 2003).

1.2 Socioeconomic Benefits

The increase in forest cover promoted by forest plantations with economic objectives and recovery of ecosystem functions is an urgent need pointed out by several authors (Ab'Sáber et al. 1990; Machado and Bacha 2002; Bacha 2008; Brancalion et al. 2011; Chaer et al. 2011; Guerra et al. 2014; Soares-Filho et al. 2014; Brazil 2017). There are numerous public and private actions in the Brazilian forestry production and conservation program aimed at the sustainability of the development of this sector: for example, the ABC Plan (Low Carbon Emissions Agriculture), and the Native Vegetation Protection Law (Law 12.651, of May 25, 2012), which defines the proportions of areas for protection of native vegetation and for forestry, agriculture, and livestock in rural properties, as well as the process of restoration of native vegetation in degraded areas. More recent legislation, called National Policy for the Recovery of Native Vegetation, was created to promote the revegetation of Brazilian biomes with mixed forest plantations and agroforestry. Moreover, other initiatives involving nongovernmental organizations, such as the Atlantic Forest Restoration Pact, bring new approaches to reconcile restoration of the biome with economic returns (Brancalion et al. 2011; Amazonas et al. 2018).

One of the great novelties of the Law 12.651 is the possibility, according to pre-established criteria, in mixing a 1:1 proportion of exotic and native species including fruit trees, in the legal reserve area (i.e., forest area of rural property intended for forest management and biodiversity conservation). This mechanism allows the formation of mixed multipurpose forests, as well as reduction of planting costs. Another relevant aspect is the legal reserve having a minimum area set at 20% in relation to the total size of the property for all regions of the country except the Amazon. In the Amazon region, legal reserve areas range around 80% (forest region), 35% (Cerrado region), and 20% (general field region). Chapter 13 details the Brazilian legal framework of multifunctional mixed forest plantations.

Mixed forest plantations provide a greater diversity of products and ecosystem services than monospecific forest plantations (Bouillet et al. 2013; Del Río et al. 2016; Dai et al. 2018; Voigtlaender et al. 2019). The mixture of two species with high timber value, with one being a N-fixer, has been the model most studied and recommended by researchers working with forest for timber production (Forrester et al. 2006; Bouillet et al. 2013; Del Río et al. 2016; Marron and Epron 2019). Pairs of mixtures of fast-growing exotic species such as *Eucalyptus* sp. or *Pinus* sp. with legumes such as *Acacia mangium* or *A. mearnsii* have the potential to offer a wide range of products in the same area, including timber, firewood, coal, tannins, resins, and essential oils. Such mixtures also contribute with the addition of ecosystem services, including the reduction of surface runoff and consequent more water infiltration, carbon sequestration, and biological fixation of N, culminating in the reduction of nitrogen fertilization use (see also Chap. 10).

Forest plantations carried out by companies on their own area or via forest out grower schemes use high-technology equipment with variation of techniques in function of the terrain slope (Malinovski et al. 2006; Gonçalves et al. 2013). The use of heavy machinery (e.g., tractors and harvesters) during planting and harvesting has restrictions because of local topography, and the alternatives for harvesting machinery in steep areas are even more expensive. Among the main limitations for the establishment of mixed forests with high technology, we cite the increase in the operational cost during harvesting. This limitation may not occur in plantations carried out in slope areas where cutting is done primarily by chainsaws. Moreover, the increase in the cost of harvesting in steep areas, in comparison to flat areas, may render unfeasible forest plantations with low value added (e.g., firewood and charcoal). Under these conditions, the planting of mixed forests with species of higher value, such as fruit trees or trees for seed production, noble wood, resins, and latex, among others, would be an interesting alternative.

The increase in income generated by the supply of both timber and non-timber products and ecosystem services (e.g., water increase, conservation of plant and animal diversity) could be much higher with mixed plantations with a greater diversity of species (Brancalion et al. 2012). Forestry production participated in the economy of 87% of Brazilian municipalities in 2017; 77.3% of revenue generation was derived from forestry (e.g., logs, firewood, and charcoal) and 22.7% from vegetable extraction (e.g., fruits, nuts, waxes, latex, resins) (IBGE 2017). Mixed forest plantations are an alternative to land use by small- and medium-sized producers who are interested in obtaining multiple forest products, and to increase ecosystem services on their properties (Brancalion et al. 2012; Brazil 2017). Species of the families Fabaceae, Myrtaceae, Arecaceae, and Lecythidaceae, among many others, can function both as a source of income and as a source of food for the fauna. Mixed plantation composed of groups of trees capable of producing wood, fruits, nuts, and extractives planted side by side allows to maintain a forest cover of sloping areas for a long term and, in this way, could reduce the surface runoff of water after rain events, and promote the infiltration of water into the soil (Gonçalves et al. 2003).

One of the main barriers to planting for restoration of native ecosystems is the high costs involved in establishment and maintenance. In some projects, high mortality may occur due to the attack of ants and competition with grasses. For example, the costs of establishment and maintenance during the first 2 years after planting native forests of the Atlantic Forest can exceed 5000 USD per hectare. These costs may be even higher if farmers request access to technical assistance, since seedlings and manpower are limited. Alternatives proposed to reduce costs of planting by generating more revenue include the mixing of *Eucalyptus* species for wood production with a relatively high-diversity (about 20–30) native species (Amazonas et al. 2018).

1.3 History of Mixed Plantations

Since the 1940s, several native and exotic forest species have been tested to select the most suitable ones for monospecific plantations. In Brazil, monocultures of *Eucalyptus* and *Pinus* have stood out in relation to native species, showing faster growth and good wood quality for multiple uses. Mixed forest plantations have been planted mainly with the objective of recovering degraded areas and restoring ecosystems (Kageyama and Castro 1989; Rodrigues et al. 2009). These plantations were planned according to the logic of ecological succession observed in natural forests, mixing groups of native and exotic species each one with different requirements for growth resources and lifetimes. The species used have little or no genetic improvement and the productivity of these types of plantations is not an important factor to consider. The use of mixed plantations for timber production is not very common in the practice of forest companies and producers.

In the last 30 years, several experiments about mixed plantations have been established in Brazil. These plantations were tested in experimental fields installed in different regions of the country in partnership with national and international research institutes and universities. Mixed forest plantations with fast-growing species with economic value, especially *Eucalyptus* × *Acacia mangium*, have recently been tested. The first experiment with *A. mangium* was set up in 1979 by EMBRAPA (Brazilian Agricultural Research Company) (Tonini et al. 2010). In 1985 the first plantation of *Acacia mangium* for genetic improvement was established and, in 1993, EMBRAPA Agrobiology established the experimental plantations that later culminated in the pioneering research center in studies of recovery of degraded areas with fast-growing leguminous species (Franco and Faria 1997; Macedo et al. 2008; Chaer et al. 2011).

From 1989 to 2000, there was a cooperation between Brazil and Germany in the studies of *A. mangium* in the North of Brazil with the project "*Studies of Human*

Impact on Forests and Floodplains in the Tropics." In 1995, the project "*Soil and Climate Zoning for the planting of fast-growing tree species in the Amazon*" was created. This project was financed under the "Pilot Program to Conserve the Brazilian Tropical Forest." The aim was to contribute to the reduction of deforestation rates in the region supplying the market with timber from areas with less legal restrictions instead of using native forests (Balieiro et al. 2018). The network of experiments was established in several units of EMBRAPA, in the states of Amazonas, Pará, Amapá, Acre, Rondônia, and Roraima. Different clones of *Eucalyptus* and *Acacia mangium* seedlings and several native species were tested.

In the last 10 years, Brazil, in cooperation with 34 countries, as France, the United States, Germany, Australia, Congo, the Netherlands, South Africa, China, Colombia, and Cuba, has developed studies specifically with *Acacia mangium* (Balieiro et al. 2018). In cooperation with the research institute CIRAD UMR Eco&Sols (La Recherche Agronomique pour le Développement), the thematic project "*Ecological intensification of eucalypt plantations by the association with nitrogen fixing tree legumes*" was approved by the Research Support Foundation of the State of São Paulo and its French counterpart, the "*Intensification écologique des écosystèmes de plantations forestières. Modélisation biophysique et évaluation socio-économique de l'association d'espèces fixatrices d'azote*," which was financed by the French National Research Agency. This project included a network of experiments installed in the southeast region, covering three states of Brazil, Minas Gerais, Rio de Janeiro, and São Paulo. The results showed that there were gains of biomass in the mixed plantations compared to *Eucalypt* monoculture when under favorable climatic conditions (hot and humid climate) for the development of *A. mangium*, low soil fertility and low water restrictions (Bouillet et al. 2013). This network of experiments has been recently expanded (since 2015) in two other Brazilian states, Tocantins and Mato Grosso. Previous studies were conducted in the Congo with similar edaphoclimatic condition as observed in the northern part of Brazil, showing a great productivity of these plantations and indicating a high potential of the eucalypt-acacia association (Bouillet et al. 2013). Chapter 2 details the studies about mixed forest plantation growth at different sites and under diverse silvicultural management.

1.4 Major Combinations of Species Already Tested in Practice and Potential

There are only few studies testing the growth of native species in mixed plantations compared to monocultures with the same species (Carvalho 1998, 2003; Machado and Bacha 2002). Carvalho (1998 and 2003) indicated the success of some of these studies to minimize the risks of pest attacks, such as the mixture of species of the Meliaceae family, as *Cedrela fissilis* and *Cabralea canjerana*, with other native or exotic fast-growing species to reduce the attacks of the cedar borer (*Hypsipyla grandella*). According to this author, species with reduced requirement of light at young age and with a large canopy when associated with fast-growing species generate stands with higher growth and straighter stem. For example, *Aspidosperma polyneuron* trees mixed with *Grevillea robusta* showed a straighter stem and 41% higher height growth than in monoculture after 16 years. Other timber species with good performance in mixed forest plantations, highlighted in Carvalho's bibliography, include the non-legumes *Cordia trichotoma, Prunus brasiliensis, Talauma ovata, Laplacea fruticosa, Luehea divaricata, Patagonula americana,* and *Tabebuia heptaphylla,* and the legumes *Anadenanthera peregrina var. falcata, Parapiptadenia rigida, Peltophorum dubium, Piptadenia gonoacantha, Piptadenia paniculata,* and *Sclerolobium paniculatum.* Additionally, potential species for timber indicated for mixed forest plantations are *Apuleia leiocarpa, Caesalpinia leiostachya, Enterolobium contortisiliquum, Hymenaea courbaril, Machaerium scleroxylon*, and *Pterogyne nitens* (Carvalho 1998). Planting of yerba mate (*Ilex paraguariensis*) in southern Brazil has been done in monoculture. In a literature review, Baggio et al. (2008) verified that there is great potential to associate this species with other natives of the southern region of Brazil, including leguminous N_2 fixers, with relevant economic gains in small properties.

The mixture of fast-growing species, including the N_2-fixing ones, has been tested in recent years in Brazil (Balieiro et al. 2002; Coelho et al. 2007; Bouillet et al. 2013; Santos et al. 2016; Soares et al. 2018). Mixed-species plantations of *Eucalyptus* and N_2-fixing *Pseudosamanea guachapele* with a 1:1 proportion were established in 1993 in the municipality of Seropédica, Rio de Janeiro state, Brazil (Balieiro et al. 2002). The authors showed that mixed stands had higher biomass production than pure plantations of each species, and despite the 10% less biomass of eucalypt in mixed than in pure stands, the efficiency of nutrient use of eucalypt increased in the consortium.

In São Paulo state, Brazil, a combination of five leguminous native trees, *Peltophorum dubium*, *Inga* sp., *Mimosa scabrella*, *Acacia polyphylla*, and *Mimosa caesalpiniaefolia*, and one exotic species, *Acacia mangium*, was tested with *Eucalyptus grandis* (Coelho et al. 2007). Each species was planted in monocultures and in consortium with *E. grandis* in commercial spacing (3 m × 3 m). The leguminous trees were planted between the plants of *E. grandis* in alternating rows with a 1:1 proportion. The study showed that interspecific competition between *E. grandis* and legumes is greater than intraspecific competition until the age of 24 months. Among the species studied, *A. mangium* was the one that best resisted to the competition with *E. grandis*.

Studies have indicated that higher yields occur mostly in mixed plantations with species of *Eucalyptus* sp. and *Acacia mangium,* and in monospecific *Eucalyptus* sp., under tropical climate (Bouillet et al. 2013). In regions of subtropical climate, mixed forest plantations of *E. grandis* and *A. mangium* are less productive than pure plantations of *E. grandis*. One of the main concerns of these authors is the high competition capacity of *E. grandis* on *A. mangium* in places where climatic conditions are optimal for the development of eucalypt and suboptimal for acacia. The same behavior was not observed in mixed plantations of the same species in an experiment in Congo (Bouillet et al. 2013). In this case, mixed plantations produced more biomass than eucalypt monocultures because they were embedded in areas characterized by nutrient-poor soils (e.g., sandy soils, deep soils, leachate), and warm and humid climates, but with low water limitation. These conditions are favorable for the growth of *A. mangium* but not optimal for eucalypt trees. Similar soil and climate conditions are found in the Brazilian Cerrado and transitional areas with the Amazon rainforest (e.g., Mato Grosso, Tocantins, and Roraima states). Santos et al. (2016) investigated the consortium between *Eucalyptus grandis* × *E. urophylla* and *A. mangium* in the municipality of Seropédica, Rio de Janeiro state, in a region with N-deficient soils and favorable climate for acacia. They also found a higher productivity of mixed plantings compared to pure eucalypt plantations.

The mixture with a 1:1 proportion of *Eucalyptus* spp. and *Acacia mearnsii* has been tested mainly in the southern part of Brazil. *A. mearnsii* has been more successful than *A. mangium* in facing eucalypt competition in this region. For example, the wood production of *A. mearnsii* in mixed stands with *E. globulus* was 77% of the production found in monospecific stand. In contrast, eucalypt wood production in the mixture was only 36% of the production found in monocultures (Soares et al. 2018). Mixed stands of *A. mearnsii* with *Eucalyptus* sp. presented similar production to monocultures (Vezzani et al. 2001; Soares et al. 2018), besides improving the nutritional status of soil and eucalypt trees (Vezzani et al. 2001).

Recent studies have suggested the use of eucalypt in consortia with many native species (20–30), in order to promote the restoration of ecosystems linked to the economic return from the sale of timber (Amazonas et al. 2018). Native species, including N-fixing legumes, established in mixture between eucalypt lines had their growth affected by eucalypt regarding their growth rates in three experimental sites with tropical climate without a dry season. The authors highlight the high capacity of interspecific competition of eucalypt and native species, reaching 75% of the basal area of pure eucalypt plantations, although with only 50% of tree density.

The planting of N_2-fixing trees is necessary for reclamation of degraded lands by agriculture and livestock or more severe situations such as mining. In a recent review, Chaer et al. (2011) described several successful studies using N_2-fixing legumes for land reclamation. The main objective of these plantings is the recuperation of the soil or substrate to provide colonization of new species in the future. A major concern today is the degraded soils of the Cerrado and Amazon region. Studies have shown that pastures cover about 62% of the deforested area of the Brazilian Legal Amazon, representing 335,700 km^2, and that the states with the highest incidence of pastures occur in Mato Grosso, Pará, and Rondônia (Almeida et al. 2016). According to EMBRAPA, half of this area is degraded, 30% is moderately degraded, and only 20% is in good condition. An alternative to recover degraded areas and improve the region's economy is through the insertion of intercropped plantations with fast-growing N_2-fixing legumes and eucalypt. The introduction of mixed acacia and eucalypt plantations is an alternative for the recovery of degraded areas and can increase the economy of small- and medium-sized farmers (Griffin et al. 2011).

1.5 Final Remarks

Brazil has millions of hectares of lands where forest plantations should be used to promote both economic and environmental gains. Forest covers promote important ecosystem services with emphasis on soil protection against erosion, silting of watercourses, and improvement of water infiltration.

Monospecific forest plantations with non-N_2-fixing species require higher fertilizer inputs and may have limited productivity on degraded soils with low nutrient and water availability. The environmental benefits of monospecific forest plantations may be more limited in these regions.

Mixed forest plantations involving the mixture between N_2-fixing and non-N_2-fixing trees have been highlighted as the most promising to sustain and/or increase the productivity of forests in regions limiting for development of monocultures. These more biodiverse plantations may be composed of two or more species used for different purposes, such as timber and non-timber products, soil protection in steep land, and recharge area of the groundwater, besides the recovery of degraded soils.

The introduction of the N_2-fixing species into eucalypt plantations, for example, is associated with improved nutrient cycling, especially nitrogen, with the addition of hundreds of kilograms of nitrogen via litterfall, root turnover, pruning of branches and leaves, and crop residues. Lower yields sometimes found in mixed forest plantations relative to monocultures are balanced by the increase in long-term sustainability.

There is need to broaden the debate on the ecosystem benefits of mixed forest plantations in relation to monocultures. Several species are promising for the composition of these more biodiverse forests, but little is known about the combinations and the edaphoclimatic conditions that permit to maximize the gains of the mixture.

References

Ab'Sáber A, Goldemberg J, Rodés L, Zulauf W (1990) Identificação de áreas para o florestamento no espaço total do Brasil. Estud. Av. 4:63–119. https://doi.org/10.1590/S0103-40141990000200005

Almeida CA, Coutinho AC, Esquerdo JCDM, Adami M, Venturieri A, Diniz CG, Dessay N, Durieux L, Gomes AR (2016) High spatial resolution land use and land cover mapping of the Brazilian Legal Amazon in 2008 using landsat-5/tm and MODIS data. Acta Amazon 46(3):291–302. https://doi.org/10.1590/1809-4392201505504

Amazonas NT, Forrester DI, Silva CC, Almeida DRA, Rodrigues RR, Brancalion PHS (2018) High diversity mixed plantations of *Eucalyptus* and native trees: An interface between production and restoration for the tropics. For Ecol Manag 417:247–256. https://doi.org/10.1016/j.foreco.2018.03.015

Bacha CJC (2008) Análise da evolução do reflorestamento no Brasil. Rev Econom Agric 55(2):5–24

Baggio AJ, Vilcahuamán LXM, Correa G (2008) Arborização da cultura da erva-mate: aspectos gerais, resultados experimentais e perspectivas. Embrapa Florestas, Colombo

Balieiro FC, Franco AA, Fontes RLF, Dias LE, Campello EFC (2002) Accumulation and distribution of aboveground biomass and nutrients under pure and mixed stands of Guachapele and *Eucalyptus*. J Plant Nutr 25:2639–2654. https://doi.org/10.1081/PLN-120015528

Balieiro FC, Tonini H, Lima RA (2018) Produção Científica Brasileira (2007-2016) sobre Acacia mangium Willd.: estado da arte e reflexões. Cad Ciên Tecnol 35(1):37–52

Boddey RM, Macedo R, Tarré RM, Ferreira E, Oliveira OC, Resende CP, Cantarutti RB, Pereira JM, Alves BJR, Urquiaga S (2004) Nitrogen cycling in Brachiaria pasture: the key to understanding the process of pasture decline. Agric. Ecosyst. Environ. 103:389–403. https://doi.org/10.1016/j.agee.2003.12.010

Bouillet JP, Laclau JP, Gonçalves JLM, Voigtlaender M, Gava JL, Leite FP, Hakamada RE, Mareschal L, Mabiala A, Tardy F, Levillain J, Deleporte P, Epron D, Nouvellon Y (2013) *Eucalyptus* and *Acacia* tree growth over entire rotation in single-and mixed-species plantations across five sites in Brazil and Congo. For Ecol Manag 301:89–101. https://doi.org/10.1016/j.foreco.2012.09.019

Brancalion PHS, Viani RAG, Strassburg BBN, Rodrigues RR (2012) Finding the money for tropical forest restoration. Unasylva 63(1):25–34

Brazil (2017) Plano Nacional de Recuperação da Vegetação Nativa (Planaveg). In: Ministério do Meio Ambiente, Ministério da Agricultura, Pecuária e Abastecimento, Ministério da Educação, Brasília

Carvalho PER (1998) Espécies nativas para fins produtivos. https://www.embrapa.br/florestas/busca-de-publicacoes/-/publicacao/307865/especies-nativas-para-fins-produtivos

Carvalho PER (2003) Coleção Espécies Arbóreas Brasileiras. In: Embrapa informação tecnológica e Embrapa Florestas, Brasília e Colombo

Chaer GM, Resende AS, Campello EFC, Faria SM, Boddey RM (2011) Nitrogen-fixing legume tree species for the reclamation of severely degraded lands in Brazil. Tree Physiol 31:139–149. https://doi.org/10.1093/treephys/tpq116

Christina M, Laclau JP, Gonçalves JLM, Jourdan C, Nouvellon Y, Bouillet JP (2011) Almost symmetrical vertical growth rates above and below ground in one of the world's most productive forests. Ecosphere 2:27–30. https://doi.org/10.1890/ES10-00158.1

Coelho SRF, Gonçalves JLM, Laclau JP, Mello SLM, Moreira RM, Silva EV (2007) Crescimento, nutrição e fixação biológica de nitrogênio em plantios mistos de eucalipto e leguminosas arbóreas. Pesquisa Agropecu Bras 42:759–768. https://doi.org/10.1590/S0100-204X2007000600001

Dai E, Zhu J, Wang X, Xi W (2018) Multiple ecosystem services of monoculture and mixed plantations: A case study of the Huitong experimental forest of Southern China. Land Use and Policy 79:717–724. https://doi.org/10.1016/j.landusepol.2018.08.014

Del Río M, Pretzsch H, Alberdi I, Bielak K, Bravo F, Brunner A, Condés S, Ducey MJ, Fonseca T, von LN, Pach M, Peric S, Perot T, Souidi Z, Spathelf P, Sterba H, Tijardovic M, Tomé M, Vallet P, Bravo-Oviedo A (2016) Characterization of the structure, dynamics, and productivity of mixed-species stands: review and perspectives. Eur J For Res 135(1):23–49. https://doi.org/10.1007/s10342-015-0927-6

Dias VP, Fernandes E (2006) Fertilizantes: uma visão global sintética. BNDES Setorial 24:97–138

Forrester DI, Bauhus J, Cowie AL, Vanclay JK (2006) Mixed-species plantations of *Eucalyptus* with nitrogen-fixing trees: a review. For Ecol Manag 233:211–230. https://doi.org/10.1016/j.foreco.2006.05.012

Franco AA, Faria SM (1997) The contribution of N2-fixing tree legumes to land reclamation and sustainability in the tropics. Soil Biol Biochem 29:897–903. https://doi.org/10.1016/S0038-0717(96)00229-5

Galloway JN (1998) The global nitrogen cycle: changes and consequences. Environ Pollut 102:15–24. https://doi.org/10.1016/S0269-7491(98)80010-9

Gonçalves JLM (1995) Recomendação de adubação para *Eucalyptus*, *Pinus* e espécies nativas da Mata Atlântica. Documentos Florestais Piracicaba 15:1–23

Gonçalves JLM, Nogueira Jr LR, Ducatti F (2003) Recuperação de solos degradados. In: Kageyama PY, Oliveira RE, Moraes LFD, Engel VL, Gandara FB (org) Restauração ecológica de ecossistemas naturais, FEPAF, Botucatu, pp 111–163

Gonçalves JLM, Alvares CA, Higa AR, Silva LD, Alfenas AC, Stahl J, Ferraz SFB, Lima WP, Brancalion PHS, Hubner A, Bouillet JP, Laclau JP, Nouvellon Y, Epron D (2013) Integrating genetic and silvicultural strategies to minimize abiotic and biotic constraints in Brazilian eucalypt plantations. For Ecol Manag 301:6–27. https://doi.org/10.1016/j.foreco.2012.12.030

Griffin AR, Midgley SJ, Bush D, Cunningham DPJ, Rinaudo AT (2011) Global uses of Australian acacias – recent trends and future prospects. Diverse distrib. 17:837–847. https://doi.org/10.1111/j.1472-4642.2011.00814

Guerra AJT, Fullen MA, Jorge COM, Alexandre ST (2014) Soil erosion and conservation in Brazil. Anu Inst Geociênc 37:81–91. https://doi.org/10.11137/2014_1_81_91

IBGE (2017) Produção da extração vegetal e da silvicultura 2017. Instituto Brasileiro de Geografia e Estatística, Rio de Janeiro

Kageyama PY, Castro CFA (1989) Sucessão secundária, estrutura genética e plantações de espécies arbóreas nativas. IPEF 41(42):83–93

Kelty MJ (2006) The role of species mixtures in plantation forestry. For Ecol Manag 233:195–204. https://doi.org/10.1016/j.foreco.2006.05.011

Laclau JP, Ranger J, Gonçalves JLM, Maquère V, Krusche AV, M'Bou AT, Nouvellon Y, Saint-André L, Bouillet JP, Piccolo MC, Deleporte P (2010) Biogeochemical cycles of nutrients in tropical *Eucalyptus* plantations: main features shown by intensive monitoring in Congo and Brazil. For Ecol Manag 259:1771–1785. https://doi.org/10.1016/j.foreco.2009.06.010

Laclau JP, Gonçalves JLM, Stape JL (2013) Perspectives for the management of eucalypt plantations under biotic and abiotic stresses. For Ecol Manag 301:1–5. https://doi.org/10.1016/j.foreco.2013.03.007

Macedo MO, Resende AS, Garcia PC, Boddey RM, Jantalia CP, Urquiaga S, Campello EFC, Franco AA (2008) Changes in soil C and N stocks and nutrient dynamics 13 years after recovery of degraded land using leguminous nitrogen-fixing trees. For Ecol Manag 255:1516–1524. https://doi.org/10.1016/j.foreco.2007.11.007

Machado JAR, Bacha CJ (2002) Análise da rentabilidade econômica dos reflorestamentos com essências nativas brasileiras: o caso do Estado de São Paulo. Rev Econ Sociol Rural 40(3):581–604. https://doi.org/10.1590/S0103-20032002000300004

Malinovski RA, Malinovski RA, Malinovski JR (2006) Yamaji FM (2006) Análise das variáveis de influência na produtividade das máquinas de colheita de madeira em função das características físicas do terreno, do povoamento e do planejamento operacional florestal. Floresta 36:169–182. https://doi.org/10.5380/rf.v36i2.6459

Marron N, Epron D (2019) Are mixed-tree plantations including a nitrogen-fixing species more productive than monocultures? For Ecol Managg 441:242–252. https://doi.org/10.1016/j.foreco.2019.03.052

Nouvellon Y, Stape JL, Le Maire G, Epron D, Gonçalves JLM, Bonnefond JM, Campoe O, Loos R, Chavez R, Bouillet JP, Laclau JP (2011) Factors controlling carbon and water balances on fast growing *Eucalyptus* plantations. In: Proceedings IUFRO 2011 Improvement and Culture of Eucalypts: Joining Silvicultural and Genetic Strategies to Minimize *Eucalyptus* Environmental Stress: from Research to Practice, Brazil, 43–46 November 2011

Paquette A, Messier C (2010) The role of plantations in managing the world's forests in the Anthropocene. Front Ecol Environ 8(1):27–34. https://doi.org/10.1890/080116

Rennenberg H, Dannenmann M, Gessler A, Kreuzwieser J, Simon J, Papen H (2009) Nitrogen balance in forest soils: nutritional limitation of plants under climate change stresses. Plant Biol. 11:4–23. https://doi.org/10.1111/j.1438-8677.2009.00241.x

Rodrigues RR, Lima RAF, Gandolfi S, Nave AG (2009) On the restoration of high diversity forests: 30 years of experience in the Brazilian Atlantic Forest. Biol Conserv 142(6):1242–1251. https://doi.org/10.1016/j.biocon.2008.12.008

Rowland L, Costa ACL, Galbraith DR, Oliveira RS, Binks OJ, Oliveira AAR, Pullen AM, Doughty CE, Metcalfe DB, Vasconcelos SS, Ferreira LV, Malhi Y, Grace J, Mencuccini M, Meir P (2015) Death from drought in tropical forests is triggered by hydraulics not carbon starvation. Nature 528:119–122. https://doi.org/10.1038/nature15539

Santos FM, Balieiro FC, Santos DH, Diniz AAR, Chaer GM (2016) Dynamics of aboveground biomass accumulation in monospecific and mixed-species plantations of Eucalyptus and Acacia on a Brazilian sandy soil. For Ecol Manag 363:86–97. https://doi.org/10.1016/j.foreco.2015.12.028

Soares-Filho B, Rojão R, Macedo M, Carneiro A, Costa W, Coe M, Rodrigues H, Alencar A (2014) Cracking Brazil's Forest Code. Science 344(6128):363–364. https://doi.org/10.1126/science.1246663

Soares GM, Silva LD, Higa AR, Simon AA, São José JFB (2018) Crescimento de *Acacia mearnsii* De Wild e *Eucalyptus globulus* Labill em monocultivos e consórcios com linhas simples e duplas de plantio. Sci For 46:571–581. https://doi.org/10.18671/scifor.v46n120.06

Stape JL, Binkley D, Ryan MG, Fonseca SRL, Takahashi EN, Silva CR, Hakamada SR, Ferreira JM, Lima AM, Gava JL, Leite FP, Silva G, Andrade H, Alves JM (2010) The Brazil *Eucalyptus* potential productivity project: influence of water, nutrients and stand uniformity on wood production. For Ecol Manag 259:1674–1684. https://doi.org/10.1016/j.foreco.2010.01.012

Tonini H, Angelo DH, Conceicao JS, Herzog FA (2010) Silvicultura da *Acacia mangium* em Roraima. In: Tonini H, HalfelD-VIeira BA, SJR S (eds) *Acacia mangium*: características e seu cultivo em Roraima. Embrapa Informação Tecnológica e Embrapa Roraima, Brasília e Boa Vista, pp 76–99

Vezzani FM, Tedesco MJ, Barros NF (2001) Alterações dos nutrientes no solo e nas plantas em consórcio de eucalipto e acácia negra. R Bras Ci Solo 25:225–231

Voigtlaender M, Brandani CB, Caldeira DRM, Tardy F, Bouillet JP, Goncalves JLM, Moreira MZ, Leite FP, Brunet D, Paula RR, Laclau JP (2019) Nitrogen cycling in monospecific and mixed-species plantations of *Acacia mangium* and *Eucalyptus* at 4 sites in Brazil. For Ecol Manag 436:56–67. https://doi.org/10.1016/j.foreco.2018.12.055

Chapter 2
Growth Patterns at Different Sites and Forest Management Systems

Carolina Braga Brandani, Felipe Martini Santos, Ivanka Rosado de Oliveira, Bruno Bordon, Maurel Bheling, Eduardo Vinicius Silva, and José Leonardo de Moraes Gonçalves

2.1 Introduction

Associating biological nitrogen-fixing trees (NFT) with non-nitrogen-fixing trees can increase biomass production of plantations (Bouillet et al. 2013; Santos et al. 2016). Nitrogen provided by biological fixation is likely the main reason for mixed-forest plantations with N_2-fixing trees being more productive than non N_2-fixing monocultures, since N plays an important role in the plant metabolism, soil-microbial activity, and cycling of other macronutrients that foster the forest growth. Hence, the introduction of N_2-fixing species in fast-growing eucalypt plantations could be a management strategy in sites where eucalypt growth is limited by N availability (Stape et al. 2010; Koutika et al. 2017; Tchichelle et al. 2017).

Decades of eucalypt breeding in Brazil, associated with adequate fertilizer inputs and weed control, have made the seedlings and clones in Brazilian plantations much more productive than N_2-fixing tree species. Therefore, the competition between eucalypt and N_2-fixing species in this scenario has differed largely from patterns observed in less productive eucalypt plantations (Laclau et al. 2008).

C. B. Brandani (✉)
College of Agricultural, Consumer and Environmental Sciences - Animal and Range Sciences, New Mexico State University, Las Cruces, NM, USA
e-mail: brandani@nmsu.edu

F. M. Santos · E. V. Silva
Federal Rural University of Rio de Janeiro, Seropédica, RJ, Brazil

I. R. de Oliveira · B. Bordon · J. L. M. Gonçalves
Department of Forest Sciences, University of São Paulo, "Luiz de Queiroz" College of Agriculture, Piracicaba, SP, Brazil

M. Bheling
Embrapa Agrosilvopastoral, Brazilian Agricultural Research Corporation, Sinop, MT, Brazil

E. J. Bran Nogueira Cardoso et al. (eds.), *Mixed Plantations of Eucalyptus and Leguminous Trees*, https://doi.org/10.1007/978-3-030-32365-3_2

The management practice aiming to supply the nutritional demand for N fertilization in eucalypt plantations has been achieved by the introduction of *Acacia mangium* in Brazilian forest plantations (Voigtlaender et al. 2012; Voigtlaender et al. 2019). Future studies seek to consolidate the increase of the productive potential of acacia in order to improve its competitiveness in terms of wood production with eucalypt in mixed stands as well as conciliate both trees' growths. Therefore, two issues are of great importance: (1) to find the best arrangements (additives and replacement designs) between trees and (2) to obtain improved acacia genetic materials that match the site conditions and woody product of interest.

Thus, it is necessary to consider that the productivity of tree plantations is a function of supply, capture, and efficiency of resources (Richards et al. 2010). In this sense, to unravel the competition for light, water, nutrients, and effects of intra- and interspecific competition on biomass partitioning between tree components (Le Maire et al. 2013) becomes fundamental, since these are the main processes influencing tree growth in mixed-forest plantations.

Our objective in this chapter is to gather a large number of data obtained in the last decade regarding above- and belowground mixed-forest growth in Brazil and to give insights into the main drivers influencing the development of *Eucalyptus* and *Acacia mangium*, including soil and climate conditions, silvicultural management, and species interactions. Additionally, we intend to provide important information for a wide range of land managers, from small farmers producing firewood to large commercial forestry companies focused on timber or pulp production, looking for sustainable mixed-forest systems.

2.2 Soil and Climatic Conditions on Stand Growth of Mixed-Forest Plantations

Climatic characteristics play a key role on the aboveground biomass production of mixed-forest plantations relative to monoculture. Acacia is well suited for the hottest and most humid sites (Atipanumpai 1989; Krisnawati et al. 2011) and, for this reason, its productivity can vary greatly according to solar radiation intensity, vapor pressure deficit, and water availability. Acacia has not been studied thoroughly for breeding characteristics as has the eucalypt, and does not offer many genotypes (hybrids, clones, genetic material) that could better adapt to specific sites. However, eucalypt has a broad option of genetic material (species and hybrids) provided by several decades of eucalypt breeding, offering different kinds of genetic materials that can be chosen to match certain climatic and edaphic conditions of each site (Gonçalves et al. 2013), in order to maximize the growth and yield of these plantations.

Although acacia has not yet achieved an exponential breeding potential, it has attracted great attention due to its physiological ability to fix atmospheric nitrogen, which benefits the soil-plant system. In the last decade, studies on the growth

Table 2.1 Main edaphic and climatic characteristics from experimental sites in Brazil and Congo

Sites	Soil type	Mean temperature (°C)	Mean air humidity (%)	Annual rainfall (mm)	Soil total N (g kg^{-1})	Soil organic matter (g kg^{-1})
Santana do Paraíso/ Brazil—Cenibra	Ferralsol	24.4	71	1240	1.7[a]	3.8[a]
Bofete/ Brazil—Suzano	Ferralsol	21.4	71	1420	0.84[a]	2.4[a]
Luiz Antônio/ Brazil—International Paper (IP)	Ferralic Arenosol	23.3	65	1420	0.64[a]	1.7[a]
Itatinga/ Brazil—(USP)	Ferralsol	19	70	1390	0.91[a]	3.5[a]
Seropédica/ Brazil	Planosol	24	81	1370	0.38[b]	0.62[b]
Pointe-Noire/ Congo	FerralicArenosol	25.7	81	1130	0.46[a]	1.14-1.71[a]

Source: Bouillet et al. (2013) and Santos et al. (2016)
[a][0–5 cm] layer
[b][0–10 cm] layer

dynamic of mixed-species plantations at different tropical and subtropical sites were carried out by a multidisciplinary thematic project entitled "Ecological intensification of eucalypt plantations through association with nitrogen-fixing leguminous tree species." The project was developed by French researchers from the *Centre de Coopération Internationale en Recherche Agronomique pour le Développement* (CIRAD), with the collaboration of Brazilian researchers from forestry companies, research institutes, and universities. A network of experiments was set up in several Brazilian soil and climatic conditions. Some of the obtained results were published by Bouillet et al. (2013) and Santos et al. (2016) (Table 2.1 and Fig. 2.1), where they evaluated *Eucalyptus* and *Acacia mangium* in pure and mixed plantations established in distinct regions of southeastern Brazil and in Congo, during a complete rotation (~6 years).

The results from Bouillet et al. (2013) showed that there were gains of aboveground biomass in the mixed plantations compared to eucalypt monoculture only in Congo. This result was attributed to the local conditions of hot and humid climate (annual averages of 27 °C and relative humidity of 80%), low soil fertility (especially with regard to N), and water restrictions. However, at the Brazilian sites, the global biomass production of the mixed plantations was not different and some sites showed inferior biomass production relative to eucalypt monoculture, which evidenced the suppression of acacia growth by the superior genotypes of *Eucalyptus* (Laclau et al. 2008; Bouillet et al. 2013). These results were attributed to the unfavorable climatic conditions of the sites located at the southeast of Brazil for the development of acacia, specially the low temperatures (ranging from 19 to 24 °C,

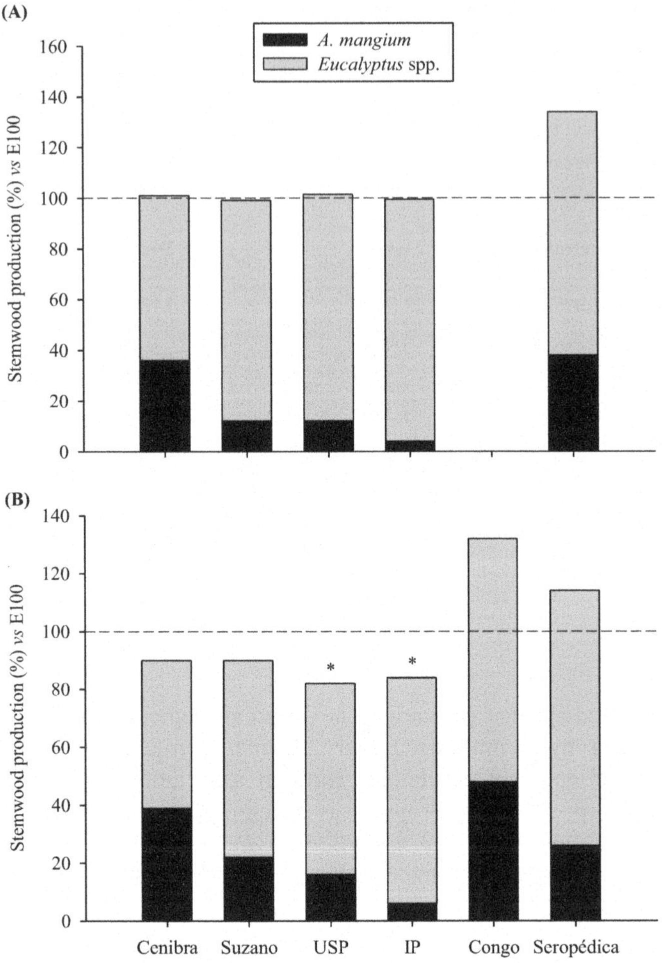

Fig. 2.1 The values of stemwood production were compared between the mixed-species *Eucalyptus* sp. and *A. mangium* 100%—E100A100 (**a**) and 50%—E50A50 (**b**) stands of the planting density composed by two species with the stemwood production of *Eucalyptus* monoculture without nitrogen fertilization (E100) in different tropical regions, site by site. Data compiled from the sites evaluated by Bouillet et al. (2013) and compared with the site in Seropédica (Santos et al. 2016). *Indicates significant statistical difference in trunk/stem production relative to control (E100). Source: Extracted from Santos et al. (2016)

throughout the year, including frost in cooler periods), and equally due to the high competition for resources in these highly productive plantations.

Contrarily, Santos et al. (2016) performed an experiment in the Southeastern Brazilian coast (in Seropédica, Rio de Janeiro state), where the climatic conditions are favorable for the development of acacia (average annual temperatures of ~25 °C, uniform rainfall distribution throughout the year, and relative humidity close to 80%). Additionally, the soil is sandy with low natural fertility, which could enhance the benefits of the biological nitrogen fixation (BNF) of *A. mangium* in consortium with

E. urophylla x grandis, according to the stress gradient hypothesis (Forrester 2014). In this study the dynamics of aboveground biomass production in monospecific and mixed stands of eucalypt and acacia over 5 years were compared with those obtained by Bouillet et al. (2013) by relativizing the overall production of the arrangements (*A. mangium* + *E. grandis*) in relation to the pure eucalypt (E100) treatment (with 1111 eucalypt ha^{-1}) of each respective site. The mixed stands with the E100A100 arrangement (with 1111 acacia +1111 eucalypt; totaling 2222 plant ha^{-1}), in all sites, produced similar amounts of stem wood biomass relative to the E100 arrangement. However, the biomass production at the Seropédica site was approximately 40% greater than that of the E100 treatment. The low relative production of *A. mangium* stem wood biomass at the sites located in the São Paulo state (Suzano, USP, and IP) may explain why the production in the E100A100 arrangement did not surpass that in the E100 treatment. These sites have similar annual average temperatures below those of the Seropédica and Cenibra sites (Santana do Paraíso, Minas Gerais state), where the contributions of *A. mangium* to the overall stem wood biomass of the population in E100A100 were considerably greater relative to the other sites.

For the E50A50 arrangement (with 555 acacia + 556 eucalypt, totaling 1111 plants ha^{-1}) a site located at Pointe-Noire, in Congo, characterized with an ideal climatic condition, was also included in this comparison. The stem wood production of E50A50 stands only exceeded that of the E100 at the Congo and Seropédica sites, although the statistical differences in both studies were not significant at the 5% level. Proportionally, *A. mangium* was more productive in Congo, followed by the Santana do Paraíso and Seropédica sites in Brazil. These sites have similar annual temperatures, above 24 °C, whereas the Seropédica and Congo sites have the poorest soils in terms of nutrients and organic matter.

In the North of Brazil (Roraima state), where the weather is characterized by a higher annual temperature (27 °C) and precipitation (2000 mm) than in the Southeast, the average productivity was 25 m^3 ha^{-1} $year^{-1}$ (Tonini 2010). There are reports of wood productivity of *Acacia mangium* from 10 to 61 m^3 ha^{-1} $year^{-1}$ obtained in the Southeast of Brazil (at the Vale do Rio Doce region), at 5.3 years of age (Silva et al. 1996), where the average annual temperature and precipitation are lower than in the Northeast of Brazil, such as in Roraima state. Plantations with adequate silvicultural management, at 3 years of age, can reach 15 m of height and 40 cm in diameter at breast height (DBH), which represents an average annual increment of 45 m^3 ha^{-1} $year^{-1}$ (Souza et al. 2004).

It is worth noting that in mixed-forest plantations, if species or sites are not complementary and correctly chosen, one species may suppress the growth of the other, which may result in less productivity than in monocultures. The success of mixed-species plantations is greatly dependent on the selection of N_2-fixing species depending on site attributes, relative growth rate of both species, and N limitation of tree growth at the site (Forrester et al. 2005; Laclau et al. 2008). It is difficult to predict which species combinations will lead to increases in productivity in mixed-forest plantations when no empirical information exists. This depends not only on the attributes of the species but also on the conditions of the site, mainly relative to water and nutrient availability, average annual temperature and precipitation, as well as weather through the year.

2.3 Silvicultural Management

Eucalyptus and *Acacia* species are chosen for their capacity to grow rapidly (the average rotation in Brazil is of ~7 years) and produce wood of excellent quality for cellulose, charcoal, and construction and fence posts. In some cases, a certain site can support high wood production with proper management, but in others, there may be serious problems.

The mixed eucalypt and acacia plantations can be established aiming at enhancing the productivity of lands that were degraded by deforestation and intensive agricultural disturbances. The development of improved forest planning and operations can increase the utilization and minimize or avoid adverse environmental effects. Some principles, aims, strategies, and silvicultural practices must be followed to develop sustainable single- and mixed-species eucalypt and acacia stands.

The success of single- and mixed-species eucalypt and acacia plantations depends directly on the soil preparation. The great advances in the last 30 years in Brazil were the understanding and abolition of burning as a way to clean the land and the adoption of conservationist techniques for soil management, which culminate in the implementation of tillage systems with minimum soil disturbance (Gonçalves et al. 2008).

Concerns regarding the conservation of natural resources and the use of post- and preemergence herbicides were the factors predisposing and permitting the adoption of minimum tillage. Weed control with herbicides was a crucial factor, since the minimum tillage system uses no ploughed land inversion (unlike the conventional system), so the weed seed bank remains on the topsoil rather than buried, which favors the infestation of plants and makes manual control operationally and economically unfeasible (Gonçalves et al. 2008). For the mixed-species eucalypt and acacia plantations, the minimum tillage has been considered as an important practice as well, since it decreases costs related to weed control, mainly because of the suppress effect of the forest residue on weed seed bank in the soil surface. A major limiting factor for plant productivity is the presence of weeds. Good practices can reduce weed infestation considerably by providing cover by crops, residues, and mulch, and by minimum soil disturbance. In forests, the problem of weeds is a great concern. The procedures for their elimination usually are carried out by combining mechanical and chemical methods, using total-action herbicides (nonselective), such as glyphosate. During the planting of forest stands, most production systems apply preemergence herbicides at a 1 m strip on the crop row. After planting, the postemergence control of weeds is performed by spraying postemergence herbicides. During application, special care must be taken to avoid the drift to leaves and stems of the cultivated plants because this can cause phytotoxic effects of reduced growth (Salgado et al. 2011).

The water availability is a critical requirement in mixed eucalypt and acacia plantations (Nouvellon et al. 2012). Soil preparation can help overcome the limitations of water resources for forest plantations in two ways. The first way is due to increased rainfall infiltration and reduced runoff, which augment the water reserve in the soil profile. The second way refers to increasing the effective soil depth when there are soil layers with physical impediment (Gonçalves et al. 2002; Stape 2002). The large infiltration is favored when compacted or hardened soil layers are disrupted, especially when forest residues are kept on the soil surface. The reduction of bulk density (resistance to soil penetration) through soil preparation in the planting row or hole facilitates root growth and, consequently, increases fertilizer use efficiency through great use of water and nutrients by adjacent seedlings. In flat and slightly undulating terrain/relief, soil preparation may consist of ripping up to a depth of 30–40 cm. Regarding soil hardening and compaction, ripping-up depth is usually about 30–35 cm and the planting hole opening, either manually or mechanically, is limited to 25–30 cm. These depths should sometimes be increased depending on the soil bulk density, for example, for soils that have a fragipan or hardpan between 50 and 80 cm, or above 80 cm, the ripping depth will be in the range of 60–90 cm, or will reach 110 cm, respectively (Gonçalves et al. 2008).

The effect of soil compaction and other soil disturbances can be severe if operations are not managed properly. Inappropriate harvesting systems have the potential to severely and adversely affect soil conditions and water availability. The effects of planting, tending, and harvesting on the physical properties of the soils will be highly dependent on the characteristics of soil and equipment used. Delimbing and debarking the stem at the stump and avoidance of fire significantly reduce nutrient exports from the aboveground biomass and soil compaction (Gonçalves et al. 2000). Rotation length is another variable that affects soil quality, because longer rotations reduce the frequency of major disturbance during harvesting, besides the decrease in wood production per unit of nutrient exported with an increase in rotation length, since the average concentration of most nutrients decreases with tree age. Hence, overall nutrient-use efficiency measured as wood production per unit of nutrient accumulated can be increased by prolonging forest rotations and using practices that lead to nutrient and organic matter retention (Gonçalves et al. 2004).

Maintenance of forest residues on soil is undoubtedly important for sustaining long-term site productivity (Rocha et al. 2016). This requires residues from the previous rotation as the main resource. Retention of forest residues helps to reduce erosion, improves water infiltration as well as moisture conservation, maintains or increases soil organic matter levels and soil microbiology indicators, as well as contributes to the nutrient cycling in the long term (Gonçalves et al. 2013; Rocha et al. 2016).

When N_2-fixing species are used to facilitate the growth of non-N_2-fixing species, they should be selected based on the rate at which they cycle nutrients in general (Bachega et al. 2016; Pereira et al. 2018) through leaf and fine-root litter and only secondarily one should consider their ability to fix N. Tree species in mixtures must also have compatible height growth dynamics to avoid the suppression of shade-intolerant plants and to reduce competition for light (Forrester et al. 2005). Mixtures should only be planted on sites where the interactions between species will increase the availability of (through facilitative interactions) or reduce competition for (through competitive reduction interactions) one of the major growth-limiting resources at that site (see Sect. 2.4 of this chapter).

Since investments in fertilizers are relatively high for most forest producers, fertilization should be combined with other silvicultural practices (i.e., soil preparation, residue management, and weed control) to reduce fertilizer demands in the short and long terms (Nambiar and Kallio 2008). Adequate nutrient supplies and balance resulting from fertilization can also improve forest vigor, and reduce the incidence of disease and the need of fungicides (Almeida et al. 2010). There are significant yield gains in response to fertilization in most forest plantations (Gonçalves et al. 2013). Regardless of weather conditions, the magnitude of the response depends on the nutritional demand of the genotype and on the availability of soil nutrients. Gains in productivity attributed to mineral fertilizers (macro- and micronutrients) are quite variable and high, but in general they represent at least 30–50% on average (Gonçalves 2011).

Nutrient cycling reduces tree dependence on net nutrient supply from soil reserves. Mobile nutrients (N, P, and K) in the plant are redistributed from the older to younger tissue, increasing efficiency for biomass production. Nutritional stages of a forest stand can be divided into before, during, and after canopy closure (Gonçalves et al. 2014). Understanding these stages and nutrient cycling is essential for the adequate planning of fertilizer application (rate, method, and time). Fertilizer recommendation should be adjusted preferably at local level to the most representative species and soil types, based on field experimentation, and should allow optimization of financial returns. Fertilization should be performed during the initial stage of tree establishment, from the planting to canopy closure. The most frequent and most significant responses to fertilizers in Brazilian soils are to N, P, K, and B. Normally, for sandy and water-deficient soils, responses to fertilizers are more common (Gonçalves et al. 2008; Gonçalves 2011).

Regarding how much fertilizer to apply, especially for fast-growing species such as acacia and eucalypt, phosphorus doses can be applied at planting, since this nutrient has low mobility in the soil and relatively low solubility. The K doses should be divided into one or two surface applications (Gonçalves et al. 2008). Doses up to 50 kg ha^{-1} of K_2O may be applied thoroughly in one single surface application, once risks of leaching are low (Maquère et al. 2005; Laclau et al. 2010). The N contribution will be initially to acacia and later will be shared with *eucalypt*. Santos et al. (2017a) showed that *A. mangium* in mixed-species plantations with eucalypt can provide ~30 kg of N ha^{-1} $year^{-1}$, only through the leaf litter deposition over 5 years of rotation, while eucalypt contribution could reach 15–20 kg N ha^{-1} $year^{-1}$ at the same period, totalizing 45–50 kg N ha^{-1} $year^{-1}$. Other studies have shown that mixed plantations of *A. mangium* and eucalypt have larger concentration of mineral N in the soil than monospecific eucalypt stands (Voigtlaender et al. 2012, Bachega et al. 2016; Tchichelle et al. 2017s; Voigtlaender et al. 2019). After mineralization of acacia litter, significant quantities of biologically fixed N become available for the eucalypt trees. This process may promote N nutrition in poor tropical soils when the fine roots of eucalypt and acacia are intermingled or are connected directly through common mycorrhizal networks (Paula et al. 2015; Pereira et al. 2018). The scenario which does not include residue burning or removal (litter, slash, and bark) and includes fertilizer application is a practical and economically proven strategy for sustaining production in the long term (Rocha et al. 2018).

Planting arrangements are one of the main factors that influence the tree growth and determine the quality and applications of the wood produced. The appropriate planting density promotes optimum growth rates and efficient plantation management.

Defining the initial spacing for forest plantations is essential because it determines the amount of resources (water and nutrients) available for each tree growth species. The planting spacing represents the number of trees in a given area and should be associated with the best way to manage and harvest the forest stand (Scolforo 1997). The recommended tree planting design is 3–4 m between rows and 1–3 m between trees, giving an initial stocking rate, usually of 1000–1800 trees per hectare.

The mixed-specie plantations can be arranged in varied designs (Figs. 2.2 and 2.3). They can be in alternating rows of each species, or even with both species interplanted in the same row. The replacement and additive series are the most common types of arrangements (Forrester et al. 2006). In replacement series, the total number of trees per stand is constant; only the proportion of each species will be changed. For example, in an area with 1111 trees per hectare 50% of the stand will be composed by one tree species (555 trees) and 50% by another species (555 trees). Recent experiments have indicated that planting trees of each species in double rows instead of alternating plants may reduce competition between eucalypt and acacia as well as facilitate the harvesting of trees at different stages of development.

In additive series, the total number of the main species of interesting (e.g., *Eucalyptus*) will be the same, while the density of the other species will vary: for example, 100% of eucalypt and 25% of another species or 100% of eucalypt and 50% of another species. These types of series allow the evaluation of the effects of the stand density and the interactions between species. Replacement and additive series have been used in mixed-species experiments to analyze the growth and the productivity of eucalypt and leguminous trees.

Based on the published mixed-species trials utilizing the replacement series design, and where mixtures were more productive than monocultures, 1:1 mixtures were one of the most productive arrangements (Binkley et al. 2003; Forrester et al. 2004). These density and proportion between two species encourage the canopy to close rapidly, reducing weed problems and improving the tree form and branching. If the goal is to produce solid timber for sawmills this density of planting is high enough to allow trees with exceptional form and vigor to be selected in a thinning plan of the stand.

During the pre-closure phase, trees tend to be more responsive to cultivation, fertilizers, and weed control. After canopy closure, intra- and interspecific competition for resources becomes strong. The density and the design of plantations directly influence the processes of facilitation, competition, and consequently tree growth (Medhurst et al. 2001). An important issue for the design of mixed plantations is the definition of the optimal spatial arrangement depending on localization (e.g., climate and type of soil) to better understand inter- and intraspecific interactions and also to facilitate the management. Wide spacing may also be used in water catchments to increase water yield in the site (Lima et al. 2012).

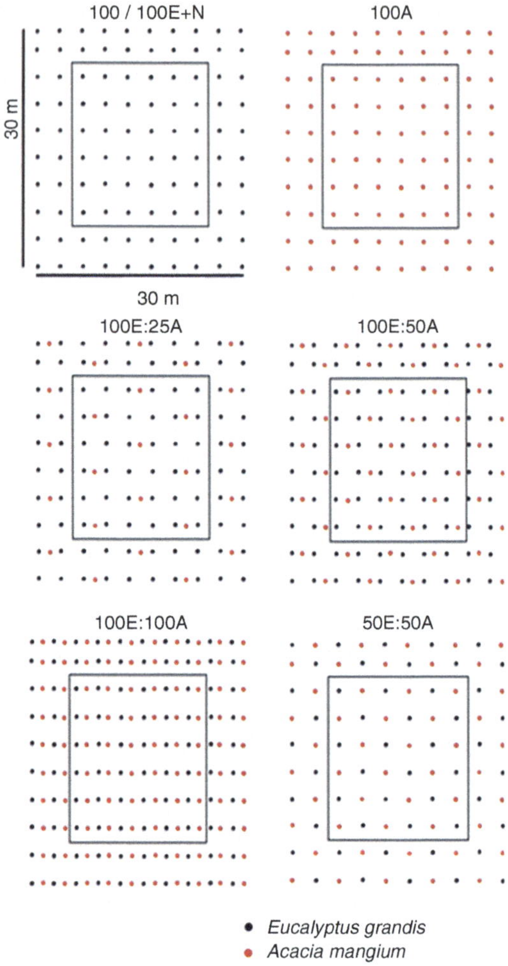

Fig. 2.2 Scheme of the different planting designs with additive and replacement series (adapted from Laclau et al. 2008)

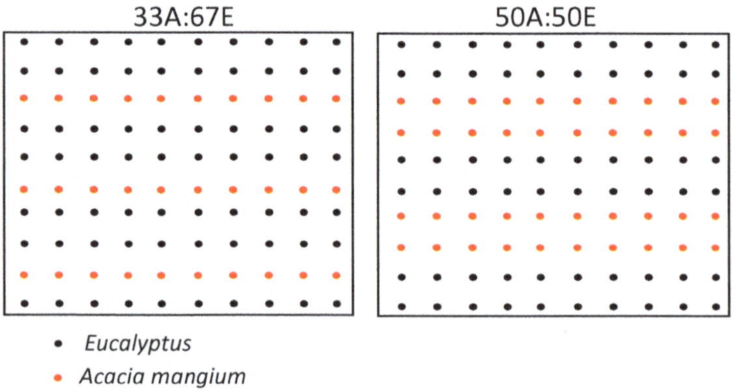

Fig. 2.3 Scheme of *Eucalyptus* and *Acacia* trees in one or double rows instead of alternating plants with two lines of *Eucalyptus* trees with one (33A:67E) or two (50A:50E) lines of *Acacia* trees

There are many different ways to manage the interactions in mixed-species plantations. These involve different designs, delayed planting of certain species, varying plant densities, or early removal and thinning of some species. Therefore, mixed-species plantations can be used to meet a wide range of economic, silvicultural, and sustainability objectives.

2.4 Intra- and Inter-specific Interactions (Competition and Facilitation Processes) in Stand Growth

In the tropics, eucalypt plantations are predominantly established in monocultures throughout successive short rotation cycles (6–8 years) (Gonçalves et al. 2013) that can result in substantial changes in soil quality and biogeochemical cycles of nutrients (Chaer and Tótola 2007; Gonçalves et al. 2004; Stape et al. 2010; Rocha et al. 2018). One of these important changes concerns the soil nitrogen balance, which generally becomes negative with multiple rotations due to combinations of high N exports from timber harvests and low doses of N fertilizers that are typically applied (Corbeels et al. 2005; Laclau et al. 2010; Rocha et al. 2018). Hence, the establishment of mixed-species plantations of eucalypt with N_2-fixing trees has been promoted due to the increase in the availability of N for eucalypt trees, making it possible to dispense the use of N fertilizers (Laclau et al. 2008; Voigtlaender et al. 2012; Voigtlaender et al. 2019). We can show other benefits of these mixtures, as increase of soil fertility and recycling of micro- and macronutrients (Balieiro et al. 2004; Forrester et al. 2005; Santos et al. 2017b), intensification of above- and belowground biomass production (Bouillet et al. 2013; Laclau et al. 2008), and increase of soil carbon (Balieiro et al. 2008; Forrester et al. 2006; Resh et al. 2002) and nitrogen stocks (Voigtlaender et al. 2012; Voigtlaender et al. 2019). In fact, these benefits have been seen in studies in which mixed plantations with leguminous trees were more productive than in monospecific forest plantations (Binkley et al. 2003; Forrester et al. 2006; Laclau et al. 2008; Bouillet et al. 2013; Santos et al. 2016). However, other studies showed productivity of mixed plantations not different from eucalypt monoculture in the same stock density (Forrester et al. 2006; Firn et al. 2007; Bouillet et al. 2013).

The benefits and processes influencing the mixed-forest growth must be better understood through concepts regarding ecological interactions between tree species in mixed-forest stands, such as competition, competitive reduction, and facilitation. The balance between competition and facilitation will cause a strong impact on the productivity and the biomass accumulation in mixed plantations.

Competition occurs when two or more species are interacting and seeking the same sources of light, water, and nutrients until a given species exerts a negative effect, such as a decrease in the growth rate or mortality of the other species, less adapted to the environmental dynamics created by the mixtures (Forrester et al. 2006; Vandermeer 1989). The competition among plants can be through the aboveground compartments when they seek light to keep photosynthetic activity (Austin et al. 1997; Hunt et al. 2006) or by belowground compartments when they are in a limiting water and nutrient uptake environment (Boyden et al. 2005; Silva et al. 2009).

When species in mixed plantings present contrasting traits the competitive reduction normally is expressed, which allows for a more efficient use of the site resources through complementary niche exploitations. This condition generally occurs through canopy and root stratification that are benefits associated to an increase in light-use efficiency and higher soil resource (i.e., water and nutrients) uptake, respectively (Vandermeer 1989). When two species with similar traits are mixed, the interspecific competition may be equal to the intraspecific competition of the same species in monoculture. However, when the species present complementary traits, the interspecific competition may be smaller relative to the intraspecific competition (Kelty 2006). The complementarity of resource uses between species is a trait that can benefit the mixed-forest productivity, which can lead to a more efficient capture of resources when compared to monocultures, as well as the reduction of competition through the niche partitioning by stratification of the canopy of two species (Hunt et al. 2006; Laclau et al. 2008). Thus, it may result in better soil exploitation without stratification of the root systems of species (Laclau et al. 2013).

Facilitation typically occurs when at least one species acts positively on the other species increasing the availability of a resource for another species, such as N_2-fixing trees increasing the availability of organic and inorganic forms of N in the ecosystem, through biological nitrogen fixation (BNF) (Vandermeer 1989; Forrester et al. 2006). Thus, when a N_2-fixing species is planted with non-N-fixing species, it is possible to improve the nutritional status of the non-fixing species and to increase the growth rates in response to transferring of biologically fixed N (Bouillet et al. 2008; Paula et al. 2015). Indirect facilitation can also happen, when the plant changes the environment, such as the faster closure of the canopy that reduces light availability for weeds (Little et al. 2002; Le Maire et al. 2013). Sometimes partial shading of a fast-growing species may be beneficial to some species.

The ecological interactions (competitive reduction and facilitation) that are favorable for the success of mixed plantations occur at the same time. However, in practice, it becomes very difficult to distinguish one from the other. Hence, they are collectively described as "complementarity" (Forrester 2014). This is particularly relevant because a more efficient capture of limiting resources may alleviate competition and contribute to enhance above- and belowground production of the mixed-species plantations (Forrester et al. 2006).

Therefore, the correct choice of species is highly important, which should be based on contrasting morphological and physiological traits, especially with respect to shade tolerance, growth rate, crown structure (i.e., leaf area density), and effective depth of the root system (Forrester et al. 2006). In addition, the spatial arrangement of species has also been defined as a key strategy for the occurrence of complementarity (Kelty 2006). Thus, the success of mixed plantations is reached when the biomass production becomes significantly higher or at least equal to that of monocultures (Forrester et al. 2005). When it happens, the complementarity effects stand out from those of interspecific competition. These factors could even contribute to an overyielding effect, in which biomass production in species mixtures exceeds the productivity of the contributing species when grown in monoculture (Bauhaus et al. 2000).

2.4.1 Ecological Interactions Change Throughout a Single Rotation

The stress-gradient hypothesis (SGh), proposed by Bertness and Callaway (1994), correlates the frequency of ecological interactions along a biotic or abiotic stress gradient. The theory predicts that under conditions of low environmental stress competition plays a more relevant role than facilitation. However, under high-stress conditions the facilitation interactions prevail in the sense of improving the neighboring habitat. This hypothesis becomes an excellent theoretical basis to explain the behavior of mixed eucalypt plantations with legume trees, especially during a complete rotation, since the balance between these interactions can be modified with the dynamics of tree growth and environmental conditions of the site (Forrester 2014).

The complementarity index represents a measure of the occurrence of ecological interactions favorable to the success of mixed plantations (competitive reduction and facilitation). This index is based on the differences between the growth (or production) at the level of the stand (Eq. (2.1)) and/or of each species (Eq. (2.2)) in the mixed stands relative to the monocultures, according to the following equations (Loreau and Hector 2001): .

$$\text{Complementary}\left(\%\right)=\frac{\text{growth}_{\text{mix}}-\text{growth}_{\text{mono}}}{\text{growth}_{\text{mono}}} \tag{2.1}$$

$$\text{Complementary}\left(\%\right)=\left(\frac{\text{yield}_{\text{mix}}}{\text{yield}_{\text{mono}}\times\text{species proportion}}-1\right) \tag{2.2}$$

where *growth* is expressed through diameter at breast height (DBH), in centimeters and *yield* is expressed by total aboveground biomass, in Mg ha^{-1}.

This index also allows inferring if these interactions change spatially or over time and what are the possible factors that are controlling these changes (see also Forrester et al. 2014).

In this chapter, we used experimental data from 5-year-old mixed and monoculture stands of *Eucalyptus urophylla x grandis* and *Acacia mangium* established in southeastern Brazil. The results revealed that complementarity interactions changed as the stands developed (Fig. 2.4). With the advancement of age, changes in the availability and uptake of water, light, and nutrients for the species occur. In the E100A100 mixture, instead of the greater competition among the trees caused by the high densification, an increase in the complementarity index was expected based on the stress-gradient hypothesis, which presupposes that facilitative interactions

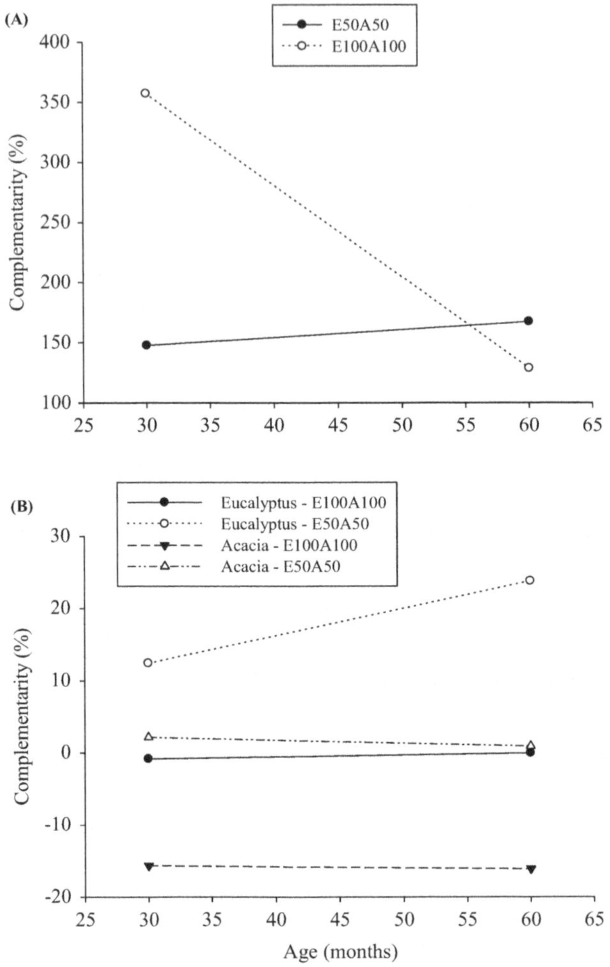

Fig. 2.4 Complementarity index at stand level (**a**) and at tree/species level (**b**) within the stand at 30 and 60 months after planting

increase, while the competition decreases with the increase of the abiotic/biotic stress, and vice versa. However, the opposite occurred because, at the stand level, complementarity decreased by almost 200% from the younger phase to the end of rotation in the E100A100 stands.

Forrester et al. (2014) reviewed studies with mixed plantations and monocultures using the same methodology and found results similar to the present study. These authors consider that complementarity may decrease as nutrient availability increases, which may explain the decrease in complementarity in the E100A100 arrangement. Thus, at the beginning of planting, the litter stock was still incipient but, with the advancement of age, there was larger deposition and release of nutrients through litter decomposition, in comparison with *E. urophylla* × *grandis* monocultures.

In contrast to E100A100, the E50A50 mixture showed a small increase in the complementarity index throughout the growth cycle. This behavior demonstrates that under this arrangement the competition between the two species was less intense. Eucalypt at E50A50 increased its complementarity index by 20%, probably due to the better growth conditions for this species, such as greater intraspecific spacing and better light utilization (i.e., greater leaf area index (LAI) than acacia, see Santos et al. (2016)). These factors may have led to increased leaf projection in the canopy and greater growth in stem diameter of eucalypt in comparison to acacia trees, which resulted in a greater allocation of C in the aboveground biomass (Forrester et al. 2014). This fact may explain why 50% of eucalypt trees, planted with 50% acacia, produced equivalent amounts of stem wood in relation to the E100 monoculture.

2.5 Aboveground Biomass

In the last decade, many studies have been carried out with different designs between eucalypt and leguminous trees, especially with *A. mangium* plantations in the southeast of Brazil (Tables 2.2 and 2.3). The following studies contributed to a better understanding of the dynamics of aboveground biomass production, as well as the main drivers and interactions for production responses (Balieiro et al. 2002; Coelho et al. 2007; Laclau et al. 2008; Nouvellon et al. 2012; Bouillet et al. 2013; Laclau et al. 2013; Le Maire et al. 2013; Epron et al. 2013; Germon et al. 2018; Santos et al. 2016; Paula et al. 2018).

Nitrogen fixed by *A. mangium* during the initial growth phase (up to 30 months) has been the key to the superior performance of the mixed stands cycle/rotation in terms of growth, stemwood biomass production, and net primary production in relation to *E. urophylla x grandis* monoculture (E100) (Santos et al. 2016). The mixed stands contributed nearly 200 kg N ha^{-1} to the soil, almost twice that of the E100 monoculture at 60 months. This great contribution can be explained mainly by the larger N richness of the acacia leaf litter (~17 g kg^{-1} for *A. mangium* vs. ~10 g kg^{-1} for *E. urophylla x grandis*), which corroborated the high BNF rates observed for acacia at the beginning. This early increase in the N levels resulted in a greater decomposition rate of eucalypt litter, as well as the biogeochemical cycling, which persists up to the mature growth phase.

In addition, canopy stratification in mixed-species stands may increase light interception as well as make them more productive than monocultures. However, this complementarity niche that occurs aboveground may not lead to an increase in stem wood biomass, if another important resource is strongly limiting tree growth. Le Maire et al. (2013), for instance, highlighted that in the *Eucalyptus grandis* W. Hill ex Maiden and *Acacia mangium* Willd. mixed-species plantations before canopy closure, the N_2-fixing trees allocated their assimilated C mainly to vertical growth in an effort to compete with *E. grandis* trees for light, because eucalypt grew faster in height. Thereafter, when *A. mangium* trees were completely dominated by

Table 2.2 Mean height, CBH, stemwood, and total aboveground biomass of harvest residues at clear cutting depending on sites (Mg ha^{-1})

Site/age	Spacing (m × m)		100E	100E + N	100A	25A:100E			50A:50E			50A:100E			100A:100E			References
			E	E	A	A	E	Total	A	E	Total	A	E	Total	A	E	Total	
Santana do Paraiso Minas Gerais 76 months	3 × 3	Height (m)	27.4	26.7	18.8	20.3	25.5	–	19.6	26.6	–	18.5	24.6	–	16.4	24.3	–	Bouillet et al. (2013)
		CBH (cm)	47.8	47.6	57.0	60.3	44.6	–	57.8	52.1	–	49.7	42.2	–	39.8	41.2	–	
		Stemwood biomass (Mg ha^{-1})	132.0	126.0	59.3	32.4	108.9	141.3	52.1	64.6	116.7	41.2	88.1	129.3	45.7	87.1	132.8	
		Total aboveground biomass (Mg ha^{-1})	153.6	146.9	82.1	46.0	127.2	173.2	74.2	75.9	150.1	59.8	103.4	163.2	68.2	102.5	170.7	
Bofete São Paulo 75 months	3 × 2	Height (m)	24.2	24.5	16.2	12.4	24.3	–	14.5	26.5	–	12.9	24.3	–	11.6	23.7	–	
		CBH (cm)	49.4	48.8	48.0	33.0	47.9	–	42.1	56.8	–	33.4	48.4	–	26.8	46.2	–	
		Stemwood biomass (Mg ha^{-1})	131.8	125.5	71.8	6.9	126.3	133.2	25.9	92.2	118.1	16.1	120.2	136.3	17.1	112.8	129.9	
		Total aboveground biomass (Mg ha^{-1})	173.0	165.0	123.7	12.7	165.6	178.3	42.9	148.2	191.1	29.5	157.7	187.2	32.3	147.9	180.3	
Luiz Antonio São Paulo 73 months	3 × 3	Height (m)	24.9	24.4	14.1	8.6	24.3	–	11.2	26.1	–	8.5	24.7	–	8.4	24.5	–	
		CBH (cm)	53.4	53.3	41.2	19.4	52.8	–	27.1	65.5	–	18.3	53.0	–	18.3	52.9	–	
		Stemwood biomass (Mg ha^{-1})	132.4	124.5	29.1	1.0	124.5	125.5	6.2	105.4	111.6	2.4	126.2	131.4	4.2	126.3	130.4	
		Total aboveground biomass (Mg ha^{-1})	159.1	149.6	44.3	2.3	149.6	151.8	10.6	127.8	138.4	5.3	152.2	160.3	9.6	151.7	161.2	

Itatinga São Paulo 72 months	3 × 3	Height (m)	24.2	24.9	15.4	9.0	25.1	–	12.6	24.8	–	9.3	24.5	–	9.1	24.0	–	Nouvellon et al. (2012), Bouillet et al. (2013), Le Maire et al. (2013)
		CBH (cm)	50.7	52.3	52.3	24.0	51.9	–	39.9	61.5	–	26.0	50.9	–	23.5	50.4	–	
		Stemwood biomass (Mg ha^{-1})	109.2	115.2	63.0	1.9	114.9	116.9	15.2	72.7	87.9	4.5	107.7	112.2	7.8	102.1	109.9	
		Total aboveground biomass (Mg ha^{-1})	136.4	143.8	108.7	3.6	143.4	147.1	24.4	99.7	124.1	8.3	134.4	142.7	14.4	127.4	141.8	
Seropédica Rio de Janeiro 60 months	3 × 3	Height (m)	16.4	20.6	13.7	–	–	–	11.0	19.9	–	–	–	–	10.5	18.2	–	Santos et al. (2016)
		CBH (cm)	45.2	51.8	43.0	–	–	–	34.5	56.2	–	–	–	–	32.9	45.2	–	
		Stemwood biomass (Mg ha^{-1})	66.3	106.9	43.5	–	–	–	17.1	58.7	75.8	–	–	–	25.0	63.6	88.6	
		Total aboveground biomass (Mg ha^{-1})	56.1	75.2	40.5	–	–	–	13.8	54.9	68.7	–	–	–	17.9	51.1	69.1	

Table 2.3 Mean height, CBH, stemwood, and total aboveground biomass of harvest residues at the mid of the rotation depending on sites (Mg ha^{-1})

Site/age	Spacing (m × m)		100E	100E + N	100A	25A:100E		50A:50E			50A:100E		100A:100E			References
			E	E	A	A	E	A	E	Total	A	E	A	E	Total	
Santana do Paraiso Minas Gerais 31 months	3 × 3	Height (m)	18.3	18.0	12.5	11.1	18.0	11.4	18.5		11.2	17.5	10.3	17.5		Bouillet et al. (2013)
		CBH (cm)	38.4	38.3	44.4	38.7	36.4	42.7	40.4		35.1	34.9	30.3	33.4		
Bofete São Paulo 38 months	3 × 2	Height (m)	19.7	19.4	12.4	9.2	19.3	10.9	20.0		9.1	18.8	8.7	18.6		
		CBH (cm)	40.3	40.3	38.9	20.4	39.3	33.1	44.9		22.8	39.5	19.0	37.5		
Luiz Antonio São Paulo 22 months	3 × 3	Height (m)	12.1	12.1	5.5	5.4	11.8	5.9	11.4		5.3	12.0	5.5	11.3		
		CBH (cm)	31.3	31.1	21.6	13.2	30.4	18.5	34.5		13.1	30.7	13.8	28.7		
Itatinga São Paulo 29 months	3 × 3	Height (m)	13.7	14.1	7.3	5.4	14.2	7.2	13.3		5.7	14.0	5.8	14.0		Laclau et al. (2008), Nouvellon et al. (2012), Bouillet et al. (2013), Le Maire et al. (2013)
		CBH (cm)	35.1	36.5	36.3	15.5	35.9	28.6	40.4		16.7	35.6	16.7	35.7		
Seropédica Rio de Janeiro 30 months	3 × 3	Height (m)	9.0	13.0	8.4	–	–	8.0	10.0		–	–	8.5	10.4		Santos et al. (2016)
		CBH (cm)	25.7	36.4	32.6	–	–	33.6	28.9		–	–	27.6	25.4		
		Stemwood biomass (Mg ha^{-1})	12.6	31.6	13.4	–	–	6.4	9.2	15.5	–	–	13.2	15.6	28.9	
		Total aboveground biomass (Mg ha^{-1})	22.3	47.7	22.1	–	–	11.07	15.6	26.6	–	–	19.8	25.0	44.8	

the *E. grandis* canopy, they invested relatively more biomass into their resource-capturing organs (i.e., leaves and fine roots) and less into stemwood production. Finally, the authors stated that *A. mangium* likely suffered from greater water stress (than in monocultures) due to competition with *E. grandis* trees.

Greater stemwood biomass production in mixed stands (E100A100) compared with the monoculture (E100) can also be explained by the improved utilization of light by two species as a function of the canopy stratification in the mixed stands (Santos et al. 2016). In this case, *E. urophylla x grandis* trees occupied the upper stratum and acacia capturing the light underutilized by *E. urophylla x grandis*, as well as by the higher density of the stand. This argument is supported by the leaf area index (LAI) of the mixed E100A100 stand in relation to that of the E100 stand, which explains the greater biomass production of the E100A100 stand. An increase in the leaf area index and the capture of photosynthetically active radiation (PAR) in mixed stands were previously reported by Nouvellon et al. (2012) for stands of *E. grandis* with *A. mangium*. The authors found that the total LAI (considering both species) was almost twice that of the eucalypt monoculture (E100); however, the increased LAI did not cause an increase in the gross primary production or wood production of mixed stands compared with the monoculture, and this result was attributed to water limitations at the site.

A net of experiments regarding replacements and additive series were replicated in three sites in the southeast of Brazil, Bofete and Luiz Antônio in São Paulo state, and Santana do Paraíso in Minas Gerais state (Bouillet et al. 2013). In the first years after planting eucalypt trees had a negative effect on acacia tree growth in all sites, as also was shown in a previous study conducted in Itatinga, located in São Paulo state (Laclau et al. 2008). However, the interspecific competition was less in São Paulo sites than in Minas Gerais probably because of more suitable climatic conditions (temperature and precipitation) in Santana do Paraíso, Minas Gerais state, for *A. mangium* development. The 50A:50E mixed stand evidenced greater eucalypt circumference growth, 52.1 cm in Santana do Paraíso, 56.8 cm in Bofete, 65.5 cm in Luiz Antônio, and 61.5 cm in Itatinga, respectively, compared to eucalypt in monocultures, either non-fertilized (47.8, 49.4, 53.4, and 50.7 cm, respectively) or fertilized with N (47.6, 48.8, 53.3, and 52.3 cm, respectively). Other replacements and additive series along the rotation at all sites evaluated in this study also showed greater growth for eucalypt (Tables 2.2 and 2.3). However, the eucalypt mean annual increment in 50A:50E was lower than in the additive series (25A:100E, 50A:100E, and 100A:100E) due to 50% less stocking density.

The same growth pattern was observed for tree growth in mixed (50A:50E and 100A:100E) and monospecific stands at Seropédica site, located in Rio de Janeiro state (Santos et al. 2016). Acacia evidenced limited competitive ability with eucalypt, with minor height and circumference growth found in the replacement treatments with greater density of eucalypt trees (Table 2.2). Eucalypt had a greater circumference growth (56.2 cm) in the replacement 50A:50E stand than in monocultures fertilized with nitrogen (51.8 cm). Even with a lower number of eucalypt trees, this arrangement resulted in an overall biomass production equal or less than the eucalypt monocultures with or without nitrogen. At both ages evaluated (30 and

60 months), the stemwood biomass production was lower in monocultures without nitrogen fertilization of both species (100A and 100E). The replacements 50A:50E and 100A:100E and eucalypt monocultures with nitrogen fertilization (100E+N) showed the highest increases in annual net primary production in comparison with monocultures of acacia and eucalypt.

These results showed the potential growth and productivity of the acacia and eucalypt mixed plantations as a result of the arrangement 50A:50E that had a better development compared to the other treatments, although with one half of the eucalypt population.

2.6 Belowground Biomass

Interactions occurring in the plant community root system can interfere with species diversity through competitive exclusion, niche partitioning, and facilitation (Schenk 2006). Some experiments have shown that plant roots interact with their biotic and abiotic environments using mechanisms that influence the availability of resources, exchange of various types of signals, and allelochemical interactions (Callaway 2002; Hierro and Callaway 2003; Semchenko et al. 2007). Some roots may detect other roots, or inert objects, and can distinguish between proper and non-proper roots. This has provided new experimental challenges to evaluate the effects of root competition on plant development (Semchenko et al. 2007). However, information on these mechanisms controlling root growth in forest environments is very limited (Kueffer et al. 2007), even more regarding belowground competition and fine-root density in mixed-species forests (Silva et al. 2011).

It is very important to understand the effects of inter- and intraspecific competition on root development to improve, e.g., the control of invasive tree species, and to model the forest dynamics (Leuschner et al. 2001; Kueffer et al. 2007) which can contribute to recommend sustainable management practices.

As shown previously, acacia growth is expected to be suppressed by eucalypt trees, depending on the region in Brazil (Bouillet et al. 2013; Santos et al. 2016). The mixtures may exploit site resources more completely through the development of a stratified canopy and soil niche separation by fine roots (Germon et al. 2018; Laclau et al. 2013; Kelty 2006; Forrester et al. 2006). In mixed plantations, the dominant species containing more fine roots, located closer to the soil surface, will have a competitive advantage over the dominated species, excluded from the resource-rich upper soil layer (Laclau et al. 2013).

Soil resources are localized along a strong vertical gradient of nutrient and water availability provided by rainfall and fertilizer application over the early growth and then throughout the biological cycle of nutrients after canopy closure. Most of the available nutrients are in the topsoil layer, particularly for forest plantations established in highly weathered tropical soils and transported by gravitational solutions (Laclau et al. 2003, 2010).

Belowground competition has been studied through the dynamics of the distribution of fine roots, but other factors that play an integral part of the interactions among species have been less studied yet, such as mycorrhizal associations (Pereira et al. 2018), which interfere in the availability of resources with low mobility (Zobel et al. 1997). Mixed plantation with eucalypt and acacia (50E50A) presented a significant increase in root colonization by arbuscular mycorrhizal fungi (*Glomus* genus) at the 0–20 and 20–50 cm soil layers, indicating a possible stimulation at superficial soil layers of the symbiosis in eucalypt roots when in consortium (Pereira et al. 2018).

Interspecific competition between *E. grandis* and leguminous trees (*Peltophorum dubium*, *Inga* sp., *Mimosa scabrella*, *Acacia polyphylla*, *Mimosa caesalpiniaefolia*, and *Acacia mangium*) was larger than intraspecific competition up to 24 months of age (Coelho et al. 2007). The *E. grandis* root system distribution relative to the distribution of the *M. scabrella* and *A. mangium* roots (the most resistant leguminous tree species to competition) in the soil profile indicated that there were different root exploration niches between species. Laclau et al. (2008) also verified that *Eucalyptus grandis* (Hill ex Maiden) and *Acacia mangium* (Willd.) tree roots occupied different niches of soils when in additive or replacement stands (Fig. 2.5).

Besides the diversified soil exploitation, fine roots showed overyielding of 27% down to 2 m of soil depth in 50A50E (445.3 g m^{-2}), when compared with 100A (352.0 g m^{-2}) and 100E (346.9 g m^{-2}) at 5 years after planting. In 50A:50E, eucalypt fine root biomass per tree was 72% greater than in 100E, whereas the opposite was found for acacia with fine root biomass per tree 17% lower than in 100A (Laclau et al. 2013). After 4 years of the replanting of the same experiment conducted by Laclau et al. (2013), total fine root biomass in 50A50E (1127 g m^{-2}) was 44% higher than in 100A (780 g m^{-2}) and 58% higher than in 100E (714 g m^{-2}) (Germon

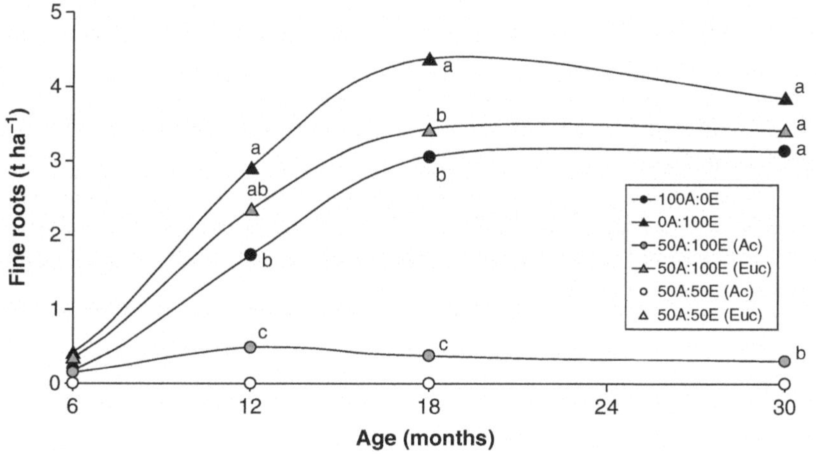

Fig. 2.5 Dynamics of fine root biomass accumulation in different additive and replacement stands between Acacia (A, Ac) and Eucalypt (E, Euc). Different letters at each age indicate significant differences in dry matter amounts ($P < 0.05$). Modified figure from Laclau et al. (2008)

et al. 2018). These results suggest that mixing acacia and eucalypt might lead to a strong fine root overyielding at very deep soil layers, increasing fine root exploration at deep soil layers compared to their respective monospecific stands, which is likely to enhance the uptake of soil resources. The authors found impressively higher concentrations of fine root biomass below 2 m of soil depth, with 50, 45, and 35%, respectively, for 100E, 50A50E, and 100A.

The reduction in acacia fine root density in 50A100E (addictive stand), when compared to the monospecific 100A0E acacia stands, in the upper soil layer at age 6–12 months after planting (Fig. 2.6), showed that belowground competition started roughly at the same time in the 50A100E treatment, due to a strong competition imposed by genetically improved eucalypt trees, relative to acacia (Bouillet et al. 2008), and because of the environmental conditions, cold for acacia trees. The highest fine root density of eucalypt in the upper soil layers occurred at 18 months of age and subsequent decreases indicated a decrease in intraspecific competition (Silva et al. 2011). Competition between eucalypt and acacia species occurred in the 50A100E stand through the horizontal distribution of the fine roots, with strong decreases of acacia roots at greater distances from the tree, whereas the eucalypt roots were not influenced by the presence of acacia. However, root competition among species in mixed treatments shifted root growth to non-favorable depths. The monospecific stands explored the soil similarly (Silva et al. 2011).

Whatever the tree stock density of eucalypt (50A100E vs. 50A50E), all authors reported a competitive exclusion of acacia fine roots from the upper soil layer (Laclau et al. 2013; Germon et al. 2018; Silva et al. 2009). According to Silva et al. (2009) and Laclau et al. (2013), the facilitation processes were weak because of the lack of aboveground transgressive overyielding in 50A50E (when the consortium production is greater than the most productive monoculture). Other factors involve the lack of influence of acacia trees on eucalypt distribution of fine roots in 50A:100E stands, besides N concentrations in eucalypt tree components, which were not significantly different from those of the 100E and 50A100E stand (Bouillet et al. 2008). However, in general, increases in soil N availability in mixed plantations of eucalypt trees planted with acacia (Tchichelle et al. 2017; Voigtlaender et al. 2012) explained the greater soil exploration by eucalypt fine roots in 50A:50E relative to 100E stands (Germon et al. 2018).

In this context, the recommendation is that comparative studies should be carried out mostly in areas highly depleted in N, where acacia trees should improve the growth of eucalypt trees through complementarity and facilitation mechanisms (Laclau et al. 2013).

2.7 Final Remarks

In this chapter we showed that mixed eucalypt plantations with acacia can increase biomass production in relation to eucalypt monocultures, especially in highly weathered soils (that are especially very poor in N) and with climatic conditions

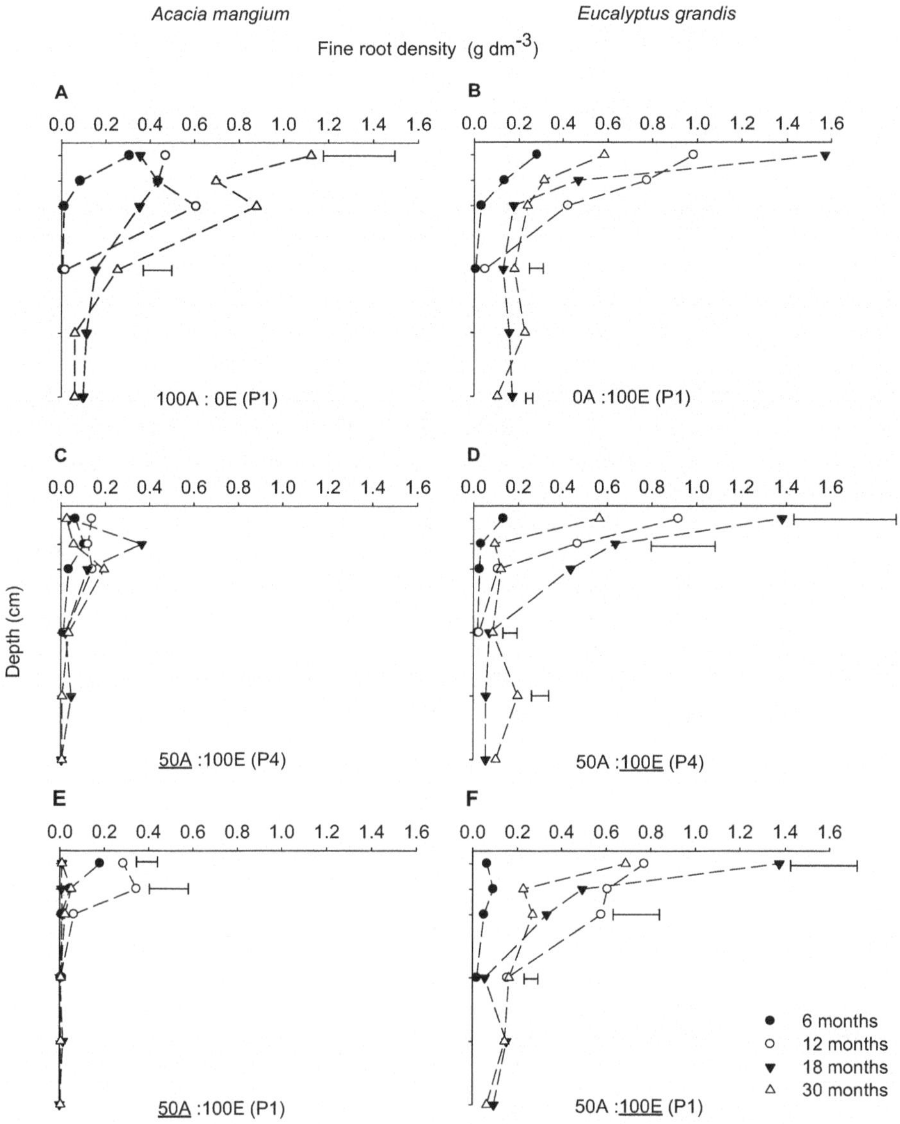

Fig. 2.6 Fine root densities according to their sampling position in the planting row: P1 position in 100A:0E close to *A. mangium* trees (**a**); P1 position in 0A:100E close to *E. grandis* trees (**b**); *A. mangium* fine roots in the P4 position in 50A:100E, close to *A. mangium* trees (**c**); *E. grandis* fine roots in the P4 position in 50A:100E, close to *A. mangium* trees (**d**); *A. mangium* fine roots in the P1 position in 50A:100E, close to *E. grandis* trees (**e**); and *E. grandis* fine roots in the P1 position in 50A:100E, close to *E. grandis* trees (**f**). The horizontal bars show the LSD values when the differences between treatments were significant ($P < 0.05$)

favorable to acacia (average annual temperatures higher than 24 °C and annual rainfall above 1000 mm). In mixed plantations, *Acacia mangium* can compete with eucalypt on equal terms or at least similar growth rates. However, for the facilitation process to occur between both species, it is necessary to develop strategic programs of *A. mangium* breeding similar to those developed for eucalypt in the last years.

Further studies should be developed to evaluate different spacing or planting arrangements that minimize interspecific competition in mixed plantations and, consequently, promote better growth and biomass production. For example, planting trees of each species in double lines instead of alternating plants could not only decrease the competition of eucalypt over acacia, but also facilitate the harvesting of trees. In addition, other questions should be answered to foster the understanding of mixed plantations in tropical conditions, such as the following: (1) What are the silvicultural practices (e.g., thinning, pruning, complementary fertilization) to ensure greater productivity? (2) Are there other legume tree species, specially native from Brazil, with productive potential that could be evaluated in the mixed planting system with eucalypt?

It is important to note that, in addition to the positive responses on growth and biomass production, mixed plantations can also result in several indirect benefits that can improve the long-term sustainability of the production system, such as increase in soil C sequestration, increase in soil N concentrations which will impact directly the dynamics of nutrient cycling, and increase in biodiversity and protection against pests and diseases, besides creating the diversification of timber and non-timber forest products, which are discussed in several chapters in this book.

References

Almeida JCR, Laclau JP, Gonçalves JLM, Ranger J, Saint-André L (2010) A positive growth response to NaCl applications in *Eucalyptus* plantations established on K-deficient soils. For Ecol Manag 259:786–1795. https://doi.org/10.1016/j.foreco.2009.08.032

Atipanumpai L (1989) *Acacia mangium*: studies on the genetic variation in ecological and physiological characteristics of fast-growing plantation tree species. Acta for Fenn 206:1–99. https://doi.org/10.14214/aff.7653

Austin MT, Brewbaker JL, Wheeler R, Fownes JH (1997) Short- rotation biomass trial of mixed and pure stands of nitrogen-fixing trees and *Eucalyptus grandis*. Aust for 60:161–168. https://doi.org/10.1080/00049158.1997.10676138

Bachega LR, Bouillet JP, Piccolo MC, Saint-Andre L, Bouvet JM, Nouvellon Y, Gonçalves JLM, Robin A, Laclau J-P (2016) Decomposition of *Eucalyptus grandis* and *Acacia mangium* leaves and fine roots in tropical conditions did not meet the Home Field Advantage hypothesis. For Ecol Manag 359:33–43. https://doi.org/10.1016/j.foreco.2015.09.026

Balieiro FC, Franco AA, Fontes RLF, Dias LE, Campello EF, Faria SM (2002) Accumulation and distribution of aboveground biomass and nutrients under pure and mixed stands of *Pseudosamanea guachapele* Dugand and *Eucalyptus grandis* W. Hill ex Maiden. J Plant Nutr 24/25:2639–2654. https://doi.org/10.1081/PLN-120015528

Balieiro FC, Franco AA, Pereira MG, Campello EFC, Dias LE, Faria SM, Alves BJR (2004) Contribution of litter and nitrogen to soil under *Pseudosamanea guachapele* and *Eucalyptus grandis* plantations. Pesquisa Agropecuária Brasileira 39:597–601. https://doi.org/10.1590/S0100-204X2004000600012

Balieiro FC, Pereira MG, Alves BJR, Resende AS, Franco AA (2008) Soil carbon and nitrogen in pasture soil reforested with *Eucalyptus* and *Pseudosamanea guachapele*. Rev Bras Cienc Solo 32:1253–1260. https://doi.org/10.1590/S0100-06832008000300033

Bauhus J, Khanna PK, Menden N (2000) Aboveground and belowground interactions in mixed plantations of *Eucalyptus globulus* and *Acacia mearnsii*. Can J For Res 30:1886–1894. https://doi.org/10.1139/cjfr-30-12-1886

Bertness M, Callaway RM (1994) Positive interactions in communities. Trends Ecol Evol 9:191–193. https://doi.org/10.1016/0169-5347(94)90088-4

Binkley D, Senock R, Bird S, Cole TG (2003) Twenty years of stand development in pure and mixed stands of *Eucalyptus saligna* and nitrogen-fixing *Falcataria moluccana*. For Ecol Manag 182:93–102. https://doi.org/10.1016/S0378-1127(03)00028-8

Bouillet JP, Laclau JP, Gonçalves JLM, Moreira MZ, Trivelin PCO, Jourdan C, Silva EV, Piccolo MC, Tsai SM, Galiana A (2008) Mixed-species plantations of *Acacia mangium* and *Eucalyptus grandis* in Brazil. 2: Nitrogen accumulation in the stands and biological N_2 fixation. For Ecol Manag 255:3918–3930. https://doi.org/10.1016/j.foreco.2007.10.050

Bouillet J-P, Laclau J-P, Gonçalves JLM, Voigtlaender M, Gava J, Leite FP, Hakamada R, Mareschal L, Mabiala A, Tardy F, Levillain J, Deleporte P, Epron D, Nouvellon Y (2013) *Eucalyptus* and *Acacia* tree growth over entire rotation in single- and mixed-species plantations across five sites in Brazil and Congo. For Ecol Manag 301:89–101. https://doi.org/10.1016/j.foreco.2012.09.019

Boyden S, Binkley D, Senock R (2005) Competition and facilitation between *Eucalyptus* and nitrogen-fixing *Falcataria* in relation to soil fertility. Ecology 86:992–1001. https://doi.org/10.1890/04-0430

Callaway RW (2002) The detection of neighbors by plants. Trends Ecol Evol 17:104–105. https://doi.org/10.1016/S0169-5347(01)02438-7

Chaer GM, Tótola MR (2007) Impact of organic residue management on soil quality indicators during replanting of eucalypt stands. Rev Bras Ciênc Solo 31:1381–1396. https://doi.org/10.1590/S0100-06832007000600016

Coelho SRF, Gonçalves JLM, Laclau JP, Mello SLM, Moreira RM, Silva EV (2007) Crescimento, nutrição e fixação biológica de nitrogênio em plantios mistos de eucalipto e leguminosas arbóreas. Pesquisa Agropec. Bras. 42:759–768. https://doi.org/10.1590/S0100-204X2007000600001

Corbeels M, McMurtrie RE, Pepper DA, Mendham DS, Grovet TS, O'Connell AM (2005) Long-term changes in productivity of eucalypt plantations under different harvest residue and nitrogen management practices: a modelling analysis. For Ecol Manag 217:1–18. https://doi.org/10.1016/j.foreco.2005.05.057

Epron D, Nouvellon Y, Mareschal L, Moreira RM, Koutika LS, Geneste B, Delgado-Rojas JS, Laclau J-P, Sola G, Gonçalves JLM, Bouillet J-P (2013) Partitioning of net primary production in eucalypt and acacia stands and in mixed-species plantations: two case-studies in contrasting tropical environments. For Ecol Manag 301:102–111. https://doi.org/10.1016/j.foreco.2012.10.034

Firn J, Erskine PD, Lamb D (2007) Woody species diversity influences productivity and soil nutrient availability in tropical plantations. Oecologia 154:521–533. https://doi.org/10.1007/s00442-007-0850-8

Forrester DI, Bauhus J, Khanna PK (2004) Growth dynamics in a mixed-species plantation of *Eucalyptus globulus* and *Acacia mearnsii*. For Ecol Manag 193:81–95. https://doi.org/10.1016/j.foreco.2004.01.024

Forrester DI, Bauhus J, Cowie AL (2005) On the success and failure of mixed-species tree plantations: Lessons learned from a model system of *Eucalyptus globulus* and *Acacia mearnsii*. For Ecol Manag 209:147–155. https://doi.org/10.1016/j.foreco.2005.01.012

Forrester DI, Bauhus J, Cowie AL, Vanclay JK (2006) Mixed-species plantations of *Eucalyptus* with nitrogen-fixing trees: A review. For Ecol Manag 233:211–230. https://doi.org/10.1016/j.foreco.2006.05.012

Forrester DI (2014) The spatial and temporal dynamics of species interactions in mixed-species forests: from pattern to process. For Ecol Manag 312:282–292. https://doi.org/10.1016/j.foreco.2013.10.003

Germon A, Guerrini I, Bordron B, Bouillet J-P, Nouvellon Y, Gonçalves J, Moreira R, Jourdan C, Laclau J-P (2018) Consequence of mixing *Acacia mangium* and *Eucalyptus grandis* trees on soil exploration by fine roots down to a depth of 17 m in a tropical planted forest. Plant Soil 424:203–220. https://doi.org/10.1007/s11104-017-3428-1

Gonçalves JLM, Stape JL, Benedetti V, Fessel VAG, Gava JL (2000) Reflexos do cultivo mínimo e intensivo do solo em sua fertilidade e nutrição das árvores. In: Gonçalves JLM, Benedetti V (eds) Nutrição e fertilização florestal. IPEF, Piracicaba, pp 1–57

Gonçalves JLM, Stape JL, Wichert MCP, Gava JL (2002) Manejo de resíduos vegetais e preparo de solo. In: Gonçalves JLM, Stape JL (eds) Conservação e cultivo de solos para plantações florestais. IPEF, Piracicaba, pp 131–204

Gonçalves JLM, Stape JL, Laclau J-P, Smethurst P, Gava JL (2004) Silvicultural effects on the productivity and wood quality of eucalypt plantations. For Ecol Manag 193:45–61. https://doi.org/10.1016/j.foreco.2004.01.022

Gonçalves JLM, Stape JL, Laclau J-P, Bouillet J-P, Ranger J (2008) Assessing the effects of early silvicultural management on long-term site productivity of fast-growing eucalypt plantations: the Brazilian experience. South For 70:105–118. https://doi.org/10.2989/SOUTH.FOR.2008.70.2.6.534

Gonçalves JLM (2011) Fertilização de plantação de eucalipto. In: Encontro Brasileiro de Silvicultura, 2. Campinas, São Paulo State, Brazil. Proceedings... Piracicaba: PTSM/IPEF/ESALQ/FUPEF, pp 85–113

Gonçalves JLM, Alvares CA, Higa AR, Silva LD, Alfenas AC, Stahl J, Ferraz SFB, Lima WP, Brancalion PHS, Hubner A, Bouillet J-P, Laclau J-P, Nouvellon Y, Epron D (2013) Integrating genetic and silvicultural strategies to minimize abiotic and biotic constraints in Brazilian eucalypt plantations. For Ecol Manag 301:6–27. https://doi.org/10.1016/j.foreco.2012.12.030

Gonçalves JLM, Duque SL, Behling M, Alvares CA (2014) Management of industrial forest plantations. In: Borges JG, Diaz-Balteiro L, McDill ME, Rodriguez LCE (eds) Managing forest ecosystems, 1st edn. Springer Netherlands, Dordrecht

Hierro JL, Callaway RM (2003) Allelopathy and exotic plant invasion. Plant Soil 256:29–39. https://doi.org/10.1126/science.1083245

Hunt MA, Battaglia M, Davidson NJ, Unwin GL (2006) Competition between plantation *Eucalyptus nitens* and *Acacia dealbata* weeds in northeastern Tasmania. For Ecol Manag 233:260–274. https://doi.org/10.1016/j.foreco.2006.05.017

Kelty MJ (2006) The role of species mixtures in plantation forestry. For Ecol Manag 233:195–204. https://doi.org/10.1016/j.foreco.2006.05.011

Koutika L-S, Tchichelle SV, Mareschal L, Epron D (2017) Nitrogen dynamics in a nutrient-poor soil under mixed-species plantations of eucalypts and acacias. Soil Biol Biochem 108:84–90. https://doi.org/10.1016/j.soilbio.2017.01.023

Krisnawati H, Kallio M, Kanninen M (2011) *Acacia mangium* Willd.: ecology, silviculture and productivity. CIFOR, Bogor

Kueffer C, Schumacher E, Fleischmann K, Edwards PJ, Dietz H (2007) Strong below-ground competition shapes tree regeneration in invasive *Cinnamomum verum* forests. J Ecol 95:273–282. https://doi.org/10.1111/j.1365-2745.2007.01213.x

Laclau J-P, Ranger J, Nzila JD, Bouillet J-P, Deleporte P (2003) Nutrient cycling in a clonal stand of *Eucalyptus* and an adjacent savanna ecosystem in Congo. 2. Chemical composition of soil solutions. For Ecol Manag 180:527–544. https://doi.org/10.1016/S0378-1127(02)00645-X

Laclau J-P, Bouillet JP, Gonçalves JLM, Silva EV, Jourdan C, Cunha MCS, Moreira MR, Saint-André L, Maquère V, Nouvellon Y, Ranger J (2008) Mixed-species plantations of *Acacia mangium* and *Eucalyptus grandis* in Brazil: 1. Growth dynamics and aboveground net primary production. For Ecol Manag 255:3905–3917. https://doi.org/10.1016/j.foreco.2007.10.049

Laclau J-P, Ranger J, Gonçalves JLM, Maquere V, Krusche AV, M'bou AT, Nouvellon Y, Saint-Andre L, Bouillet J-P, Piccolo MC, Deleporte P (2010) Biogeochemical cycles of nutrients in tropical eucalypt plantations. Main features shown by intensive monitoring in Congo and Brazil. For Ecol Manag 259:1771–1785. https://doi.org/10.1016/j.foreco.2009.06.010

Laclau JP, Nouvellon Y, Reine C, Gonçalves JLM, Krushe AV, Jourdan C, Le Maire G, Bouillet JP (2013) Mixing *Eucalyptus* and *Acacia* trees leads to fine root over-yielding and vertical segregation between species. Oecologia 172:903–913. https://doi.org/10.1007/s00442-012-2526-2

Le Maire G, Nouvellon Y, Christina M, Ponzoni F, Gonçalves JLM, Bouillet J-P, Laclau J-P (2013) Tree and stand light use efficiencies over a full rotation of single- and mixed-species *Eucalyptus grandis* and *Acacia mangium* plantations. For Ecol Manag 288:31–42. https://doi.org/10.1016/j.foreco.2012.03.005

Leuschner C, Hertel D, Coners H, Büttner V (2001) Root competition between beech and oak: a hypothesis. Oecologia 126:276–284. https://doi.org/10.1007/s004420000507

Lima WP, Laprovitera R, Ferraz SFB, Rodrigues CB, Silva MM (2012) Forest plantations and water consumption: a strategy for hydrosolidarity. Int J Forest Res 2012:8p. https://doi.org/10.1155/2012/908465

Little KM, Schumann AW, Noble AD (2002) Performance of a *Eucalyptus grandis* × *E. camaldulensis* hybrid clone as influenced by a cowpea cover crop. For Ecol Manag 168:43–52. https://doi.org/10.1016/S0378-1127(01)00728-9

Loreau M, Hector A (2001) Partitioning selection and complementarity in biodiversity experiments. Nature 412:72–76. https://doi.org/10.1038/35083573

Maquère V, Laclau J-P, Gonçalves JLM, Krushe AV, Piccolo MC, Gine MF, Ranger J (2005) Influence of fertilizer inputs on soil solution chemistry in eucalypt plantations established on Brazilian sandy soils. In: Proceedings of the workshop on management of tropical sandy soils for sustainable agriculture: a holistic approach for sustainable development of problem soils in the tropics. International Union of Soil Science, Khon Kaen, Khon Kaen, pp 466–471

Medhurst J, Beadle C, Neilsen W (2001) Early-age and later-age thinning affects growth, dominance, and intraspecific competition in *Eucalyptus nitens* plantations. Can J For Res 31(2):187–197. https://doi.org/10.1139/cjfr-31-2-187

Nambiar EKS, Kallio MH (2008) Increasing and sustaining productivity in tropical forest plantations: making a difference through cooperative research partnership. In: Nambiar EKS (ed) Site management and productivity in tropical plantation forests: workshop proceedings, 22–26 November 2004, Piracicaba, Brazil, and 6–9 November, Bogor, Indonesia. Center for International Forestry Research, Bogor, pp 205–227

Nouvellon Y, Ranger J, Bouillet J-P (2012) Introducing *Acacia mangium* trees in Eu*calyptus grandis* plantations: consequences for soil organic matter stocks and nitrogen mineralization. Plant Soil 352:99–111. https://doi.org/10.1007/s11104-011-0982-9

Paula RR, Bouillet J-P, Trivelin PCO, Zeller B, Gonçalves JLM, Nouvellon Y, Bouvet JM, Plassard C, Laclau J-P (2015) Evidence of short-term belowground transfer of nitrogen from *Acacia mangium* to *Eucalyptus grandis* trees in a tropical planted forest. Soil Biol Biochem 91:99–108. https://doi.org/10.1016/j.soilbio.2015.08.017

Paula RR, Bouillet J-P, Gonçalves JLM, Trivelin PC, Balieiro FC, Nouvellon Y, Oliveira JC, Deus JC, Bordron B, Laclau J-P (2018) Nitrogen fixation rate of *Acacia mangium* Wild at mid rotation in Brazil is higher in mixed plantations with *Eucalyptus grandis* Hill ex Maiden than in monocultures. Ann Forest Sci 75:14–14. https://doi.org/10.1007/s13595-018-0695-9

Pereira APA, Santana MC, Bonfim JA, Mescolotti DL, Cardoso EJBN (2018) Digging deeper to study the distribution of mycorrhizal arbuscular fungi along the soil profile in pure and mixed *Eucalyptus grandis* and *Acacia mangium* plantations. Appl Soil Ecol 128:1–11. https://doi.org/10.1016/j.apsoil.2018.03.015

Resh SC, Binkley D, Parrotta JA (2002) Greater soil carbon sequestration under nitrogen - fixing trees compared with *Eucalyptus* species. Ecosystems 5:217–231. https://doi.org/10.1007/s10021-001-0067-3

Richards AE, Forrester DI, Bauhus J, Scherer-Lorenzen M (2010) The influence of mixed tree plantations on the nutrition of individual species: a review. Tree Physiol 30(9):1192–1208. https://doi.org/10.1093/treephys/tpq035

Rocha JHT, Gonçalves JLM, Gava JL, Godinho TO, Melo EASC, Bazani JH, Hubner A, Arthur Junior JC, Wichert MP (2016) Forest residue maintenance increased the wood productivity of a Eucalyptus plantation over two short rotations. For Ecol and Manage 379:1–10. https://doi.org/10.1016/j.foreco.2016.07.042

Rocha JHT, Gonçalves JLM, Brandani CB, Ferraz AV, Franci AF, Marques ERG, Arthur Junior JC, Hubner A (2018) Forest residue removal decreases soil quality and affects wood productivity even with high rates of fertilizer application. For Ecol Manag 430:188–195. https://doi.org/10.1016/j.foreco.2018.08.010

Salgado TP, Alves PLCA, Kuva MA, Takahashi EN, Dias TCS, Lemes LN (2011) Sintomas de intoxicação inicial de *Eucalyptus* proporcionados por subdoses de glyphosate aplicadas no caule ou nas folhas. Planta Daninha 29:913–922. https://doi.org/10.1590/S0100-83582011000400022

Santos FM, Balieiro FC, Ataíde DHS, Diniz AR, Chaer GM (2016) Dynamics of aboveground biomass accumulation in monospecific and mixed-species plantations of *Eucalyptus* and *Acacia* on a Brazilian sandy soil. For Ecol Manag 363:86–97. https://doi.org/10.1016/j.foreco.2015.12.028

Santos FM, Chaer GM, Diniz AR, Balieiro FC (2017a) Nutrient cycling over five years of mixed-species plantations of *Eucalyptus* and *Acacia* on a sandy tropical soil. For Ecol Manag 384:110–121. https://doi.org/10.1016/j.foreco.2016.10.041

Santos FM, Balieiro FC, Fontes MA, Chaer GM (2017b) Understanding the enhanced litter decomposition of mixed-species plantations of *Eucalyptus* and *Acacia mangium*. Plant Soil 423:1/2) 1–1/2)15. https://doi.org/10.1007/s11104-017-3491-7

Schenk HJ (2006) Root competition: beyond resource depletion. J Ecol 94:725–739. https://doi.org/10.1111/j.1365-2745.2006.01124.x

Scolforo JRS (1997) Manejo florestal. Universidade Federal de Lavras; Fundação de Apoio ao Ensino, Pesquisa e Extensão, Lavras, p 433

Semchenko M, Hutchings MJ, John EA (2007) Challenging the tragedy of the commons in root competition: confounding effects of neighbor presence and substrate volume. J Ecol 95:252–260. https://doi.org/10.1111/j.1365-2745.2007.01210.x

Silva FP, Borges RCG, Pires IE (1996) Avaliação de procedências de *Acacia mangium* Willd, aos 63 meses de idade, no Vale do Rio Doce-MG. Revista Árvore 20(3):299–308

Silva EV, Gonçalves JLM, Coelho SR, Moreira RM, Mello SLM, Bouillet J-P, Jourdan C, Laclau J-P (2009) Dynamics of fine root distribution after establishment of monospecific and mixed-species plantations of *Eucalyptus grandis* and *Acacia mangium*. Plant Soil 325:305–318. https://doi.org/10.1007/s11104-009-9980-6

Silva EV, Bouillet J-P, Gonçalves JLM, Abreu Junior CH, Trevelin PCO, Hinsinger P, Jourdan C, Nouvellon Y, Stape JL, Laclau J-P (2011) Functional specialization of *Eucalyptus* fine roots: contrasting potential uptake rates for nitrogen, potassium and calcium tracers at varying soil depths. Funct Ecol 25:996–1006. https://doi.org/10.1111/j.1365-2435.2011.01867.x

Souza CR, Rossi LMB, Azevedo CP, Lima RMB (2004) Comportamento da *Acacia mangium* e de clones de *Eucalyptus grandis* x *E. urophylla* em plantios experimentais na Amazônia Central. Sci Forest 65:95–101

Stape JL (2002) Production ecology of clonal *Eucalyptus* plantations in northeastern Brazil. PhD thesis. Colorado State University, Ft. Collins, p 225

Stape JL, Binkley D, Ryan MG, Fonseca S, Loose RA, Takahashi EN, Silva CR, Silva SR, Hakamada RE, Ferreira JMDA, Lima AMN, Gava JL, Leite FP, Andrade HB, Alves JM, Silva GGC, Azevedo MR (2010) The Brazil *Eucalyptus* potential productivity project: influence of water, nutrients and stand uniformity on wood production. For Ecol Manag 259:1684–1694. https://doi.org/10.1016/j.foreco.2010.01.012

Tchichelle SV, Epron D, Mialoundama F, Koutika LS, Harmand J-M, Bouillet J-P, Mareschal L (2017) Differences in nitrogen cycling and soil mineralisation between a eucalypt plantation and a mixed eucalypt and *Acacia mangium* plantation on a sandy tropical soil. Southern Forests 79(1):1–8. https://doi.org/10.2989/20702620.2016.1221702

Tonini H (2010) Características em plantios e propriedades da madeira de *Acacia mangium*. In: Tonini H, Halfeld-Viera BA, Silva SJR (eds) *Acacia mangium*: Características e seu cultivo em Roraima. Embrapa Informação Tecnológica, Brasília, p 63

Vandermeer J (1989) The ecology of intercropping. Cambridge University Press, New York

Voigtlaender M, Laclau J-P, Gonçalves JLM, Piccolo MC, Moreira MZ, Nouvellon Y, Ranger J, Bouillet J-P (2012) Introducing *Acacia mangium* trees in *Eucalyptus grandis* plantations: Consequences for soil organic matter stocks and nitrogen mineralization. Plant Soil 352:99–111. https://doi.org/10.1007/s11104-011-0982-9

Voigtlaender M, Brandani CB, Caldeira DRM, Tardy F, Bouillet J-P, Goncalves JLM, Moreira MZ, Leite FP, Brunet D, Paula RR, Laclau J-P (2019) Nitrogen cycling in monospecific and mixed-species plantations of *Acacia mangium* and *Eucalyptus* at 4 sites in Brazil. For Ecol Manag 43:56–67. https://doi.org/10.1016/j.foreco.2018.12.055

Zobel M, Moora M, Haukioja E (1997) Plant coexistence in the interactive environment: Arbuscular mycorrhiza should not be out of mind. Oikos 78:202–208. https://doi.org/10.2307/3545818

Chapter 3
Nutrient Cycling in Mixed-Forest Plantations

José Henrique Tertulino Rocha, José Leonardo de Moraes Gonçalves, and Alexandre de Vicente Ferraz

3.1 Introduction

Nutrient cycling in forests was defined by Attiwill and Adams 1993 as the range of natural processes that govern the availability of nutrients for the forest trees, as well as the interactions between plants and soil in the uptake and return of nutrients, microbial interactions in which nutrients are transformed between organic and inorganic forms, and balance between input and output of nutrients. Thus, nutrient cycling is a term used to cover all the pathways and processes by which nutrients enter, leave, and move within forest ecosystems.

In planted or managed natural forest for wood production the nutrient cycle is open once a large amount of nutrients is removed with harvest and in some places large amounts of nutrients are applied through fertilizers. The magnitude of the nutrient output and, consequently, the dependence on fertilizer application increase with the management intensity. In Brazil, most of the wood consumed and exported comes from planted forest managed in short rotation (5–7 years) with high productivity (from 20 to 80 m^3 ha^{-1} $year^{-1}$). The main genus planted is *Eucalyptus*. This system of production is highly efficient and highly productive, but highly dependent on fertilizer application. This dependency is intensified because of highly weathered soils, poor in nutrients or plantation established at steep sites susceptive to soil erosion. As commented in other chapters an alternative to reduce the dependence of fertilizer is the introduction of nitrogen-fixing trees (NFT) into eucalypt plantations.

J. H. T. Rocha (✉)
Agriculture and Forest Engineering College (FAEF), Garça, SP, Brazil

J. L. M. Gonçalves
Department of Forest Sciences, University of São Paulo, "Luiz de Queiroz" College of Agriculture, Piracicaba, SP, Brazil

A. V. Ferraz
Institute of Forest Science and Research (IPEF), Piracicaba, SP, Brazil

E. J. Bran Nogueira Cardoso et al. (eds.), *Mixed Plantations of Eucalyptus and Leguminous Trees*, https://doi.org/10.1007/978-3-030-32365-3_3

In natural ecosystems, the wood productivity is lower and there is no dependence on fertilizer application; on the other hand, in eucalypt plantation, the high wood productivity increases the dependence in terms of fertilizer application. As suggested in this book, perhaps the sustainable alternative is somewhere in between the natural forest and the traditional monospecific eucalypt plantation. Thus, the main goal of this chapter is to compare the nutrient cycling in mixed forest (mainly *Acacia* with *Eucalyptus*) with monospecific plantation and natural forest (Atlantic Forest and Cerrado—Brazilian savannah). Nutrient cycling can be divided into six main stages, as suggested by Attiwill and Adams (1993). They are (1) inputs of nutrients by rain, dust, biological fixation, and parental rock weathering; (2) uptake and accumulation of nutrients by trees; (3) outputs of nutrients by leaching and gaseous forms and in harvested material; (4) internal redistribution of nutrients within and among plants; (5) return of nutrients from plant to soil; and (6) decomposition of the forest floor and nutrient mineralization. Some of these subjects are presented in more details in Chaps. 4 and 6.

3.2 Nutrient Inputs

In tropical planted forest, fertilizer application is frequently the main nutrient input into the system, but atmospheric deposition, biological N_2 fixation, and parental rock weathering can also play an important role. Biological N_2 fixation (BNF) is discussed in another chapter. In this chapter, we discuss the role of atmospheric deposition and parental rock weathering.

3.2.1 Atmospheric Deposition

In forest plantations the importance of atmospheric deposition increases with the annual deposition rate and with the length of the rotation (Ranger and Turpault 1999; du Toit et al. 2014). The main sources of nutrients contributing to atmospheric deposition are mineral and marine aerosols, wildfires, industrial activity, combustion of fossil fuel, and agricultural activity (Wieder et al. 2016; Lequy et al. 2014; Nyaga et al. 2013).

The amount of nutrients deposited is highly dependent on the source, and highly variable in the spatial and temporal scale. In the literature, we found references to annual depositions of N, P, K, Ca, Mg, and S ranging from 1 to 10, 0.1 to 5, 1 to 25, 1 to 30, 0.3 to 3, and 1 to 10 kg ha^{-1} $year^{-1}$, respectively (Table 3.1). The deposition reduced exponentially with the distance from the emission center. Unlike N and S deposition, the K, Ca, and Mg depositions occur more concentrated around the emission center (Wieder et al. 2016; Nyaga et al. 2013). Wieder et al. (2016) found small Ca and Mg deposition rates 69 km from the emission center, while for N and S a small deposition rate was present even in the most distantly assessed point (251 km).

Table 3.1 Nutrient deposition by rainfall (kg ha^{-1} year^{-1})

Location	N	P	K	Ca	Mg	S	Source
Alberta, Canada	1.2	0.1		9.0	2.0	5.0	Wieder et al. (2016)
Northeastern France		0.5	1.1	0.8	0.3		Lequy et al. (2014)
West coast of South Africa	3.9	0.1	3.0	15	5		Nyaga et al. (2013)
São Paulo, Brazil	4.1		3.6	9.3	1.5		Laclau et al. (2010)
Kondi, Congo	5.4		6.2	7.3	3.1		Laclau et al. (2010)
São Paulo, Brazil	4.2		4.4	7.4	2		Vital et al. (1999)
Rio de Janeiro, Brazil	15.5						de Souza et al. (2017)
Average	**3.8**	**0.2**	**3.7**	**8.1**	**2.3**	**5.0**	

3.2.2 *Rock Weathering*

Rock weathering rates are difficult to be quantified and frequently low in relation to the rotation scale. Many methods have been proposed, and despite the good relationship among the results, the accuracy of absolute data is uncertain (Hodson and Langan 1999; Klaminder et al. 2011; Koseva et al. 2010; Ouimet and Duchesne 2005; Whitfield et al. 2006, 2011). Generally, in highly weathered soils, with low levels of primary minerals, nutrient inputs by weathering are effectively negligible (Melo et al. 2005).

In young and shallow soils, rich in primary minerals and where trees are grown in long rotations, this nutrient input can be important for nutrient supply to the stand. Starr and Lindroos (2006) assessed the rate of Ca and Mg released by weathering in a soil chronosequence ranging from 340 to 5279 years of age in Finland with the same parent material under *Pinus sylvestris* forests. They found releases of around 2.0 and 0.6 kg ha^{-1} year^{-1} of Ca and Mg, respectively, in the youngest soil, and releases of 0.4 and 0.2 kg ha^{-1} year^{-1} of Ca and Mg in older soil. There was a drastic reduction in the Ca and Mg release up to soil ages of 1000 years followed by stabilization thereafter. Under an 80-year-old *P. sylvestris* forest in Finland, Starr et al. (2014) found weathering rates (1.8, 0.5, and 0.7 kg ha^{-1} year^{-1} of Ca, Mg, and K, respectively) which almost equaled the leaching rate. They reported that the quantities of exchangeable cations at the 0–40 cm soil layer are equivalent to approximately 30 years of weathering and the quantities accumulated in the aboveground biomass are equivalent to almost 50 years of weathering. These data sets indicate that in some sites weathering on its own is sufficient to supply the K, Ca, and Mg requirements of the trees. However, in Oxisols in tropical climate, the release of K, Ca, and Mg is very close to zero (Melo et al. 2005).

3.3 Nutrient Uptake and Accumulation

The nutrient uptake and consequent accumulation in the biomass are linearly related with the growth rate in planted forest, but are also related with the site nutrient availability (Gonçalves et al. 2014; Rocha et al. 2019). In natural unmanaged forest on steady state, the uptake rate is equal to the nutrient return to soil and the amount of nutrients accumulated in the biomass is proportional to the biomass stock.

The amount of nutrients accumulated in the biomass is equally affected by the species composition. Santos et al. (2017), comparing mixed-species and monospecific plantations of *Acacia mangium* and *Eucalyptus* (hybrid between *E. urophylla* and *E. grandis*), found 354 t ha^{-1} of aboveground biomass and 268 kg ha^{-1} of N accumulated in this biomass in monospecific eucalypt at 5-year-old stands. At the same site and age, these authors found, in the monospecific acacia plantation, 107 t ha^{-1} of aboveground biomass and 186 kg ha^{-1} of N accumulated. The overall N concentration of the aboveground biomass of eucalypt was 2.19 g kg^{-1} and of acacia was 2.44 g kg^{-1}. If the productivity of both stands were the same, the N accumulation in the acacia aboveground biomass would be 11% bigger than in the eucalypt plantation. When we look at the overall P concentration in the aboveground biomass for the same productivity, the eucalypt monospecific plantation accumulates around 33% less P than acacia monospecific plantation. Due to genetic differences among the species, when we mix acacia and eucalypt in a plantation, there is an increase in the N, P, K, Ca, and Mg content per ton of biomass produced (Santos et al. 2017).

The mixture of *A. mangium* with *Eucalyptus* plantation increases the fine root biomass and consequently the soil exploration, especially in deep soil layers (see also Chap. 2). Germon et al. (2018) studying soil exploration by fine roots down to a depth of 17 m found an increase of 58% in the root biomass when eucalypt was mixed with *acacia* (50%E 50%A), when compared with monospecific *Eucalypt* plantation. Beyond the root biomass, they also found an increase of 50% in the root specific area (cm^2 g^{-1}) of acacia in mixed plantations when compared with acacia in monospecific plantation. In this study, the root of eucalypt dominated the upper soil layer and "forced" acacia to increase the root density in deeper soil layers. The root front of the monospecific acacia plantation was down to 12 m while under mixed plantation acacia roots reached 17 m.

3.4 Nutrient Outputs

The harvest output increased linearly with stand productivity and with harvest intensity. Rocha et al. (2019), based on 45 stands, estimated the nutrient harvest output for three levels of productivity and two levels of harvest intensity for monospecific eucalypt plantations (Table 3.2). Santos et al. (2017) assessed the harvest outputs of 5-year-old monospecific and mixed plantations of eucalypt with acacia in Rio de Janeiro state, Brazil. The productivity of plantations was 110, 50, and 80 t ha^{-1} of stem wood, when comparing monospecific eucalypt, acacia, and mixed plantation, respectively. The nutrient harvest outputs of the monospecific *eucalypt* plantation were higher than those in the mixed plantation due to the higher productivity (Table 3.3).

Table 3.2 Nutrient outputs by harvesting[a] in eucalypt plantations (with and without bark) in rotations of 7 years and mean annual increment (MAI) ranging from 30 to 50 m^3 ha^{-1} $year^{-1}$

	MAI (m^3 ha^{-1} $year^{-1}$)		
	30	40	50
Nutrient	kg ha^{-1}		
Wood with bark			
N	198	264	330
P	41	54	67
K	116	155	194
Ca	202	270	338
Mg	23	31	39
S	37	49	61
Wood			
N	168	224	280
P	32	42	53
K	66	88	110
Ca	83	110	138
Mg	12	16	20
S	34	45	56

[a]Adapted from Rocha et al. (2019)

Table 3.3 Biomass and nutrient outputs[a] by harvest of a monospecific eucalypt plantation (hybrid between *E. urophylla* and *E. grandis*—100E), a monospecific *Acacia mangium* plantation (100A), and a mixed eucalypt with acacia plantation (50E50A), all 5 years old, harvested in the system of only stem and full tree

	Stem			Full tree		
Biomass/nutrient	100E	100A	50E50A	100E	100A	50E50A
Biomass (t ha^{-1})	110	50	80	123	63	95
N (kg ha^{-1})	120	62	92	269	187	232
P (kg ha^{-1})	13	10	12	23	23	24
K (kg ha^{-1})	98	43	74	189	135	165
Ca (kg ha^{-1})	84	56	76	130	125	137
Mg (kg ha^{-1})	19	13	17	40	39	41

[a]Adapted from Santos et al. (2017)

Beyond harvest outputs, other nutrient losses can be significant in forest plantation. The soil loss by erosion under eucalypt plantation managed by minimum tillage is low, ranging from 0 to 2 t ha^{-1} $year^{-1}$ and being influenced by the age and management of the plantation (Martins et al. 2003; Silva et al. 2011). The soil loss under acacia plantation is also low, around 1 t ha^{-1} $year^{-1}$ (Barros et al. 2009). Due to the depth of the root system and the low deep-water drainage, nutrient leaching under forest plantation is negligible (Laclau et al. 2013; Christina et al. 2017). Ammonia volatilization in forest plantations in Brazil is also negligible, because these plantations are established normally on acidic soils.

3.5 Nutrient Redistribution Within and Among Plants

Nutrient redistribution or biochemical nutrient cycling is a well-known process in deciduous trees as well as in evergreen trees. The nutrients differ greatly in their mobility. Calcium, for example, is considered immobile, because it is a structural element, while K is highly mobile due to being a nonstructural element. Some authors found that under conditions of high nutrient availability, the retranslocation tends to be reduced (Boerner 1984; Pugnaire and Chapin 1993; Andrews et al. 1999), but others found no nutrient retranslocation (Millard and Proe 1993).

Among species, the N retranslocation rate is higher in eucalypt trees, K and P retranslocation rate is higher in acacia trees, and the Mg retranslocation rate is equal in both species (Santos et al. 2017). These authors found no difference in the retranslocation rate of both species, when comparing mixed with monospecific plantations. Since the K and P retranslocation rates are higher in acacia trees, the introduction of this species in monospecific eucalypt plantations can increase the nutrient retranslocation (Table 3.4).

Beyond the nutrient retranslocation within the trees, the nutrient retranslocation among trees can play an important role in the nutrition of mixed plantations, especially when there are NFTs. Paula et al. (2015), using ^{15}N, found transference from acacia to eucalypt trees 5 days after application among trees 6 m away from each other in a mixed plantation located in Itatinga, Brazil. These authors concluded that the transference belowground may provide a significant amount of N requirement of the tree close to NFT. This transference may be direct, when roots of eucalypt and acacia are connected by mycorrhizal network, or indirect, by root exudation of N compounds (See also Chap. 6).

3.6 Return of Nutrients from Plants to Soil

Monospecific eucalypt plantation returns to soil on average 5.6 t ha^{-1} $year^{-1}$ of litter (Table 3.5). The litterfall rate normally increases until 3 to 4 years of age and stabilizes or shows a little reduction afterwards (Rocha 2017). This litterfall rate results in a return to the soil of around 45, 2, 16, 40, 12, and 5 kg ha^{-1} $year^{-1}$ of N, P, K, Ca, Mg, and S, respectively. When compared with the native Atlantic Forest these amounts are markedly lower, especially for the nutrients, which indicates a lower nutrient concentration in the eucalypt litterfall (Table 3.5).

In mixed plantations, there is an increase in the total amount of nutrients deposited on the soil, especially N and P. This higher nutrient deposition is a result of higher nutrient concentration in the litter and of a higher litterfall rate (Table 3.5). These findings indicate that the introduction of acacia into monospecific eucalypt plantation accelerates and increases the nutrient cycling as also evidenced by Binkley (1992) and Forrester et al. (2005).

Table 3.4 Nutrient retranslocation rate[a] of *Eucalyptus* (hybrid between *E. urophylla* and *E. grandis*) and *Acacia mangium* trees at 30 and 60 months after planting

	N	P	K	Mg
Age (month)	%			
Eucalyptus				
30	77	68	61	34
60	51	70	70	43
A. mangium				
30	62	84	74	46
60	45	83	81	37

[a]Adapted from Santos et al. (2017)

Table 3.5 Litterfall rate and amount of nutrients deposited on the soil by litterfall in monospecific eucalypt plantation, Natural Forests, and in a trial which compares monospecific eucalypt plantation (100% eucalypt) with mixed-species plantation (50% eucalypt, 50% acacia)

			N	P	K	Ca	Mg	S	
Species	Age (year)	Mass ($t\ ha^{-1}\ year^{-1}$)	$kg\ ha^{-1}\ year^{-1}$						Source [a]
Monospecific eucalypt plantation									
E. grandis and hybrid[b]	1–9	5.6[c] (3.8–7.8) [14]	44.0 (24.0–83.5) [14]	1.9 (0.9–5.1) [14]	15.8 (4.4–44.2) [14]	39.4 (11.2–84.0) [14]	11.7 (7.0–16.2) [13]	4.9 (2.5–8.1) [6]	1, 2, 3, 4, 5, 6, 7, and 8
Eucalypt with acacia trials									
100% Eucalypt	2–6	8.5 (5.0–11.5) [8]	49.5 (30.0–62.0) [8]	5.3 (1.8–8.8) [4]	15.6 [1]	30.0 [1]	8.8 [1]	–	9, 10, 11, and 12
50% Eucalypt 50% Acacia	2–6	8.7 (6.1–11.0) [8]	70.7 (63.0–80.0) [8]	6.2 (1.7–10.7) [4]	18.8 [1]	33.2 [1]	9.0 [1]	–	9, 10, 11, and 12
Natural Forest									
Atlantic Forest		9.1 (6.3–13.0) [10]	169.6 (122–218.9) [10]	5.9 (1.6–11.6) [10]	44.3 (11.7–67.7) [10]	148.2 (88.9–231.1) [9]	25.8 (11.0–38.7) [9]	13.6 (13.5–13.6) [2]	6, 13, 14, 15, 16, 17, 18, and 19
Cerrado		3,8 (2,1–7,8) [4]	34.4 (12.7–64.7) [4]	2.1 (0.4–4.7) [4]	6.3 (2.3–12.5) [4]	14.6 (4.7–26.5) [4]	5.2 (1.9–10.9) [4]	0.7 [1]	19 and 20

[a]1—Gonçalves et al. (2000), 2—Zaia and Gama-Rodrigues (2004), 3—Cunha et al. (2005), 4—Ferraz (2009), 5—Silva (2006), 6—Gama-Rodrigues and Barros (2002), 7—Silva et al. (2013), 8—Rocha (2017), 9—Voigtlaender et al. (2019), 10—Koutika et al. (2014), 11—Santos et al. (2016), 12—Santos et al. (2017), 13—Vital et al. (2004), 14—Pinto et al. (2009), 15—Pimenta et al. (2011), 16—Godinho et al. (2013), 17—Domingos et al. (1997), 18—Pereira et al. (2008), 19—Toledo et al. (2002), 20—Nardoto et al. (2006)

[b]Hybrid between *E. grandis* and *E. urophylla*

[c]Average, followed by the amplitude between parentheses and followed by the number of sites plus the number of years assessed between brackets

Table 3.6 Litterfall, litter layer, decomposition rate (k), half lifetime, and decomposition time of 95% of the litter in monospecific eucalypt plantation and in Natural Forests

Species	Age (year)	Litterfall $t\ ha^{-1}\ year^{-1}$	Layer ($t\ ha^{-1}$)	*k*	Decomposition time 50% (year)	Decomposition time 95% (year)	Source[a]
Monospecific eucalypt plantation							
E. grandis and hybrid[b]	1-9	5.6[b] (3.8–7.8) [10]	11.6 (3.9–23.7) [10]	0.63 (0.23–1.2) [10]	1.46 (0.58–2.97) [10]	6,31 (2,50–12,84) [10]	1, 2, 3, 4, and 5
Natural Forest							
Atlantic Forest		8.5 (6.3–10,6) [8]	6.0 (3.4–10.1) [8]	1.53 (0.93–2.45) [8]	0.49 (0.28–0.74) [8]	2.10 (1.22–3.22) [8]	5, 6, 7, 8, 9, and 10

[a]1—Zaia and Gama-Rodrigues (2004), 2—Cunha et al. (2005), 3—Ferraz (2009), 4—Gonçalves et al. (2000), 5—Gama-Rodrigues and Barros (2002), 6—Vital et al. (2004), 7—Morellato (1992); 8—Pinto et al. (2009), 9—Pimenta et al. (2011), 10—Godinho et al. (2013)

[b]Average, followed by the amplitude between parentheses and followed by the number of sites plus the number of years assessed between brackets

3.7 Decomposition of Forest Litter

We will discuss litter layer decomposition in detail in the next chapter. In this topic, we will be comparing only the litter decomposition in eucalypt stands with the natural vegetation. Under monospecific eucalypt plantation the litterfall and litter layer rates are around 5.5 $t\ ha^{-1}\ year^{-1}$ and 11.6 $t\ ha^{-1}$, and, under Atlantic Forest, 8.6 $t\ ha^{-1}\ year^{-1}$ and 6.0 $t\ ha^{-1}$, respectively. The decomposition rate (k) of the Atlantic Forest litter is 2.4 times greater than the k of eucalypt plantation (Table 3.6).

When NFTs are mixed with eucalypt despite an increase in the litter N and P concentration and a reduction in the concentration of phenol, the *k* does not necessarily increase (Bachega et al. 2016). A large increase in the N mineralization under NFT in monospecific or mixed plantations was detected (Voigtlaender et al. 2012, 2019). On the other hand, as discussed in Chap. 4, changes in decomposition rates could be site specific.

3.8 Conclusion

The introduction of NFT, such as *Acacia mangium*, at monospecific eucalypt stands can improve the capacity of the trees in obtaining nutrients, mainly due to the atmospheric N_2 fixation and by the wider soil exploration. The NFT also accelerates and increases nutrient cycling and contributes to a large return of nutrients to soil by litterfall, increasing the topsoil nutrient availability. The N mineralization increases greatly. Thus, the dependence of mixed plantations on nitrogen fertilizer application is lower. More studies need to be incentivized, encompassing other NFT species.

The concentration of some nutrients in the acacia biomass is higher than that in eucalypt biomass. If mixed plantations reach the same productivity of monospecific eucalypt plantation, an increase in the nutrient harvest output can occur.

References

Andrews JA, Siccama TG, Vogt KA (1999) The effect of soil nutrient availability on retranslocation of Ca, Mg and K from senescing sapwood in Atlantic white cedar. Plant Soil 208(1):117–123. https://doi.org/10.1023/A:1004512317397

Attiwill PM, Adams MA (1993) Nutrient cycling in forests. New Phytol 124:561–582. https://doi.org/10.1111/j.1469-8137.1993.tb03847.x

Bachega LR, Bouillet JP, Piccolo MC, Saint-Andre L, Bouvet JM, Nouvellon Y, Gonçalves JLM, Robin A, Laclau J-P (2016) Decomposition of *Eucalyptus grandis* and *Acacia mangium* leaves and fine roots in tropical conditions did not meet the Home Field Advantage hypothesis. For Ecol Manag 359:33–43. https://doi.org/10.1016/j.foreco.2015.09.026

Barros LD, do Vale JF, Schaefer C, Mourão M (2009) Soil and water losses in *Acacia mangium* wild plantations and natural savanna in Roraima, northern Amazon. Rev Bras Ciênc Solo 33(2):447-454

Binkley D (1992) Mixtures of nitrogen-fixing and non-nitrogen-fixing tree species. In: Cannell M, Malcolm D, Robertson P (eds) The ecology of mixed-species stands of trees. Blackwell Scientific, Oxford, pp 99–123

Boerner REJ (1984) Foliar nutrient dynamics and nutrient use efficiency of 4 deciduous tree species in relation to site fertility. J Appl Ecol 21:1029–1040. https://doi.org/10.2307/2405065

Christina M, Nouvellon Y, Laclau J-P, Stape JL, Bouillet J-P, Lambais GR, le Maire G (2017) Importance of deep water uptake in tropical eucalypt forest. Functecol 31:509–519. https://doi.org/10.1111/1365-2435.12727

Cunha GM, Gama-Rodrigues AC, Costa G (2005) Nutrient cycling in a eucalypt plantation (*Eucalyptus grandis* W. Hill ex Maiden) in Northern Rio de Janeiro State. Rev Árvore 29:353–363. https://doi.org/10.1590/S0100-67622005000300002

Domingos M, Moraes RM, Vuono YS, Anselmo CE (1997) Produção de serapilheira e retorno de nutrientes em um trecho de Mata Atlântica secundária, na Reserva Biológica de Paranapiacaba, SP. Rev Bras Bot 20(1):91–96. https://doi.org/10.1590/S0100-84041997000100009

de Souza PA, de Mello WZ, da Silva JJN, Renato de A. R. Rodrigues, da Conceição MCG (2017) Atmospheric wet, dry and bulk deposition of inorganic nitrogen in the Rio de Janeiro State. Revista Virtual de Química 9(5):2052–2066

du Toit B, Gush MB, Pryke JS, Samways MJ, Dovey SB (2014) Ecological impacts of biomass production at stand and landscape levels. In: Seifert T (ed) Bioenergy from wood, Chapter 10. Springer, Dordrecht, pp 211–236

Ferraz AV (2009) Ciclagem de nutrientes e metais pesados em plantios de *Eucalyptus grandis* adubados com lodos de esgoto produzidos em diferentes estações de tratamento da região metropolitana de São Paulo. In: Recursos Florestais. ESALQ - USP, p 122

Forrester DI, Bauhus J, Cowie AL (2005) Nutrient cycling in a mixed-species plantation of *Eucalyptus globulus* and *Acacia mearnsii*. Can J Res 35:2942–2950. https://doi.org/10.1139/x05-214

Gama-Rodrigues AC, Barros NF (2002) Ciclagem de nutrientes em floresta natural e em plantios de eucalipto e de dandá no sudeste da Bahia, Brasil. Rev Árvore 26:193–207

Germon A, Guerrini IA, Bordron B, Bouillet J-P, Nouvellon Y, GoncalvesJLM JC, Paula RR, Laclau J-P (2018) Consequences of mixing *Acacia mangium* and *Eucalyptus grandis* trees on soil exploration by fine-root down to a depth of 17 m. Plant Soil 424:203–220. https://doi.org/10.1007/s11104-017-3428-1

Godinho TD, Caldeira MVW, Caliman JP, Prezotti LC, Watzlawicks LF, Azevedo HCA, Rocha JHT (2013) Biomass, macronutrients and organic carbon in the litter in a section of submontane seasonal semideciduous forest, ES. Sci For 41(97):131–144

Gonçalves JLM, Stape JL, Benedetti V, Fessel VAG, Gava JL (2000) Reflexos do cultivo mínimo e intensivo do solo em sua fertilidade e na nutrição das árvores. In: Gonçalves JLM, Benedetti V (eds) Nutrição e Fertilização Florestal. IPEF, Piracicaba, pp 1–58

Gonçalves JLM, Alvares CA, Behling M, Alves JM, Pizzi GT, Angeli A (2014) Productivity of eucalypt plantations managed under high forest and coppice systems, depending on edaphoclimatic factors. Sci Forest 42(103):411–419

Hodson ME, Langan SJ (1999) Considerations of uncertainty in setting critical loads of acidity of soils: the role of weathering rate determination. Environ Pollut106:73-81. doi:https://doi.org/10.1016/S0269-7491(99)00058-5

Klaminder J, Lucas RW, Futter MN, Bishop KH, Kohler SJ, Egnell G, Laudon H (2011) Silicate mineral weathering rate estimates: Are they precise enough to be useful when predicting the recovery of nutrient pools after harvesting? For Ecol Manag 261:1–9. https://doi.org/10.1016/j.foreco.2010.09.040

Koseva IS, Watmough SA, Aherne J (2010) Estimating base cation weathering rates in Canadian forest soils using a simple texture-based model. Biogeochemistry 101:183–196. https://doi.org/10.1007/s10533-010-9506-6

Koutika LS, Epron D, Bouillet JP, Mareschal L (2014) Changes in N and C concentrations, soil acidity and P availability in tropical mixed acacia and eucalypt plantations on a nutrient-poor sandy soil. Plant Soil 379:205–216. https://doi.org/10.1007/s11104-014-2047-3

Laclau J-P, Ranger J, Gonçalves JLM, Maquere V, Krusche AV, M'Bou AT, Nouvellon Y, Saint-Andre L, Bouillet J-P, Piccolo MC, Deleporte P (2010) Biogeochemical cycles of nutrients in tropical *Eucalyptus* plantations Main features shown by intensive monitoring in Congo and Brazil. For Ecol Manag 259:1771–1785. https://doi.org/10.1016/j.foreco.2009.06.010

Laclau J-P, Da Silva EA, Lambais RG, Bernoux M, Le Maire G, Stape JL, Bouillet J-P, Gonçalves JLM, Jourdan C, Nouvellon Y (2013) Dynamics of soil exploration by fine roots down to a depth of 10 m throughout the entire rotation in *Eucalyptus grandis* plantations. Front Plant Sci 4:243. https://doi.org/10.3389/fpls.2013.00243

Lequy E, Calvaruso C, Conil S, Turpault MP (2014) Atmospheric particulate deposition in temperate deciduous forest ecosystems: Interactions with the canopy and nutrient inputs in two beech stands of Northeastern France. Sci Total Environ 487:206–215. https://doi.org/10.1016/j.scitotenv.2014.04.028

Martins SG, Silva MLN, Curi N, Ferreira MM, Fonseca S, Marques J (2003) Soil and water losses by erosion in forest ecosystems in the region of Aracruz, state of Espírito Santo, Brazil. Rev Bras Ciênc Solo 27:395–403. https://doi.org/10.1590/S0100-06832003000300001

Melo VD, Correa GF, RibeiroAN MPA (2005) Kinetics of potassium and magnesium release from clay minerals of soils in the Triangulo Mineiro Region, Minas Gerais State, Brazil. Rev Bras Ciênc Solo 29(4):533–545. https://doi.org/10.1590/S0100-06832005000400006

Millard P, Proe MF (1993) Nitrogen uptake, partitioning and internal cycling in Picea sitchensis (Bong.) Carr. as influenced by nitrogen supply. New Phytol 125:113–119. https://doi.org/10.1111/j.1469-8137.1993.tb03869.x

Morellato LPC (1992) Nutrient cycling in two south-east Brazilian forests.1. litterfall and litter standing crop. J Trop Ecol 8:205–215

Nardoto GB, Bustamante MMD, Pinto AS, Klink CA (2006) Nutrient use efficiency at ecosystem and species level in savanna areas of Central Brazil and impacts of fire. J Trop Ecol 22:191–201. https://doi.org/10.1017/S0266467405002865

Nyaga JM, Cramer MD, Neff JC (2013) Atmospheric nutrient deposition to the west coast of South Africa. Atmos Environ 81:625–632. https://doi.org/10.1016/j.atmosenv.2013.09.021

Ouimet R, Duchesne L (2005) Base cation mineral weathering and total release rates from soils in three calibrated forest watersheds on the Canadian Boreal Shield. Can J Soil Sci 85:245–260. https://doi.org/10.4141/S04-061

Paula RR, Bouillet J-P, Trivelin PCO, Zeller B, Goncalves JLM, Nouvellon Y, Bouvet JM, Plassard C, Laclau J-P (2015) Evidence of short-term belowground transfer of nitrogen from *Acacia mangium* to *Eucalyptus grandis* trees in a tropical planted forest. Soil Biol Biochem 91:99–108. https://doi.org/10.1016/j.soilbio.2015.08.017

Pereira MG, Menezes LFT, Schultz N (2008) Aporte e decomposição da serapilheira na Floresta Atlântica, Ilha da Marambaia, Mangaratiba, RJ. Ciênc Florestal 18(4):443–454. https://doi.org/10.5902/19805098428

Pimenta JA, Rossi LB, Domingues TJM, Cavalheiro AL, Bianchini E (2011) Litter production and nutrient cycling in a reforested area and a seasonal semideciduous forest in southern Brazil. Acta Bot Bras 25:53–57. https://doi.org/10.1590/S0102-33062011000100008

Pinto SI, Martins SV, Barros NF, Teixeira DHC (2009) Nutrient cycling in two sites of semideciduous forest in mata do Paraíso forest reserve in Viçosa, MG, Brazil. Rev Árvore 33(4):653–663

Pugnaire FI, Chapin FS (1993) Controls over nutrient resorption from leaves of evergreen Mediterranean species. Ecology 74:124–129. https://doi.org/10.2307/1939507

Ranger J, Turpault MP (1999) Input-output nutrient budgets as a diagnostic tool for sustainable forest management. For Ecol Manag 122:139–154. https://doi.org/10.1016/S0378-1127(99)00038-9

Rocha JHT (2017) Manejo de resíduos florestais e deficiência nutricional em duas rotações de cultivo de eucalipto. In: Recursos Florestais Escola Superior de Agricultura "Luiz de Queiroz". Universidade de São Paulo, Piracicaba, p 173

Rocha JHT, du Toit B, Gonçalves JLM (2019) Ca and Mg nutrition and its application in E*ucalyptus* and P*inus* plantations. Forest Ecololgy Management 442:63–78. https://doi.org/10.1016/j.foreco.2019.03.062

Santos FM, Balieiro FD, Ataide DHD, Diniz AR, Chaer GM (2016) Dynamics of aboveground biomass accumulation in monospecific and mixed-species plantations of *Eucalyptus* and *Acacia* on a Brazilian sandy soil. For Ecol Manag 363:86–97. https://doi.org/10.1016/j.foreco.2015.12.028

Santos FM, Chaer GM, Diniz AR, Balieiro FD (2017) Nutrient cycling over five years of mixed-species plantations of *Eucalyptus* and *Acacia* on a sandy tropical soil. For Ecol Manag 384:110–121. https://doi.org/10.1016/j.foreco.2016.10.041

Silva PHM (2006) Produção de madeira, ciclagem de nutrientes e fertilidade do solo em plantios de *Eucalyptus grandis*, após aplicação de lodo de esgoto. In: Ciências Florestais. ESALQ-USP, Piracicaba, p 118

Silva MA, Silva MLN, Curi N, Avanzi JC, Leite FP (2011) Management systems in the eucalypt forest plantations and the soil and water losses in vale do rio doce, MG state. Ciênc Florest 21:765–776. https://doi.org/10.5902/198050984520

Silva PHM, Poggiani F, Libardi PL, Gonçalves AN (2013) Fertilizer management of eucalypt plantations on sandy soil in Brazil: Initial growth and nutrient cycling. For Ecol Manag 301:67–78. https://doi.org/10.1016/j.foreco.2012.10.033

Starr M, Lindroos AJ (2006) Changes in the rate of release of Ca and Mg and normative mineralogy due to weathering along a 5300-year chronosequence of boreal forest soils. Geoderma 133:269–280. https://doi.org/10.1016/j.geoderma.2005.07.013

Starr M, Lindroos AJ, Ukonmaanaho L (2014) Weathering release rates of base cations from soils within a boreal forested catchment: variation and comparison to deposition, litterfall and leaching fluxes. Environ Earth Sci 72:5101–5111. https://doi.org/10.1007/s12665-014-3381-8

Toledo LO, Pereira MG, Menezes CEG (2002) Produção de serapilheira e transferência de nutrientes em florestas secundárias localizadas na região de Pinheiral, RJ. Ciên Florestal 12(2):9–16

Vital ART, Guerrini IA, Franken WK, Fonseca RCB (2004) Produção de serapilheira e ciclagem de nutrientes de uma Floresta Estacional Semidecidual em zona ripária. Rev Árvore 8:793–800. https://doi.org/10.1590/S0100-67622004000600004

Vital ART, Lima WP, Camargo FRA (1999) Efeitos do corte raso de plantação de *Eucalyptus* sobre o balanço hídrico, a qualidade da água e as perdas de solo e de nutrientes em uma microbacia no vale do Paraíba, SP. Sci For 55:5–16

Voigtlaender M, Laclau J-P, Gonçalves JLM, Piccolo MC, Moreira MZ, Nouvellon Y, Ranger J, Bouillet J-P (2012) Introducing Acacia mangium trees in Eucalyptus grandis plantations: Consequences for soil organic matter stocks and nitrogen mineralization. Plant Soil 352:99–111. https://doi.org/10.1007/s11104-011-0982-9

Voigtlaender M, Brandani CB, Caldeira DRM, Tardy F, Bouillet J-P, Gonçalves JLM, Moreira MZ, Leite FP, Brunet D, Paula RR, Laclau J-P (2019) Nitrogen cycling in monospecific and mixed-species plantations of Acacia mangium and Eucalyptus at 4 sites in Brazil. For Ecol Manag 43:56–67. https://doi.org/10.1016/j.foreco.2018.12.055

Whitfield CJ, Watmough SA, Aherne J, Dillon PJ (2006) A comparison of weathering rates for acid-sensitive catchments in Nova Scotia, Canada and their impact on critical load calculations. Geoderma 136:899–911. https://doi.org/10.1016/j.geoderma.2006.06.004

Whitfield CJ, Watmough SA, Aherne J (2011) Evaluation of elemental depletion weathering rate estimation methods on acid-sensitive soils of northeastern Alberta, Canada. Geoderma 166:189–197. https://doi.org/10.1016/j.geoderma.2011.07.029

Wieder RK, Vile MA, Albright CM, Scott KD, Vitt DH, Quinn JC, Burke-Scoll M (2016) Effects of altered atmospheric nutrient deposition from Alberta oil sands development on Sphagnum fuscum growth and C, N and S accumulation in peat. Biogeochemistry 129:1–19. https://doi.org/10.1007/s10533-016-0216-6

Zaia FC, Gama-Rodrigues AC (2004) Nutrient cycling and balance in eucalypt plantation systems in north of Rio de Janeiro state, Brazil. Rev Bras Ciênc Solo 28:843–852

Chapter 4
Litter Decomposition and Soil Carbon Stocks in Mixed Plantations of *Eucalyptus* spp. and Nitrogen-Fixing Trees

Fabiano de Carvalho Balieiro, Fernando Vieira Cesário, and Felipe Martini Santos

4.1 Introduction

Increased demand for forest products around the world has contributed to the growth of planted forest areas over the last few decades (FAO 2015). The annual growth rate of planted forests in the tropics has been 2.5% per year, corresponding to an increase of around 20 million hectares from 1990 to 2015 (Payn et al. 2015). In Brazil, the planted forest generates and offers a huge diversity of products, especially pulp, paper, charcoal, sawn wood, and plywood, among others. Brazil currently has 7.84 million hectares of planted trees and its planted tree industry is responsible for 91% of the wood produced for industrial purposes in the country and 6.2% of the Brazilian gross domestic product; it is one of the industries with the greatest potential to help build a green economy (IBA 2017). Most of these planted forests are monocultures of *Eucalyptus* spp. (72%) and *Pinus* spp. (20%), but other species can also attend the internal and external market, such as *Acacia mangium* and *Acacia mearnsii* that together occupy 2.0% of the planted forest area. Other species include *Hevea brasiliensis* (2.9%), *Schizolobium amazonicun* (1.1%), and *Tectona grandis* (1.1%) (IBA 2017).

Despite the success of the Brazilian forestry agribusiness, most of these forest plantations occupy extensive areas of marginal soils, previously managed soils with low fertility or at some stage of degradation (Gonçalves et al. 2013). The sus-

F. de Carvalho Balieiro (✉)
EMBRAPA Soils, Brazilian Agricultural Research Corporation, Rio de Janeiro, RJ, Brazil
e-mail: fabiano.balieiro@embrapa.br

F. V. Cesário
Fluminense Federal University, Niterói, RJ, Brazil

F. M. Santos
Federal Rural University of Rio de Janeiro, Seropédica, RJ, Brazil

E. J. Bran Nogueira Cardoso et al. (eds.), *Mixed Plantations of Eucalyptus and Leguminous Trees*, https://doi.org/10.1007/978-3-030-32365-3_4

tainability of these plantations could be compromised in the medium and long terms, because they have been managed under short rotations and a fertilization regime, often far from high nutrient exports due to timber harvests, especially for P, K, and N (Santana et al. 2008; Laclau et al. 2010; Bouillet et al. 2013; Gonçalves et al. 2013).

Soil organic matter (SOM) in tropical planted or native forests has many functions that affect soil properties and processes. It plays a role of the nutrient reserve, and energy source for animals, plants, and microorganisms since its soil minerals are not great suppliers of mineral nutrients due to the high degree of weathering (Sanchez 1979; Maquere et al. 2008; Quesada et al. 2010). In addition, SOM features the greatest cation exchange capacity (CEC) of these soils (Senesi and Loffredo 1999; Oorts et al. 2000; Motta et al. 2002), which prevents essential nutrients from being easily lost by leaching and, consequently, not utilized by forests. Instead, the SOM allows such nutrients to be exchanged easily with the soil solution and to be taken up by plants. SOM participates in soil aggregation and consequently in soil pore space formation (Tisdall and Oades 1982; Denef et al. 2001; Denef and Six 2005). With an expressive and balanced pore space (i.e., macro- and micropores) the roots exploit the soil profile much better, water infiltration is facilitated, and plants get benefit from increased soil water storage (Doran and Parkin 1994; Franzluebbers 2002).

Poorly managed planted forests stimulate decomposition of SOM and C loss. This poor management includes excessive use of machines, with high traffic intensity in soil preparation or harvesting, in addition to removing or burning crop residues from the harvested area and excessive soil disturbance during preparation, which comprises the provision of numerous soil ecosystem services and the sustainability of these forests (Chaer and Tótola 2007; Dominati et al. 2010; Gonçalves et al. 2013; Jesus et al. 2015; and Chap. 10).

The rates of SOM and litter decomposition are clearly influenced by many factors, including temperature, air humidity, soil moisture content, soil microbial community, and litter quality. The last one may be described by lignin, polyphenols, carbon, and nutrient concentrations (especially N and P) or ratios of these (Attiwill and Adams 1993; Hättenschwiler et al. 2011). The N_2-fixing species often have higher N concentrations and decompose more rapidly. Thus, mixtures of *Eucalyptus* litter with more readily decomposable and more nutrient-rich litter may enhance the litter decomposition rates (Briones and Ineson 1996). Gartner and Cardon (2004) in a meta-analysis found synergistic effects in the majority of mixed-litter decomposition studies (47.5%), while antagonistic effects (19.1%) were less frequent. However, some expected improvements due to the introduction of N_2-fixing trees in litter decomposition, such as microbial activity and nutrient release (especially N), seem to be conditioned by the structural quality of the residues, the N:P ratio of litter, and the integration with microorganisms, driven by the increase in diversity of the plant community (Forrester et al. 2006; Rachid et al. 2015; Santos et al. 2018).

This chapter presents and discusses some data about changes in key soil processes, litter decomposition, and C stabilization (i.e., humification) arising from the introduction of N_2-fixing legumes in mixed *Eucalyptus* spp. plantations. We must emphasize the dynamics of C in tropical soils, especially Brazilian soils, and of the understanding of the main drivers of C stock in mixed-forest soil. Many researchers believe that one of the legacies of mixed plantations is the increase of soil C stocks, and at the end of this chapter, some of the practices and challenges to overcome obstacles of C storage will be introduced.

4.2 Carbon Assimilation and Partitioning in Forest Plantation

Understandings of carbon dynamics in forests involve knowledge of the biotic and abiotic factors that interfere in the growth of species that colonize a particular site, as well as where assimilated C is allocated (Cannell 1989; Davidson and Hirsch 2001). According to Cannell (1989), the CO_2 conversion efficiency in forest biomass is variable and associated with factors such as earth geometry, geographic location of the plantations, CO_2 diffusion rate for chloroplasts, canopy interception capacity (I_o), respiration rate, and proportion of the different aboveground tree compartments.

Carbon accumulation in a forest occurs in several aboveground and belowground compartments. In the case of woody tissues such as stem, thick roots, and branches, this accumulation may last for years or decades. On the other hand, in labile tissues such as leaves, flowers, and fine roots, after the senescence C will return to the atmosphere in days or weeks via decomposition (Landsberg and Gower 1997; Fearnside 2000; Schlesinger and Lichter 2001; Nouvellon et al. 2012; Bachega et al. 2016).

In general, about half of all CO_2 annually fixed in terrestrial biomass via photosynthesis (~120 Pg C y^{-1}) is respired by plants (~60 Pg C y^{-1}) and the other half is respired heterotrophically (Janzen 2004). In other words, if not disturbed, the C reservoirs shall remain constant in these environments, including the soil compartment. Because soil is the largest active C compartment of the terrestrial ecosystem (~1500–2000 Pg) and is in direct connection with other environmental components (i.e., atmosphere, hydrosphere, pedosphere, and biosphere), land use and climate changes have led to the loss of soil capacity to provide varied ecosystem services associated to SOM loss.

The net primary production (NPP) of forests increases with the reduction of latitude because there is a higher incidence of global radiation, higher evapotranspiration, and, consequently, higher cloud formation and rainfall in tropical regions (Ometto 1981). Reducing the vapor saturation deficit in these regions increases stomatal conductance and rate of carbon fixation of plant species (Sands and Mulligan 1990; Novais and Barros 1997; Whitehead 1998). Recent papers with *Eucalyptus* in Brazil have shown that NPP of the genus is determined preferentially by local water availability and, to a lesser extent, by soil fertility (Reis et al. 1985; Stape et al. 2004, 2010; Rigatto et al. 2005; Balieiro et al. 2008). The importance of water supply for *Eucalyptus grandis* and *E. urophylla* was highlighted by Stape et al. (2004), when evaluating 14 sites with a significant productivity gradient (9.7–39.1 Mg ha^{-1} $year^{-1}$). According to the authors, the NPP of the sites with intermediate productivity (i.e., average of 16 Mg ha^{-1} $year^{-1}$) was 46% higher than NPP of the sites with low productivity (i.e., average of 11.9 Mg ha^{-1} $year^{-1}$), where 32% of this variation was related to the site of water supply. Although the productivity difference between the high and medium productivity sites was 72%, one-third of this difference was related to the availability of soil water.

Similarly, when measured by the average monthly increment of the stem, the productivity of eucalypt was directly related to precipitation in the Rio Doce Basin, in Minas Gerais (Souza et al. 2006). According to the authors, for each increase of 100 mm in the total precipitation within a year, there was an average increase of 0.45 m^3 ha^{-1} $month^{-1}$, while 100 mm reduction affects this increase in 0.64 m^3 ha^{-1} $month^{-1}$. Rigatto et al. (2005) also found very high correlations between the height of *Pinus taeda* plants and the available water in eight different forest sites. Stape et al. (2010) across a large edaphoclimatic gradient (~1000 km) and eight regions found that fertilization beyond the current operational rates did not increase the growth of clonal *Eucalyptus* plantations, whereas irrigation raised growth about 30% (to 30.6 Mg ha^{-1} $year^{-1}$).

At mixed-forest plantations in Brazilian and Congolese conditions, where the set of trials with *Eucalyptus* and *Acacia mangium* (*Acacia*) were developed, the complementary interactions in the mixed stands of acacia and eucalypt led to significant biomass production. However, the NPP was higher than in monocultures without N fertilization only where appropriate climate conditions (i.e., temperature, high humidity, and rainfall) for *Acacia* and poor soil occur (Bouillet et al. 2013; Santos et al. 2016; Voigtlaender et al. 2019). Under milder and drier climate, *Acacia* cannot compete with *Eucalyptus*. For E50:A50 arrangement, and comparing six different sites (five in Brazil and one in Congo), Santos et al. (2016) after Bouillet et al. (2013) found that stemwood production only exceeded E100 ones at the Congo and Seropédica sites, although the differences in both studies were not significant at the 5% level (Santos et al. 2016).

Using the C budget approach to quantify growth, C uptake, and C partitioning in pure and mixed plantations with the same stocking density of *E. grandis* and *A. mangium* plantations, Nouvellon et al. (2012) developed an interesting work with C allocation of mixed plantation in São Paulo state. According to the results, the production in mixed plantation is lower than in *Eucalyptus* due to the lower gross primary production and net primary production values, as well as shifts in C allocation from above- to belowground and from growth to litter production. However, the pattern seems to be site specific due to the two contrasting tropical site environments (i.e., Brazilian and Congolese). Epron et al. (2013) compared Brazilian and Congolese site, and found that mixed-species plantations at the Brazilian site had a lower stand of wood biomass and aboveground net primary production (ANPP) without change in total belowground C fluxes (TBCF). In contrast, the mixed-species plantations overyielded the monocultures at the Congolese site, which led to higher standing wood biomass at the harvest. The NPP partitioning of the mixed plantations shifted towards aboveground growth at Kissoko (Congo site) and towards belowground growth at Itatinga (Brazil).

4.3 Litterfall and Nutrient Deposition in Pure and Mixed-Forest Plantations

As Chap. 3 explores this subject, we briefly discuss the main nutrient cycling alterations during the presence of N_2-fixing trees in *Eucalyptus* plantations.

In general, the amount of litter deposited by a given species or forest follows the same pattern of NPP (Fraser et al. 2015). Studies of litterfall are essential for the sustainability of planted forests since litter management significantly changes the biogeochemistry of these ecosystems and has already known consequences on plant growth, soil quality, and climate (Bowen and Nambiar 1984; Bonan 2008; Ponge 2013; Berg 2018). The C accumulation pattern and partitioning in the different tissues of plants, including the litterfall per unit area, depend on the density, arrangement, and interaction between the species (Reis et al. 1985; Leite et al. 1998; Nouvellon et al. 2012; Epron et al. 2013; Laclau et al. 2013).

The amount and quality of the material deposited on forest soils are related to not only genetics (i.e., planted species), structural, and aging factors of the plantation, but also climatic variables and soil type (Bernhard Reversat 1996; Landsberg and Gower 1997; Laclau et al. 2010; Voigtlaender et al. 2019). Soil fertility and climate significantly shape these two variables (i.e., quantity and quality of leaf litter) (Reis and Barros 1990; Negi and Sharma 1996; Stape et al. 2010). In soils with low fertility, it is natural for species to use nutrients more efficiently than nutrient-richer areas (Novais and Barros 1997; Malhi et al. 2006; Laclau et al. 2010). Therefore, it is common to find more nutrients being recycled in litter in soils of better fertility or without nutritional limitation. However, if there is water deficit during forest development, the water deficit induces stomata closure and, consequently, reduction in carbon fixation; that is, the plant will not succeed to express its potential and efficiency of nutrient use (Novais and Barros 1997; Epron et al. 2009; Stape et al. 2010).

Based on some studies about N deposition via litterfall in pure and mixed plantations, it was found that the litterfall in N_2-fixing leguminous plantation is 65% higher (in average) than that in the monocultures of *Eucalyptus* (without N). In addition, combining such legumes with *Eucalyptus* can promote significant increase, even 42% higher, which evidences the N input increase with the introduction of a leguminous in a mixed plantation (Fig. 4.1).

The presence of *A. mangium* in adult mixed plantations (i.e., >5 years) of five sites in Brazil also intensified the contribution of N via litterfall (Table 4.1), although the deposition pattern was different in the sites analyzed. In four sites, *Acacia* showed higher deposition in comparison to *Eucalyptus* without N fertilization,

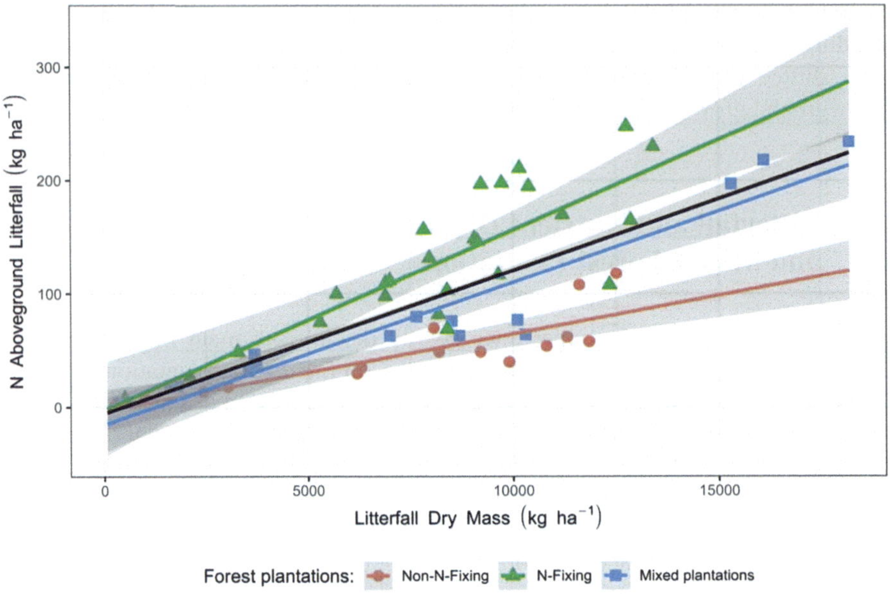

Fig. 4.1 Scatterplot between litterfall dry mass and N aboveground litterfall (Kg ha^{-1}) among N_2-fixing, non-N_2-fixing, and mixed plantations. Data from $n = 53$ selected studies (see Tables 4.1 and 4.2). The black line represents the mean regression line independent of plantations, that is, the overall regression line ($R^2 = 0.60$, $p < 0.001$). The regression line for red (circle) for non-N_2-fixing species, blue (square) for mixed plantations (E50:A50), and green (triangle) for N_2-fixing species are given. The shadows represent the standard error around lines for non-N_2-fixing, N_2-fixing, and mixed plantations only

while the other one showed similar deposition. This trend correlated with the climatic adaptation of the *A. mangium*.

In Brazilian sandy soils, the addition of 120 kg ha^{-1} of N via fertilizer in the *Eucalyptus* monoculture increased the N via litterfall around 10 kg ha^{-1} above the monoculture without N fertilization (final phase of the first rotation). On the other hand, in mixed plantations with half of the plant density replaced by *Acacia mangium* (E50:A50) and double density (E100:A100), the contributions were 110 and 79 kg ha^{-1} higher than in the monocultures without N fertilization, respectively. In younger plantations, the contribution of N inputs via litterfall was lower. But in the first and second rotations and up to 33 months, the contributions for pure *Acacia* or mixed plantations are significantly higher than in the *Eucalyptus* monocultures (without N fertilization) (Santos et al. 2016; Tchichelle et al. 2017).

Most studies referred in Table 4.1 do not provide information about deposition of other nutrients, so it is not possible to imply whether mixed plantations improve the contributions of these elements. On the other hand, since litterfall in mixed plantations is superior to monocultures, it is believed that the overall contribution is higher, as observed for N (Fig. 4.2a). The variability of P and N deposition data (Fig. 4.2)

Table 4.1 Litterfall mass and nutrient deposition associated (kg ha^{-1} ano^{-1}) under pure and mixed-forest plantations of N_2-fixing and non-N_2-fixing trees in different countries around the world*

Species (proportion)	Country	Rainfall	Temp.	Soil (texture) or	Age	Mass	N	P	K	Ca	Mg	Reference
	State	mm	°C	substrate	years	kg ha^{-1} $year^{-1}$						
Pseudosamanea guachapele	Brazil (RJ)	1500	25	Planosol (sandy)	5	7,968	132	7	34	119	51	Froufe (1999)
P. guachapele + *E. grandis*	Brazil (RJ)	1500	25	Planosol (sandy)	5	7,653	80	5	20	71	20	Froufe (1999)
A. mangium	Brazil (RJ)	1500	25	Planosol (sandy)	5	12,854	165	6	57	96	23	Froufe (1999)
Eucalyptus grandis	Brazil (RJ)	1500	25	Planosol (sandy)	5	9,903	40	6	29	84	21	Froufe (1999)
E. urograndis-	Brazil (RJ)	1370	24	Planosol (sandy)	2.5	500	4	0.3	2	5	1	Santos et al. (2016)
E. urograndis + N	Brazil (RJ)	1370	24	Planosol (sandy)	2.5	1,000	8	0.6	4	9	2	Santos et al. (2016)
A. mangium + *E. urograndis* (100:100)	Brazil (RJ)	1370	24	Planosol (sandy)	2.5	1,800	19	0.6	6	13	3	Santos et al. (2016)
A. mangium + *E. urograndis* (50:50)	Brazil (RJ)	1370	24	Planosol (sandy)	2.5	1,620	19	0.6	5	12	3	Santos et al. (2016)
A. mangium	Brazil (RJ)	1370	24	Planosol (sandy)	2.5	2,100	27	0.4	7	13	3	Santos et al. (2016)
E. urograndis	Brazil (RJ)	1370	24	Planosol (sandy)	5	11,600	108	4	37	69	0	Santos et al. (2016)
E. urograndis + N	Brazil (RJ)	1370	24	Planosol (sandy)	5	12,500	118	4	38	74	21	Santos et al. (2016)
A. mangium + *E. urograndis* (100:100)	Brazil (RJ)	1370	24	Planosol (sandy)	5	16,100	218	5	51	89	24	Santos et al. (2016)

(continued)

Table 4.1 (continued)

Species (proportion)	Country	Rainfall	Temp.	Soil (texture) or substrate	Age	Mass	N	P	K	Ca	Mg	Reference
	State	mm	°C		years	kg ha^{-1} year^{-1}						
A. mangium + *E. urograndis* (50:50)	Brazil (RJ)	1370	24	Planosol (sandy)	5	15,300	197	4	47	83	22	Santos et al. (2016)
A. mangium	Brazil (SP)	1390	24	Planosol (sandy)	5	13,400	230	3	52	61	15	Santos et al. (2016)
A. mangium	Brazil (SP)	1390	19	Ferralsol (loam)	6	6,890	98		–	–	–	Voigtlaender et al. (2019)
A. mangium + *E. grandis* (50:50)	Brazil (SP)	1390	19	Ferralsol (loam)	6	8,693	63		–	–	–	Voigtlaender et al. (2019)
E. grandis	Brazil (SP)	1390	19	Ferralsol (loam)	6	9,201	49		–	–	–	Voigtlaender et al. (2019)
A. mangium	Brazil (SP)	1420	21	Ferralsol (loam)	6.3	8,394	103		–	–	–	Voigtlaender et al. (2019)
A. mangium + *E. grandis* (50:50)	Brazil (SP)	1420	21	Ferralsol (loam)	6.3	10,107	77		–	–	–	Voigtlaender et al. (2019)
E. grandis	Brazil (SP)	1420	21	Ferralsol (loam)	6.3	11,302	62		–	–	–	Voigtlaender et al. (2019)
A. mangium	Brazil (SP)	1420	23	Arenosol (sandy)	6.1	5,297	75		–	–	–	Voigtlaender et al. (2019)
A. mangium + *E. urograndis* (50:50)	Brazil (SP)	1420	23	Arenosol (sandy)	6.1	10,296	64		–	–	–	Voigtlaender et al. (2019)
E. urograndis	Brazil (SP)	1420	23	Arenosol (sandy)	6.1	10,804	54		–	–	–	Voigtlaender et al. (2019)
A. mangium	Brazil (SP)	1240	24	Ferralsol (clayed)	6.3	8,185	82		–	–	–	Voigtlaender et al. (2019)
A. mangium + *E. urograndis* (50:50)	Brazil (SP)	1240	24	Ferralsol (clayed)	6.3	8,504	76		–	–	–	Voigtlaender et al. (2019)

E. urograndis	Brazil (SP)	1240	24	Ferralsol (clayed)	6.3	8,195	49		–	–	–	Voigtlaender et al. (2019)
P. guachapele	Brazil (SP)	1250	24	Planosol (sandy)	7	12,750	248	–	–	–	–	Balieiro et al. (2004)
E. grandis	Brazil (RJ)	1250	24	Planosol (sandy)	7	11,840	58	–	–	–	–	Balieiro et al. (2004)
E. globulus	Australia	1009	16	Acrisol (sandy loam)	8.8	2,430	14	0.5	–	–	–	Forrester et al. (2013)
A. mearnsii	Australia	1009	16	Acrisol (sandy loam)	8.8	3,290	49	1	–	–	–	Forrester et al. (2013)
E. globulus + *A. mearnsii*	Australia	1009	16	Acrisol (sandy loam)	8.8	3,560	32	1	–	–	–	Forrester et al. (2013)
E. globulus + *A. mearnsii*	Australia	1009	16	Acrisol (sandy loam)	8.8	3,740	39	1	–	–	–	Forrester et al. (2013)
E. globulus + *A. mearnsii*	Australia	1009	16	Acrisol (sandy loam)	8.8	3,690	47	1	–	–	–	Forrester et al. (2013)
E. globulus + N	Australia	1009	16	Acrisol (sandy loam)	8.8	3,050	18	1	–	–	–	Forrester et al. (2013)
E. saligna	Hawaii	4600	21	Andosol	4	8,070	70	5	4	61	10	Binkley et al. (1992)
Albizia falcataria	Hawaii	4600	21	Andosol	4	12,340	108	6	6	86	14	Binkley et al. (1992)
E. saligna + *A. falcataria* (50:50)	Hawaii	4600	21	Andosol	4	18,200	234	9	7	72	19	Binkley et al. (1992)
A. mangium	Congo	1200	25	Arenosol (sandy)	7	7,000	112	10	–	–	–	Koutika et al. (2014)
A. mangium + *E. urograndis* (50:50)	Congo	1200	25	Arenosol (sandy)	7	7,000	63	11	–	–	–	Koutika et al. (2014)
E. urograndis	Congo	1200	25	Arenosol (sandy)	7	6,200	30	9	–	–	–	Koutika et al. (2014)

(continued)

Table 4.1 (continued)

Species (proportion)	Country	Rainfall	Temp.	Soil (texture) or substrate	Age	Mass	N	P	K	Ca	Mg	Reference
	State	mm	°C		years	kg ha^{-1} year^{-1}						
A. mangium (2nd rotation)	Congo	1200	25	Arenosol (sandy)	2	5,000	388					Tchichelle et al. (2017)
A. mangium + *E. urograndis* (50:50) (2sd rotation)	Congo	1200	25	Arenosol (sandy)	2	3,300	306					Tchichelle et al. (2017)
E. urograndis (2nd rotation)	Congo	1200	25	Arenosol (sandy)	2	2,700	89					Tchichelle et al. (2017)
P. guachapele + E. grandis – M (50:50)	Brazil (RJ)	1250	24	Planosol (sandy)	7	66	–	–	–	–	–	Balieiro et al. (2004)
Leguminous tree plantations												
A. mangium	Brazil (RJ)	1005	25	Planosol (sandy)	4	10,155	211	12	28	80	24	Andrade et al. (2000)
A. holosericea	Brazil (RJ)	1005	25	Planosol (sandy)	4	9,062	149	6	20	83	17	Andrade et al. (2000)
Mimosa caesalpiniaefolia	Brazil (RJ)	1005	25	Planosol (sandy)	4	9,132	147	4	21	60	11	Andrade et al. (2000)
Sclerolobium paniculatum–P	Brazil (AP)	2100	27	Ferralsol (sandy loam)	9	9,646	117	3	6	26	10	Mochiutti et al. (2006)
M. caesalpiniaefolia	Brazil (PE)	1300	26	Acrisol (sandy loam)	10	7,830	157	10	55	115	26	Ferreira et al. (2007)
Degraded lands												
Mimosa caesalpiniaefolia	Brazil (RJ)	1005	25	Gravel extraction pit	10	11,200	170	7	31	190	40	Costa et al. (1998)
M. caesalpiniaefolia + *Gliricidia sepium*	Brazil (RJ)	1005	25	Gravel extraction pit	10		100	4	18	110	25	Costa et al. (1998)
Gliricidia sepium	Brazil (RJ)	1005	25	Gravel extraction pit	10	5,700	100	4	18	100	32	Costa et al. (1998)

Acacia auriculiformis	Brazil (RJ)	1005	25	Gravel extraction pit	10	6,900	110	5	19	120	29	Costa et al. (1998)
M. caesalpiniaefolia/A. auriculiformis	Brazil (RJ)	1005	25	Gravel extraction pit	10		130	5	24	150	29	Costa et al. (1998)
M. caesalpiniaefolia	Brazil (MA)	1900	27	Red mud deposits	3	9,727	198	–	–	–	–	Fortes (2000)
M. acutistipula	Brazil (MA)	1900	27	Red mud deposits	3	9,221	197	–	–	–	–	Fortes (2000)
A. mangium	Brazil (MA)	1900	27	Red mud deposits	3	10,376	195	–	–	–	–	Fortes (2000)
E. camaldulensis	Brazil (RJ)	1020	24	Clay extraction pit	4	6,300	35	3	19	45	10	Silva et al. (2015)
A. mangium	Brazil (RJ)	1020	24	Clay extraction pit	4	8,400	69	4	18	31	9	Silva et al. (2015)

*Most of the information in the table is short rotation (1 year). When data were not available in tables or supplementary materials within the papers, the "web plot digitizer program" was used to extract the graphical data (https://automeris.io/WebPlotDigitizer/)

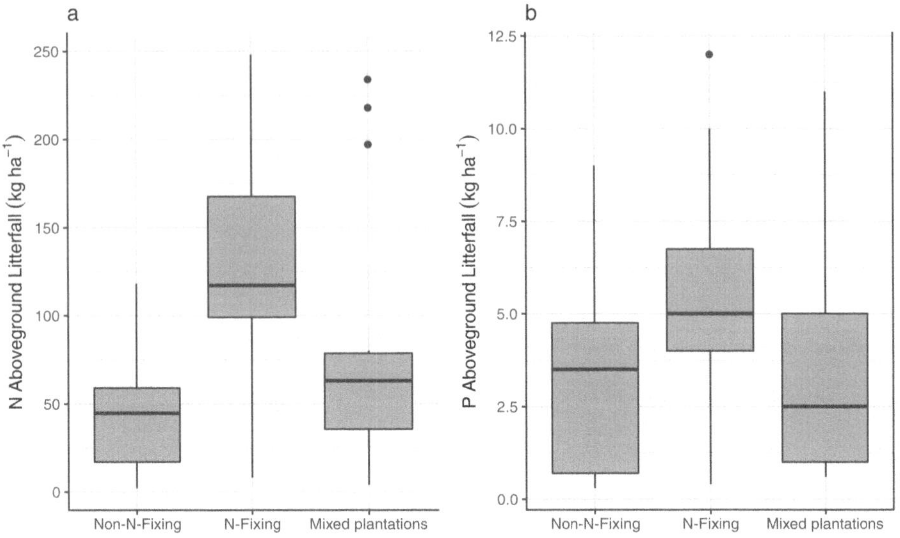

Fig. 4.2 Box plots of N (**a**) and P (**b**) in aboveground litterfall (Kg ha^{-1}) under N_2-fixing and non-N_2-fixing monocultures and mixed plantations of both. Data from n = 58 selected studies (see Tables 4.1 and 4.2). Black points represent outliers

is mostly due to the evaluated species, age of plantations, and variations in the soil types studied that present different mineralogy and fertility, as well as different patterns of growth and accumulation in plants.

4.4 Litter Decomposition Under Pure and Mixed-Forest Plantation

The logic of non-N_2-fixing and N_2-fixing species' mixed plantations is to promote forest sustainability through complementarity and competitive reduction interactions among species. The "extra" supply of N provided by legume trees via biological nitrogen fixation and its redistribution offer throughout the system an intensification of leaf litter decomposition, N mineralization, and transferring between fixing and non-fixing species. Collectively, they are key processes to improve the growth and yield of the mixed forests, eliminating or using less N fertilizers (Forrester et al. 2006; Kaye et al. 2000; Koutika and Richardson 2019).

In general, decomposition of forest residues is driven by the litter chemical attributes, environmental conditions, and surface area/volume ratio of the residue (Landsberg and Gower 1997; Gholz et al. 2000; Laclau et al. 2010). Although the N deposition is high in pure and mixed plantations with N_2-fixing trees (Fig. 4.1), this chemical characteristic is not a guarantee of higher litter decomposition, especially under tropical conditions. For instance, *A. mangium* in sandy and loamy textured

Table 4.2 Changes in soil C stocks (Mg ha^{-1}) under plantations of *Eucalyptus* spp. in comparison to native vegetation as a function of rotation, soil texture, and biome evaluated

Depth	*n* observations; comments	Max	Min	Median	SD	Mean*
		Carbon stock budget in pure stands of *Eucalyptus* related to natural vegetation Mg ha^{-1}				
0–20 cm	50	20.0	−20.9	−1.8	10.0	−1.5
20–40 cm	39	42.0	−25.0	−1.9	15.9	0.3
		Carbon stock change induced by time rotations Mg ha^{-1}				
0–20 cm	13; 1st rotation	19.0	−20.7	1.2	11.6	−2.3
	15; 2nd rotation	20.0	−20.9	3.0	11.5	2.3
	21; not reported	14.0	−18.2	−4.0	7.3	−3.8
20–40 cm	9; 1st rotation	24.0	−21.2	5.0	16.3	−1.0
	14; 2nd rotation	42.0	−25.0	−0.6	20.6	3.1
	14; not reported	20.0	−21.0	−1.8	10.0	−1.5
		Carbon stock change induced by soil texture Mg ha^{-1}				
0–20 cm	10; sandy	7.8	−10.0	−1.6	6.9	−1.0
	17; clayed	19.0	−18.2	0.0	11.5	1.2
	22; not reported	20.0	−20.9	−2.2	9.9	−3.9
20–40 cm	12; sandy	24.0	−12.0	−1.5	10.7	2.5
	13; clayed	32.8	−21.0	1.0	15.5	2.6
	12; not reported	42.0	−25.0	−9.1	20.3	−4.4
		Carbon stock change induced by biome Mg ha^{-1}				
0–20 cm	13; Cerrado	19.0	−10.0	1.0	8.7	3.5
	22; Atlantic rainforest	15.0	−21.0	−6.5	10.3	−5.2
	12; Pampa	20.0	−13.0	−3.5	9.4	−0.3
20–40 cm	10; Cerrado	33.0	−12.0	2.5	13.6	6.4
	18; Atlantic rainforest	14.0	−23.0	−4.5	12.1	−5.3
	9; Pampa	42.0	−25.0	10.0	21.6	5.6

Adapted from Fialho and Zinn (2012); * In average, *Eucalyptus* spp. plantations do not affect the soil organic carbon stocks in Brazil (t-test at $p < 0.05$ were used in all comparasions)

soils shows lower leaf and fine root decomposition in comparison to *Eucalyptus*, although it has high N leaf concentration (Balieiro et al. 2004; Bachega et al. 2016; Doughty et al. 2018). Besides, *A. mangium* has high internal cycling of P (Balieiro et al. 2004; Bachega et al. 2016; Doughty et al. 2018), providing a litter with low P concentration. Furthermore, *A. mangium* has a more recalcitrant leaf litter, lignin rich with low contents of nonstructural carbohydrates or low-molecular-weight phenols and P concentration (Santos et al. 2017). Thus, the microbiota faces a harsh condition to decompose its residues. The concept of "decomposer starvation," proposed by Hättenschwiler et al. (2011), reveals a syndrome of poor C litter in tropical rainforests which could be applied to pure *A. mangium* plantations in oligotrophic soils (e.g., Arenosol). According to these authors "*in the neotropical rainforest, natural selection favored a leaf litter trait that leads to starvation-inhibition of decomposers, thereby increasing the tree ability to compete for the uptake of highly limiting nutrients, P in particular, via mycorrhizal associations*." Other authors who have also observed a decrease of P in the soil in areas in mixed plantations claim that this decrease comes from the P uptake by *Eucalyptus* to maintain the N:P stoichiometry of their leaves (Koutika et al. 2014).

It is worth noting that legume trees have a different demand for P as a function of symbiosis with diazotrophic bacteria (Giller and Cadisch 1995; Vadez et al. 1995). Therefore, different strategies are required for capturing this element, such as acidification of the rhizosphere that improves the solubilization of less soluble forms of P (Raven et al. 1990) and root exudation of acid phosphatases and organic acids that leads to PO_4 desorption from the soil matrix and improves P availability (Vance 2001; Venterink 2011). These strategies are paramount in tropical soils since they are mainly able to fix P by ligand exchange and occupation of P sorption sites (Bhatti et al. 1998; de Campos et al. 2016). This is possibly due to the high content of aluminum and iron oxides in the Oxisols and Ultisols typically found in moist tropical sites (Leal and Velloso 1973; Lloyd et al. 2001).

Land-use history may also affect the litter decomposition. Under Ferralsols, previously managed with *Eucalyptus* (for 60 years), Bachega et al. (2016) detected that early decomposition of leaves and fine roots of *Acacia mangium* was markedly slower than that of *Eucalyptus* residues, despite higher N and P concentrations in both tissues of *Acacia* (respectively, 1.9 and 1.5 times higher for leaves and 2.9 and 3.3 times for roots). The lower values of C:N and C:P ratios were associated to litter decomposition rates of *Acacia*, but the authors did not confirm the home field advantage (HFA) that states which plants create a specialization of local decomposer communities of their litter. For *Acacia*, authors claim that the time since the start of the first rotation was not long enough to allow the decomposers to become specialized for its residues. In contrast, several researches have demonstrated that in the short term the litter and soil bacterial and fungi communities are very specific for both (Rachid 2013; Rachid et al. 2015; Bini et al. 2013).

As was pointed out in the perspective of the HFA theory, our research group studied the litter leaf decomposition of both species (i.e., *Acacia* and *Eucalyptus*) in a sandy soil. Our team observed a distinct HFA for decomposition litter from *Acacia* and *Eucalyptus*, with *Acacia* decomposing in the home stands faster than under *Eucalyptus* stands. In contrast, litter from *Eucalyptus* decomposed faster under *Acacia* stands. Additionally, the litter of each material showed very distinct fungal communities and did not change in function of time, and the local of incubation did not influence the microbial community (Rachid 2013). Higher diversity and lower dominance of fungi were reported in litter from *Acacia* and mixed plantations during the decomposition period (180d), regardless of the place at which they were established (Rachid 2013). By using infrared spectroscopy, it was found that *Eucalyptus* litter during the decomposition under *Acacia* stands remain more proteinaceous material than the initial residue, and the migration of N to *Eucalyptus* litter may be considered. On the other hand, the *Acacia* litter did not change its chemical composition during the incubation (Novotny et al. 2013). These results demonstrate the importance of biological N_2 fixed for the *Eucalyptus* litter decomposition, and suggests that the most diverse fungi community is essential for mixed litter decomposition allowing N mobility. Nonetheless, it is interesting to note that the higher P concentration in *Eucalyptus* litter acts simultaneously and synergistically with the decomposition process in the mixed plantation (Santos et al. 2017). In addition, the C:N ratio alone seems not to be the main predictor of soil organic and litter decomposition (Cotrufo et al. 2013; Lehmann and Kleber 2015; Berg 2018). Some predictors such as N:P, lignin:N, and lignin:P, when possible, should be analyzed collectively.

The N:P ratio of the litter trait is often cited as an essential drive of litter decomposition (Güsewell 2004; Bakker et al. 2011). However, differences in the N:P ratio observed in aboveground biomass and leaf litter usually reflect even more significant changes in the available N:P ratio of soil (Güsewell 2004). Figure 4.3 demonstrates the broad range of the N:P ratio (mass) of the litter in the papers analyzed and, consequently, about edaphic conditions of the studies. Also, there is a trend towards increasing this relation in mixed plantations: non-N_2-fixing (N:P–14) < N_2 fixing (N:P–25) < mixed (N:P–32). Although the contribution of N via litterfall is higher for leguminous plantations, followed by mixed plantings compared to *Eucalyptus* plantations, for P there is a lower

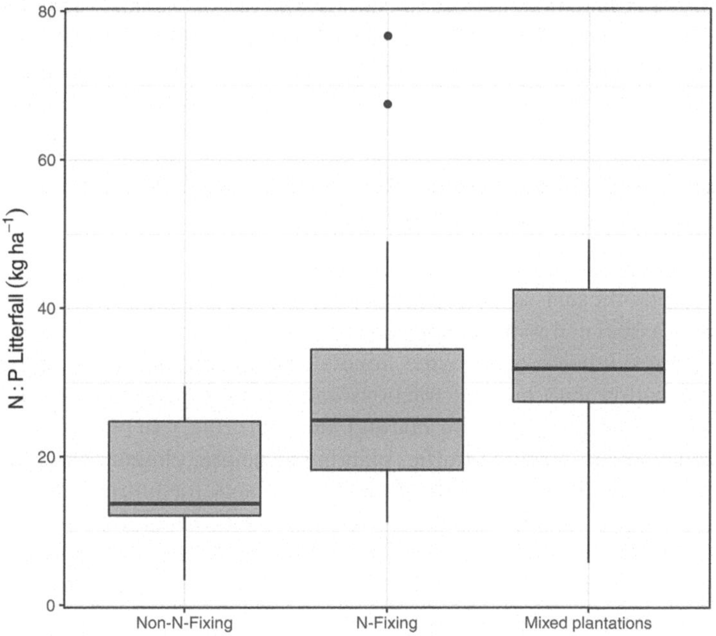

Fig. 4.3 Box plots of N:P relationship in aboveground litterfall (Kg ha^{-1}) among N_2-fixing, non-N_2-fixing, and mixed plantations. Data from $n = 58$ selected studies (see Tables 4.1 and 4.2). Black points represent outliers

recycling trend in mixed planting conditions (Fig. 4.2b). This finding corroborates the productivity of these plantations and the accumulation of P in the aboveground biomass of the species, besides the edaphic limitations in the P supply. In other words, although P is being absorbed more efficiently in these plantations, the plants begin to drain this P for internal use. This process decreases the availability of P for the soil in the medium and long terms (Sanchez 1979; Binkley et al. 2000) and requires attentive management with phosphate fertilization in future rotations.

Climate change and especially temperature rise may also affect the chemical characteristics of leaf litter with effects on decomposition. Trees growing in high-temperature environments have a reduction in the N content of the leaves and an increase in the rate of carboxylation, which results in increased structural and nonstructural carbon levels in the leaves (e.g., glucose, sucrose, fructose) (Güsewell 2004; Pandey et al. 2015) and a recalcitrant leaf litter. These effects can be intensified in the tropics, where the soils are acidic and poor (i.e., with minerals with high adsorption capacity of P). In addition, the presence of leguminous species with high internal P recycling, such as *A. mangium* (Balieiro et al. 2005; Inagaki et al. 2011; Santos et al. 2017), may impair the decomposition process with unfavorable stoichiometric ratios in the leaf litter (e.g., C:N, C:P, or N:P) (Güsewell 2004). Therefore, changes in leaf litter quality may be reflected by changes in soil enzymatic activity (Fanin and Bertrand 2016; Santos et al. 2017), stabilization of C stocks (Fisk et al. 2015; Castellano et al. 2015), emission of CO_2 through microbial respiration (Zhou et al. 2013, 2015), and nutrient mineralization rates (e.g., P, N, and S) (Marklein et al. 2016; van Huysen et al. 2016). All these aspects are feedbacks of climate change and productivity of forest systems (i.e., plantations or natural forests) (Bonan 2008).

4.5 Are Soil Carbon Stocks Really Higher in Mixed Plantations than in Monocultures?

In general, soils present several mechanisms related to the protection of organic matter that is associated with chemical complexation (e.g., polymerization, humification, organic synthesis, and organo-mineral interaction) and physical protection (e.g., complexation with mineral fractions and degree of aggregation) (Feller and Beare 1997; Sollins et al. 1996; Roscoe and Machado 2002). In planted forests, crop residue management, machine traffic, planting structure, climate, and planted and understory species all were identified as determinants for nutrient dynamics and degree of C storage in soils (Bernhard Reversat 1996; Binkley et al. 2000; Kaye et al. 2000; Resh et al. 2002; Qiao et al. 2014; Jesus et al. 2015). With the possibility of managing the communities of soil bacteria and fungi in mixed *Eucalyptus* and *A. mangium* plantations (Rachid et al. 2013, 2015; Bini et al. 2013), it is also reasonable to think about changes in belowground interactions (i.e., belowground competition for nutrients and water). These consequences are still poorly studied in soil C dynamics, although Sokol and Bradford (2019) have claimed that belowground

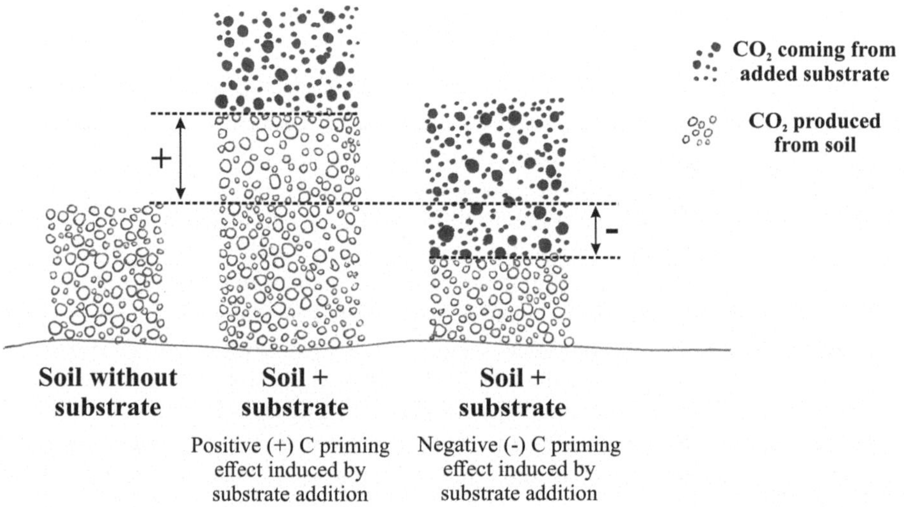

Fig. 4.4 Schematization of the priming effect—nonadditive interactions between decomposition of the added substrate and of soil organic matter (SOM): (+) acceleration of SOM decomposition—positive effect; (−) retardation of SOM decomposition—negative priming effect (adapted from Kuzyakov et al. (2000))

inputs provided by the microbial formation pathway form mineral-stabilized soil C are more effective than aboveground inputs, partly due to the higher efficiency of formation by the rhizosphere microbial community relative to the bulk soil community. In mixed plantations of *A. mangium* and *Eucalyptus*, there might be a more favorable environment for soil C stabilization, because we observed a complementarity in terms of nutrient deposition (e.g., N and P deposition) (Santos et al. 2017, 2018), soil bacterial and fungi communities (Rachid et al. 2013, 2015), and fine root growth (segregated) in soil profile (Silva et al. 2009; Laclau et al. 2013).

The dynamics of soil organic matter from natural and planted forests, including mixed plantations, have been studied from the viewpoint of the "priming effect" by several authors (Resh et al. 2002; Balieiro et al. 2008; Hoosbeek and Scarascia-Mugnozza 2009; Forrester et al. 2013; Koutika et al. 2014). The priming effects were defined by Kuzyakov et al. (2000) as the "*strong short-term changes in the turnover of soil organic matter caused by comparatively moderate treatments of the soil.*" Such changes might be the input of organic or mineral fertilizer to the soil, exudation of the organic substances by roots, and simple mechanical treatment of soil or its dry and wet cycles. Figure 4.4 presents the schematic representation of the priming effect. Under an ecological point of view, the negative priming effects have a much greater significance than positive ones, but the direction of these changes depends mostly on the nutrient status of the soil and the C:N ratio of the active SOM pool (i.e., labile organic matter) (Kuzyakov et al. 2000).

4.5.1 Soil Carbon Stocks in Pure Plantations of Eucalyptus

The impacts of *Eucalyptus* on soil carbon stocks vary according to many factors, including the land-use history, previous crop, post-logging residue management, climate, and spatial variability of soil attributes (Forrester et al. 2006; Chaer and Tótola 2007; Balieiro et al. 2008; Gonçalves et al. 2013). Under a broader point of view, in a recent meta-analysis, Fialho and Zinn (2014) compiled data on the organic soil C stocks using 50 observations for depths between 0 and 20 cm and 39 between 0 and 40 cm of studies in Brazil in paired plots (i.e., plantations and natural forests) aiming to evaluate the impact of native vegetation conversion on plantations of *Eucalyptus*. The authors verified that, on average, the net effect of the conversion is null; that is, it does not damage the original C stocks of the soil, although losses and gains are related to local site conditions (Table 4.2). The authors conclude that this null effect, even after considering the rotation time, texture, and biome, suggests that other factors may control the direction and intensity of changes in soil C stocks in Eucalyptus plantations and point to the productivity, techniques of soil preparation, soil type, and management as essential factors in this evaluation.

Cook et al. (2016) studied the effects of *Eucalyptus* plantations on soil carbon stocks, 0–30 cm deep, over two decades, in 306 operational eucalypt plantations across a 1200 km gradient in Brazil. The study included two tropical states (Bahia and Espírito Santo) and one subtropical state (São Paulo), and resulted in the findings that the size and rates of change in soil C stocks were due to different factors. These factors include the history of the site, soil order, clay content, seasonal precipitation (especially dry season), and mean annual temperature. In general, across all sites, the soil C showed a slight decrease (-0.22 ± 0.05 Mg ha^{-1} year^{-1}) from the original sampling that ranged in approximately 3–4 rotations, but in subtropical regions the stocks remained the same (0.06 Mg ha^{-1} year^{-1}).

Maquere et al. (2008) studied the impact of different land uses (i.e., savanna, pasture, and *Eucalyptus saligna* plantations) and a management (i.e., 60 years under short rotation vs. 60 years under continuous growth) on soil carbon and nitrogen stocks. The authors found significant soil carbon increases (approximately 25%) with *Eucalyptus* under short rotation management when compared to Cerrado native vegetation, whereas soil carbon stocks in the continuous forest plantation increased by 15% in relation to Cerrado vegetation. In the same biome and in degraded pastures Lima et al. (2006) reported that afforestation of former degraded pasture land leads to increased C storage in the soil in the short term (30 years). They observed carbon sequestration rates up to 0.57 Mg C ha^{-1} year^{-1} by *Eucalyptus* afforestation.

Finally, it is evident in the studies that pure *Eucalyptus* plantations present a potential for soil C increase, especially if preceded by degraded pastures. However, local abiotic conditions and forest management seem to be the significant constraints to the effective increase of C in soil. On the other hand, the introduction of legume trees associated with diazotrophic bacteria to *Eucalyptus* monocultures has gained expression in recent years. The increase of soil C stocks so far had always been related to these plantations but, as will be seen below, this phenomenon cannot be generalized, mainly for the Brazilian tropics.

4.5.2 Soil C Stocks Under Mixed Plantations of N_2-Fixing and Non-N_2-Fixing Plantations

Several authors reported higher soil C accumulation of N_2-fixing species over non-N_2-fixing ones (Binkley et al. 1992; Kaye et al. 2000; Resh et al. 2002). However, when a small or no variation is detected in soil C stocks, changes in soil morphological and molecular levels of soil organic matter have also been observed (Kindel et al. 2003; Voigtlaender 2012; Koutika et al. 2014; Santana et al. 2015; Voigtlaender et al. 2019), and this may have consequences for the nutrient cycling in these plantations.

Resh et al. (2002), using isotopic techniques, detected that 55% of the soil C enriching the N_2-fixing plantations came from the retention of native C, in a comparison between the C soil stocks of N_2-fixing species, as *Albizia falcataria, Leucaena leucocephala*, and *Casuarina equisetifolia*, and the soil C stocks of *Eucalyptus saligna* in four tropical soils, two Andisols and one Vertisol and Entisol. They reported that the native C stock was derived from old sugarcane plantations and pastures with metabolic assimilation of C_4, while the other 45% came from C stabilized and derived from recently allocated C_3 from the trees, showing a clear negative priming effect. In addition, 62% of the explanation of the stabilization of native C (C_4) occurred due to the accumulation of N in the soil of the legume trees and *Casuarina* plantations. Very similar results were reported by Kaye et al. (2000) in Andisols from Hawaii, where *Albizia* monocultures had 2.3 Mg ha^{-1} more N and 20 Mg ha^{-1} more C than Eucalyptus monocultures in the 50 cm topsoil. In Acrisols (Brown Dermasols) from Australia, Forrester et al. (2013) reported a difference greater than 15 Mg ha^{-1} between N_2-fixing and non-N_2-fixing trees in the first 30 cm topsoil. However, in Arenosols from Congo and Brazil, a lower gain in soil carbon storage was detected in the N_2-fixing *Acacia mangium* plantations at 7 and 5 years after planting in comparison to monocultive of *Eucalyptus*, respectively (Koutika et al. 2014; Rocha et al. 2019). In the *Pseudosamanea guachapele* plantations, the C storage was even much lower than that found under *Eucalyptus grandis* plantations in southeastern Brazil, in the same soils (Balieiro et al. 2008), a fact that was justified by the authors by high contribution of N (~250 kg ha^{-1} y^{-1}) and other nutrients cycled through litterfall, resulting in a rapid decomposition of the litter and SOM. In is worth mentioning here that the pure legume tree stands studied in Brazil (*Acacia mangium* and *Pseudosamanea guachapele*) had no similar growth in comparison to *Eucalyptus*, contrary to the plantations of *Falcataria moluccana* (*Albizia falcataria*) in Hawaii that present higher aboveground biomass than Eucalyptus saligna (Kaye et al. 2000).

Climate and soil characteristics can control soil C stocks, but the aboveground biomass and NPP are also drivers of C stocks in various environments (Lal 2005; Cornwell et al. 2008; Qiao et al. 2014; Nottingham et al. 2015; Lange et al. 2015). With the literature data consulted, it is not possible to state that C stocks are a function of aboveground biomass production since the soil C augmentation pattern is not very different between monocultures and mixed ones.

Under mixed plantations and Brazilian conditions, mixed plantations presented aboveground biomass higher than monocultures only in sites with poor soils (for

Acacia mangium and *Pseudosamanea guachapele*) (Santos et al. 2017; Balieiro et al. 2010). However, despite the soil C stocks in N_2-fixing plantations did not differ statistically from *Eucalyptus* monocultures (without N) (Balieiro et al. 2008; Rocha et al. 2019), under mixed plantations were detected higher rates of C sequestration (up to 1.44 Mg ha^{-1} y^{-1}), in comparison to *Eucalyptus* monocultives (without N fertilization), demonstrating that for oligotrophic soils the mixed plantation is an great alternative to improve soil quality.

In Australia, Forrester et al. (2013) studied mixed plantations of *Eucalyptus globulus* and *Acacia mearnsii*. They detected a linear relationship between aboveground production and soil C, but not at an N_2-fixing proportion, as found by Kaye et al. (2000), who observed higher aboveground biomass production in mixed plantations and lower in monocultures. In both papers, the increase in C soil stocks in mixed plantations is justified by the negative priming effect of the mixed litter, that is, the maintenance of the C originated from previous crops (sugarcane) and by the addition and stabilization of C derivatives from the trees.

A significant contrast found among these plantations seems to be the pattern of C allocation in the belowground biomass, because Hawaiian monocultures accumulated more C in the belowground than mixed plantations, contrary to what was observed in Australia (Forrester et al. 2013; Binkley and Ryan 1998). In Hawaii, intraspecific competition intensified the use of soil resources, while interspecific competition in Australia further pressured soil resources, culminating in belowground investments to compensate for distinct constraints. In monocultures, the accumulation patterns of the species can lead to the storage for the aboveground biomass of critical elements for soil C storage; in contrast, and because of the complementarity of niches and facilitation (in both directions), mixed plantations can more efficiently cycle nutrients and better use available water. As previously mentioned (Sokol and Bradford 2019) and pointed out by other authors (Schmidt et al. 2011; Nouvellon et al. 2012), the change in the pattern of C allocation in plantations may be due to the large C stock drive to the soil.

The differential input of N by N_2-fixing species appears not to be the major reason for increasing soil C storage (Forrester et al. 2013). The complementarity of niches (above- and belowground) (Tilman 1999; Tilman et al. 2001; Forrester et al. 2006), soil microbial sharing and increased fungal richness (Rachid et al. 2013, 2015), and stimulus to grow and microbial activity (Bini et al. 2013; Pereira et al. 2017; Santos et al. 2017) seem to prove that belowground dynamics commands soil C storage (Sokol and Bradford 2019), but studies in this sense need to be performed for mixed plantations.

The difference in the magnitude of C stocks of soils has a great association with mineralogy and texture (Feller and Beare 1997; Hassink 1997), hence the variations presented in Table 4.3. In general, Andisols from Hawaii have high specific surface phyllosilicate clays and naturally higher SOM contents due to the complexes formed with Al (i.e., noncrystalline Al hydroxide and Al-insoluble organic complexes) and allophanes (Feller and Beare 1997). Even in soils with a recognized SOM stability, the biologically fixed N input was responsible for a significant increase of the C and N stocks, showing the benefit of the N_2-fixing species in that environment, which presents a nonexistent water deficit (i.e., 4500 mm of annual precipitation).

Table 4.3 Soil C stocks (Mg ha^{-1}) in pure and mixed plantations of eucalyptus and N-fixing trees

Species	Country	Age years	Non-N-fixing Proportion	N-fixing	Soil layer cm	Carbon stock Mg ha^{-1}	N:P	Soil classification	Reference
Eucalyptus saligna	Hawaii	16	100	0	0–50	123.4[a]		Andosol	Kaye et al. (2000)
E. saligna + *Albizia falcataria*	Hawaii	16	75	25	0–50	129.7[a]		Andosol	Kaye et al. (2000)
E. saligna + *Albizia falcataria*	Hawaii	16	66	34	0–50	132[a]		Andosol	Kaye et al. (2000)
E. saligna + *Albizia falcataria*	Hawaii	16	50	50	0–50	136.1[a]		Andosol	Kaye et al. (2000)
E. saligna + *Albizia falcataria*	Hawaii	15	34	66	0–50	140.1[a]		Andosol	Kaye et al. (2000)
Albizia falcataria	Hawaii	15	0	100	0–50	148.8[a]		Andosol	Kaye et al. (2000)
E. globulus	Australia	8.8	100	0	0–30	66.3	30[b]	Acrisol	Forrester et al. (2013)
E. globulus + *Acacia mearnsii*	Australia	8.8	75	25	0–30	77.9	46[b]	Acrisol	Forrester et al. (2013)
E. globulus + *Acacia mearnsii*	Australia	8.8	50	50	0–30	84.5	49[b]	Acrisol	Forrester et al. (2013)
E. globulus + *Acacia mearnsii*	Australia		25	75	0–30	86.1	57[b]	Acrisol	Forrester et al. (2013)
Acacia mearnsii	Australia		0	100	0–30	82.8	66[b]	Acrisol	Forrester et al. (2013)
E. urograndis	Brazil	5	100	0	0–40	20.0	27[b]	Planosol	Rocha et al. (2019)); Santos et al. (2017)
A. mangium + *E. urograndis*	Brazil	5	50	50	0–40	22.4	49[b]	Planosol	Rocha et al. (2019); Santos et al. (2017)
A. mangium	Brazil	5	0	100	0–40	22.1	77[b]	Planosol	Rocha et al. (2019); Santos et al. (2017)
E. grandis	Brazil	7	100	0	0–40	17.2	7[c]	Planosol	Balieiro et al. (2008); Froufe (1999)
A. mangium + *E. grandis*	Brazil	7	50	50	0–40	23.8	16[c]	Planosol	Balieiro et al. (2008); Froufe (1999)
A. mangium	Brazil	7	0	100	0–40	14.2	19[c]	Planosol	Balieiro et al. (2008); Froufe (1999)
E. urograndis	Congo	7	100	0	0–25	15.9	62[d]	Arenosol	Koutika et al. (2014)
A. mangium + *E. urograndis*	Congo	7	50	50	0–25	17.8	75[d]	Arenosol	Koutika et al. (2014)

(continued)

Table 4.3 (continued)

Species	Country	Age years	Non-N-fixing Proportion	N-fixing	Soil layer cm	Carbon stock Mg ha^{-1}	N:P	Soil classification	Reference
A. mangium	Congo	7	0	100	0–25	16.7	59[d]	Arenosol	Koutika et al. (2014)
E. grandis (Bofete)	Brazil	5	100	0	0–15	27.6	—	Ferralsol	Voigtlaender et al. (2019)
A. mangium + *E. grandis* (Bofete)	Brazil	5	50	50	0–15	29.6	—	Ferralsol	Voigtlaender et al. (2019)
A. mangium (Bofete)	Brazil	5	0	100	0–15	26.7	—	Ferralsol	Voigtlaender et al. (2019)
E. urograndis (Luiz Antônio)	Brazil	6	100	0	0–15	22.2	—	FerralicArenosol	Voigtlaender et al. (2019)
A. mangium + *E. urograndis* (Luiz Antônio)	Brazil	6	50	50	0–15	24.1	—	FerralicArenosol	Voigtlaender et al. (2019)
A. mangium (Luiz Antônio)	Brazil	6	0	100	0–15	21.6	—	FerralicArenosol	Voigtlaender et al. (2019)
E. urograndis (Santana do Paraíso)	Brazil	6	100	0	0–15	35.1	—	Ferralsol	Voigtlaender et al. (2019)
A. mangium + *E. urograndis* (Santana do Paraíso)	Brazil	6	50	50	0–15	34.5	—	Ferralsol	Voigtlaender et al. (2019)
A. mangium (Santana do Paraíso)	Brazil	6	0	100	0–15	38.7	—	Ferralsol	Voigtlaender et al. (2019)
Casuarina equisetifolia	Porto Rico	4	0	100	0–40	65.1	191	Andosol	Parrota (1999)
Eucalyptus robusta	Porto Rico	4	100	0	0–40	75.7	124	Andosol	Parrota (1999)
Leucaena leucocephala	Porto Rico	4	0	100	0–40	75.6	150	Andosol	Parrota (1999)
C. equisetifolia + *E. robusta*	Porto Rico	4	50	50	0–40	65.9	133	Andosol	Parrota (1999)
C. equisetifolia + *L. leucocephala*	Porto Rico	4	50	50	0–40	56.6	179	Andosol	Parrota (1999)
E. robusta L. leucocephala	Porto Rico	4	50	50	0–40	61.7	146	Andosol	Parrota (1999)

[a]Values estimated from models adjusted by authors for monocultures and mixed stands
[b]N:P ratio estimated using the total of N and P (kg ha^{-1} y^{-1}) deposited by litterfall under each plantation
[c]N:P ratio estimated using N and P stocked (kg ha^{-1}) in soil (0–40 cm)
[d]N:P ratio calculated from N and P stocks cited by authors
[e]We classified isohyperthmeric typic troposamments as Andosol, with reservations

Thus, the feedback effect of the N input and NPP (Kaye et al. 2000), associated with the fact that N favors, in consonance with high fertility and humification of soil organic matter (Kirkby et al. 2011) justifies such results for Hawaii. On the other hand, the Ferralsols, in which the Brazilian plantations were established, are highly weathered and acidic and have low fertility, which contrast with the high physical aggregate stability that confers protection of soil C (Feller and Beare 1997; Silva et al. 2013). Thus, even under tropical conditions, plantings associated with N_2-fixing species can store significant amounts of C in the soil. It is interesting to note that even under a single soil class (Ferralsols), large variations in soil C stocks occur in plantations, due to variations in soil clay content and climate (Voigtlaender 2012; Bouillet et al. 2013; Voigtlaender et al. 2019). The Arenosols of Brazil and Congo, from which the main results of this class originated, are in turn naturally lower in amounts of C due to accelerated biochemical dynamics and reduced SOM protection, as well as poor structure and aggregation (Feller and Beare 1997; Zinn et al. 2002). However, it is a fact that even among the Arenosols, small variations in the clay contents of these soils can confer significant changes in the C stocks, as well as capping of the quartz grains by kaolinite and hematite or goethite type clays that increase the formation possibilities of organo-mineral complexes and C storage of these soils (Scheidegger et al. 1993; Donagemma et al. 2008).

The storage of C in the soil of mixed plantations is also due to the rate of decomposition of the deposited residues, which determines how fast the C is incorporated into the mineral phase. By comparing monocultures, Kaye et al. (2000), Resh et al. (2002), Forrester et al. (2013), Koutika et al. (2014), and Pereira et al. (2017) detected greater incorporation of C in soil under N_2-fixing species. However, considering the literature on mixed plantations the C stock in soil was higher but do not show differences in relation to monocultures of Eucalyptus and legume trees (Fig. 4.5) due to higher variability of data. In addition, C stocks occur preferentially in the more labile compartments of SOM fractions (Bini et al. 2013; Koutika and Mareschal 2017; Pereira et al. 2017).

These authors compared the C content of the microbial biomass as well as C and N of the soil organic fraction (physically fractionated between 2000 and 75 μm) and found that, in the four treatments, Eucalyptus without N fertilization, Eucalyptus + N fertilization, Eucalyptus + *Acacia mangium* (E50:A50), or pure *A. mangium* plantations, they differed and were significantly higher in plantings of *Acacia* or in mixed plantations, confirming positive changes in microbial indicators and increases in concentration and nutrient cycling in Ferralsols. When analyzing Table 4.3, it is possible to detect that, with the exception of allophanic soils, the storage of C in poor soils follows the N:P ratio up to a certain limit. This implies that the management of these plantations in oxidic soils with high P adsorption capacity (Lloyd et al. 2001) or sandy soil with low SOM and P reserves should require, in the short and medium terms, a special attention to the management of P fertilizer.

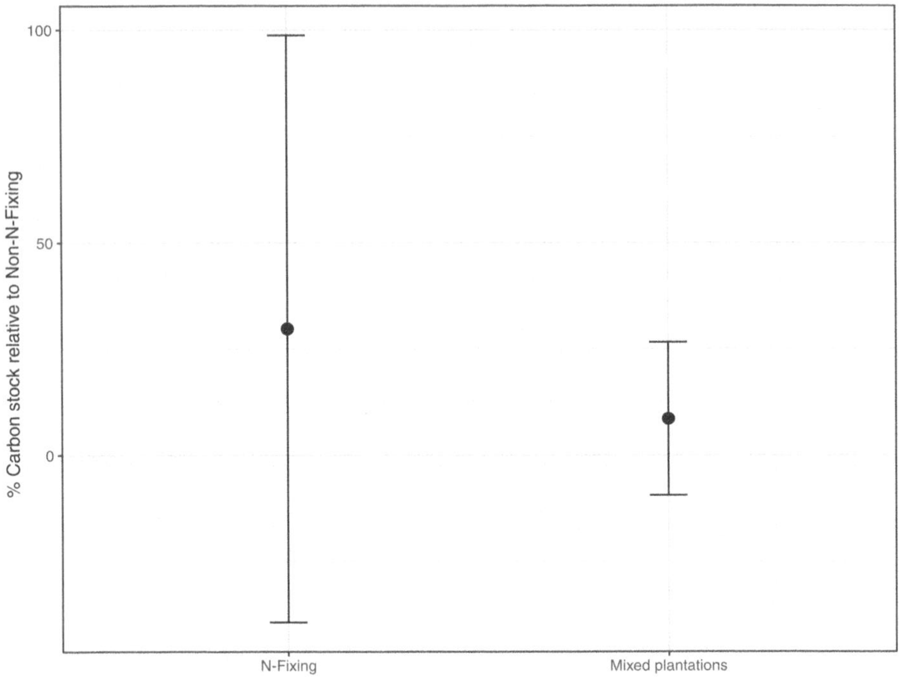

Fig. 4.5 Mean (●) and SD (whiskers) of carbon stock differences (%) of N_2-fixing and mixed plantations relative to non-N_2-fixing species plantations (Eucalyptus) ($n = 29$)

4.6 How to Improve the Soil C Stocks in Mixed-Species Plantations Under Tropical Conditions?

Based on the reviewed literature and data collected, it can be suggested that soil C stocks in planted forests can be increased with improved management practices to overcome environmental restraints and with a broader logic of available biological resources.

Planted forests in the tropics and mainly in Brazil occupy marginal soils, with a history of intense use. Thus, these soils present low fertility, high acidity, Al saturation, and adsorption capacity of P and S (Leal and Velloso 1973; Motta et al. 2002). However, huge differences in these attributes occur within the tropical climatic zone (Sanche 1997), and may partially justify changes in species responses under monoculture and mixed conditions described in this chapter. Although the responses of the forest plantations to liming and fertilization of the soil are small, the non-replacement of the exported nutrients leads to a decrease in the nutritional capital of the soil and the future commitment of these forests. The focus of foresters shall be on the construction of soil fertility under planted forests, with the adoption of practices and technologies that potentiate the above- and belowground growth of species. For the P case specifically, although most are in the non-labile form in soil, considered reserves exist in labile and moderately labile fractions (up to 7.7 and 15.5 Tg, respectively) (Withers et al. 2018) in areas under higher fertilizer regimes. Especially when combined with no till, correction of acidity, and adoption of the 4R principles of nutrient management (right rate, right time, right place, and right form) (INPI 2012) the stocks of soil natural capital related migth be at least maintained.

In Brazilian Eucalyptus plantations, accumulation in the aboveground biomass of N, P, K, Ca, and Mg in 100 Mg is in the range of 238–298 kg for N, 16–29 kg for P, 123–236 kg for K, 176–590 kg for Ca, and 40–92 kg for Mg. From these totals, 35% of N, 29% of P, 36% of K, and 21% of Ca and Mg are potentially exportable by wood (Santana et al. 2008). In the average 7-year rotations for Eucalyptus, Gonçalves et al. (2013) recommend up to 2 Mg ha^{-1} of lime, 60–80 kg ha^{-1} of N, 60–80 kg ha^{-1} of P_2O_5, and 140–160 kg ha^{-1} of K_2O. For trace elements around 1–5 kg ha^{-1} of B are applied, depending on the local water deficit, and 1 kg ha^{-1} of Cu and Zn. Roughly, the natural nutritional capital of the soil (Sanche 1997; Dominati et al. 2010) seems to be compromised, with detrimental consequences to the organic and inorganic nutrient reserves of these forests.

Managing mixed plantings or rotations with legumes that associate with bacteria that fix N_2 from the atmosphere will lead to substantial changes in fertilizer management given the changes in biogeochemical cycles of nutrients and differentiated nutrient exports (Chap. 3). Legumes have a differentiated demand for P in relation to non-N_2-fixing species (Vádez et al. 1995; Inagaki et al. 2009, Venterink 2011). *Acacia mangium*, the species most studied under Brazilian conditions, presents high absorption of P in the seedling and adult phase (Inagaki et al. 2011; Santos et al. 2017), and it is very efficient in recycling internally this absorbed P, so that it deposits litter with a high N:P ratio (Inagaki et al. 2010; Santos et al. 2018). Since plants and microorganisms compete strongly for soil P, affecting key soil biological processes and growth of tropical forests (Wieder et al. 2009; Hättenschwiler et al. 2011), strategies to reduce species competition for soil P or to increase complementarity in recycling due to P deposition should be pursued jointly with the adoption of management practices that favor the permanence of crop residues on the soil and SOM.

The recommendations of crop fertilization in Brazil are based on response curves of crop production (shoot biomass or stem in the case of Eucalyptus or its productivity) in relation to the applied fertilizer dose or its availability (i.e., concentration) in the soil (Cantarutti et al. 2007). Little or no attention has been given to changes in the C allocated to roots by planted forests and their relationship to fertility management and soil C stocks. Based on current hypotheses that the belowground inputs provided by the microbial formation pathway form mineral-stabilized soil C more efficiently than aboveground inputs partly due to the greater efficiency of formation by the rhizosphere microbial community relative to the bulk soil community (Sokol and Bradford 2019), it can be considered that, by managing the soil chemical limitations (i.e., acidity, P, N, S, Zn concentration, among others), there will be higher plant growth and consequently root growth, which together will increase soil C stocks. Recent works (Kirkby et al. 2011, 2014) also demonstrate the possibility of altering the potential for soil C accumulation and sequestration of soils from the introduction of nutrients to the soil, correcting the stoichiometry among C, N, P, and S of the heavy fraction of organic matter (i.e., humidified). The formation of a "new-fine fraction of soil organic matter," the most stable C component in soil, increased threefold by increasing the residues with supplementary nutrients, which in other words implies to say that we can manipulate the nutrients in favor of C sequestration and restoration of the fertility. In other words, although *Eucalyptus* or *Acacia* (or other leguminous species) is tolerant to acidity and high saturation of Al, it does not mean that the practice of soil fertilizing and correction shall not be stimulated. On the other hand, gains in productivity in oxidic soils under the correct management of soil fertility are always accompanied by increased biomass and microbial activity (i.e., basal respiration and enzymatic activity) and, consequently, increases in SOM levels.

4.7 Final Considerations

As shown in this chapter, it seems clear that the establishment of mixed plantations of *Eucalyptus* and *Acacia mangium* represents an alternative to increase soil C stocks in marginal lands. However, the success of these plantations is not the guarantee to increase C stocks. It is necessary to take into account the climatic and soil conditions of the site for the intercropped species, to perform conservationist practices of soil preparation, residue management, and conservative harvesting practices. Better results were obtained where exist appropriated climate condition for Acacia and oligotrophic soils (e.g. sandy soils).

Finding a new set of other species (including N_2-fixing and non-N_2-fixing) that have ecological and economic interest seems to be the main future challenge, although this is not such a simple task. It could begin through a list of priority legume tree species with litter traits that, together with eucalypt, can promote the increase of the global stand biomass production (and also belowground) and improvements in soil properties and functions. It is interesting to note that C sequestration is only an ecosystem service that the soil can provide and that the mix of non-N_2-fixing with N_2-fixing trees may provide many other ecosystem services, such as flood mitigation; greenhouse gas regulation; filtration and recycling of nutrients; and biodiversity preservation among others. In Chap. 10, these services are presented and discussed.

New researches should also elucidate under which arrangements and trees density could provide positive ecological interactions (complementarity and competitive reduction) that can increase the biomass production (above- and belowground) and accelerate litter decomposition and soil C sequestration.

Finally, due to the close relationship between the N and C cycle, it has been reported that as a result of the presence of N_2-fixing trees in *Eucalyptus* plantations positive changes in soil N stocks are more pronounced than for C (Bernhard Reversat 1996; Voigtlaender 2012; Rachid et al. 2013; Voigtlaender et al. 2019; Rocha et al.

2019). These general results confirm that the presence of legumes in *Eucalyptus* plantation might contribute to reducing the need for mineral N fertilization in the long term and increasing the soil nutritional capital. New mixtures of species and arrangements of mixed plantations in combination with nutrient management could enhance C sequestration and produce a more stable organic matter (humus).

Acknowledgments To the National Council for Scientific and Technological Development (CNPq) for the Productivity Grant given to the first author. We also thank the Fundação de Amparo à Pesquisa do Estado de São Paulo (FAPESP) and the Fundação de Amparo à Pesquisa do Estado do Rio de Janeiro (FAPERJ) for the financial support of our research, the field work and laboratory analyses and to Coordination of Superior Level Staff Improvement (CAPES) for the grants to the students involved.

References

Andrade AG, Costa GS, Faria SM (2000) Deposição e decomposição da serapilheira em povoamentos de Mimosa caesalpiniifolia, Acacia mangium e Acacia holosericea com quatro anos de idade em planossolo. Revista Brasileira de Ciência do Solo 24(4):777–785

Assis PCR, Stone LF, Medeiros JC et al (2015) Atributos físicos do solo em sistemas de integração lavoura-pecuária-floresta. Rev Bras Eng Agr Amb 19:309–316

Attiwill PM, Adams MA (1993) Nutrient cycling in forests. New Phytol 124:561–582

Bachega LR, Bouillet J-P, de Cássia Piccolo M et al (2016) Decomposition of Eucalyptus grandis and Acacia mangium leaves and fine roots in tropical conditions did not meet the Home Field Advantage hypothesis. Forest Ecol Manag 359:33–43

Bakker MA, Carreño-Rocabado G, Poorter L (2011) Leaf economics traits predict litter decomposition of tropical plants and differ among land use types. Funct Ecol 25:473–483

Balieiro F d C, Dias LE, Franco AA et al (2005) Acúmulo de nutrientes na parte aérea, na serapilheira acumulada sobre o solo e decomposição de filódios de *Acacia mangium* Willd. Cienc Florest 14:59–65

Balieiro F d C, Franco AA, Pereira MG et al (2004) Contribution of litter and nitrogen to soil under *Pseudosamanea guachapele* and *Eucalyptus grandis* plantations. Pesqui Agropecu Bras 39:597–601

Balieiro F d C, Pereira MG, Alves BJR et al (2008) Soil carbon and nitrogen in pasture soil reforested with eucalyptus and guachapele. Rev Bras Cienc Solo 32:1253–1260

Balieiro FC, Franco AA, Pereira MG, Campello EF, Faria SM, Dias LE (2010) Alves BJR Acúmulo e distribuição de biomassa e nutrientes na parte aérea de *Pseudosamanea guachapele* e *Eucalyptus grandis* em consórcio e monocultivos. Embrapa Solos, Rio de Janeiro. Boletim de Pesquisa

Berg B (2018) Decomposing litter, limit values, humus accumulation, locally and regionally. Appl Soil Ecol 123:494–508

Bernhard Reversat F (1996) Nitrogen cycling in tree plantations grown on a poor sandy savanna soil in Congo. Appl Soil Ecol 4:161–172

Bhatti JS, Comerford NB, Johnston CT (1998) Influence of oxalate and soil organic matter on sorption and desorption of phosphate onto a spodic horizon. Soil Sci Soc Am J 62:1089–1095

Bini D, Figueiredo AF, da Silva MCP et al (2013) Microbial biomass and activity in litter during the initial development of pure and mixed plantations of *Eucalyptus grandis* and *Acacia mangium*. Rev Bras Cienc Solo 37:76–85

Binkley D, Dunkin KA, DeBell D, Ryan MG (1992) Production and nutrient cycling in mixed plantations of eucalyptus and albizia in Hawaii. For Sci 38:393–408

Binkley D, Ryan MG (1998) Net primary production and nutrient cycling in replicated stands of *Eucalyptus saligna* and *Albizia falcataria*. For Ecol Manag 112:79–85

Binkley D, Giardina C, Bashkin MA (2000) Soil phosphorus pools and supply under the influence of *Eucalyptus saligna* and nitrogen-fixing *Albizia facaltaria*. For Ecol Manag 128:241–247

Bonan GB (2008) Forests and climate change: forcings, feedbacks, and the climate benefits of forests. Science 320:1444–1449. https://doi.org/10.1126/science.1155121

Bouillet J-P, Laclau J-P, Gonçalves JL de M et al (2013) Eucalyptus and Acacia tree growth over entire rotation in single- and mixed-species plantations across five sites in Brazil and Congo. For Ecol Manag 301:89–101. https://doi.org/10.1016/j.foreco.2012.09.019

Bowen GD, Nambiar ES (1984) Nutrition of plantation forests. Academic, London

Briones MJI, Ineson P (1996) Decomposition of eucalyptus leaves in litter mixtures. Soil Biol Biochem 28:1381–1388

Cannell MGR (1989) Physiological basis of wood production: A review. Scand J Forest Res 4:459–490. https://doi.org/10.1080/02827588909382582

Cantarutti RB, Barros ND, Martinez HEP, Novais RF (2007) Avaliação da fertilidade do solo e recomendação de fertilizantes. Fertilidade do solo. Sociedade Brasileira de Ciência do Solo, Viçosa, pp 769–850

Castellano MJ, Mueller KE, Olk DC et al (2015) Integrating plant litter quality, soil organic matter stabilization, and the carbon saturation concept. Glob Change Biol 21:3200–3209. https://doi.org/10.1111/gcb.12982

Chaer GM, Tótola MR (2007) Impact of organic residue management on soil quality indicators during replanting of eucalypt stands. Rev Bras Cienc Solo 31:1381–1396. https://doi.org/10.1590/S0100-06832007000600016

Cook RL, Binkley D, Stape JL (2016) *Eucalyptus* plantation effects on soil carbon after 20 years and three rotations in Brazil. For Ecol Manag 359:92–98

Cornwell WK, Cornelissen JHC, Amatangelo K et al (2008) Plant species traits are the predominant control on litter decomposition rates within biomes worldwide. Ecol Lett 11:1065–1071. https://doi.org/10.1111/j.1461-0248.2008.01219.x

Costa GS, de Andrade AG, de Faria SM (1998) Aporte de nutrientes pela serapilheira de Mimosa caesalpiniifolia (Sabiá) com seis anos de idade. In: 3 Simpósio Nacional de Recuperação de Áreas Degradadas

Costa GS, da Gama-Rodrigues AC, Cunha GdM (2005) Decomposição e liberação de nutrientes da serapilheira foliar em povoamentos de Eucalyptus grandis no norte fluminense. Revista Árvore 29(4):563–570

Cotrufo MF, Wallenstein MD, Boot CM et al (2013) The Microbial Efficiency-Matrix Stabilization (MEMS) framework integrates plant litter decomposition with soil organic matter stabilization: do labile plant inputs form stable soil organic matter? Glob Change Biol 19:988–995. https://doi.org/10.1111/gcb.12113

de Campos M, Antonangelo JA, Alleoni LRF (2016) Phosphorus sorption index in humid tropical soils. Soil Till Res 156:110–118. https://doi.org/10.1016/j.still.2015.09.020

Davidson EA, Hirsch AI (2001) Fertile forest experiments. Nature 411:431–433

Denef K, Six J (2005) Clay mineralogy determines the importance of biological versus abiotic processes for macroaggregate formation and stabilization. Eur J Soil Sci 56:469–479

Denef K, Six J, Paustian K, Merckx R (2001) Importance of macroaggregate dynamics in controlling soil carbon stabilization: short-term effects of physical disturbance induced by dry-wet cycles. Soil Biol Biochem 33:2145–2153

Dominati E, Patterson M, Mackay A (2010) A framework for classifying and quantifying the natural capital and ecosystem services of soils. Ecol Econ 69:1858–1868. https://doi.org/10.1016/j.ecolecon.2010.05.002

Donagemma GK, Ruiz HA, Alvarez VVH et al (2008) Fósforo remanescente em argila e silte retirados de Latossolos após pré-tratamentos na análise textural. Rev Bras Cienc Solo 32:1785–1791. https://doi.org/10.1590/S0100-06832008000400043

Doran JW, Parkin TB (1994) Defining and assessing soil quality. In: Doran JW, Coleman DC, Bezdicek DF, Stewart BA (eds) Defining soil quality for a sustainable environment, SSSA

Special Publication, vol 35. Soil Science Society of America/The American Society of Agronomy, Madison, pp 1–21

Doughty CE, Santos-Andrade PE, Shenkin A et al (2018) Tropical forest leaves may darken in response to climate change. Nature Ecol Evol 2(12):1918. https://doi.org/10.1038/s41559-018-0716-y

Epron D, Marsden C, ThongoM'Bou A et al (2009) Soil carbon dynamics following afforestation of a tropical savannah with Eucalyptus in Congo. Plant Soil 323:309–322. https://doi.org/10.1007/s11104-009-9939-7

Epron D, Nouvellon Y, Mareschal L et al (2013) Partitioning of net primary production in Eucalyptus and Acacia stands and in mixed-species plantations: Two case-studies in contrasting tropical environments. Forest Ecol Manag 301:102–111. https://doi.org/10.1016/j.foreco.2012.10.034

Fanin N, Bertrand I (2016) Aboveground litter quality is a better predictor than belowground microbial communities when estimating carbon mineralization along a land-use gradient. Soil Biol Biochem 94:48–60. https://doi.org/10.1016/j.soilbio.2015.11.007

Food and Agriculture Organization – FAO (2015) Global forest resources assessment 2015: How have the world's forests changed? FAO, Rome, p 253

Fearnside PM (2000) Global warming and tropical land-use change: greenhouse gas emissions from biomass burning, decomposition and soils in forest conversion, shifting cultivation and secondary vegetation. Clim Chang 46:115–158. https://doi.org/10.1023/A:1005569915357

Feller C, Beare MH (1997) Physical control of soil organic matter dynamics in the tropics. Geoderma 79:69–116. https://doi.org/10.1016/S0016-7061(97)00039-6

Ferreira RLC, Lira Junior M d A, da Rocha MS, dos Santos MVF, Lira M d A, Barreto LP (2007) Deposição e acúmulo de matéria seca e nutrientes em serapilheira em um bosque de sabiá (Mimosa caesalpiniifolia Benth.). Revista Árvore 31(1):7–12

Fialho RC, Zinn YL (2014) Changes in soil organic carbon under eucalyptus plantations in Brazil: a comparative analysis. Land Degrad Develop 25:428–437. https://doi.org/10.1002/ldr.2158

Fisk M, Santangelo S, Minick K (2015) Carbon mineralization is promoted by phosphorus and reduced by nitrogen addition in the organic horizon of northern hardwood forests. Soil Biol Biochem 81:212–218. https://doi.org/10.1016/j.soilbio.2014.11.022

Forrester DI, Cowie AL, Bauhus J et al (2006) Effects of changing the supply of nitrogen and phosphorus on growth and interactions between *Eucalyptus globulus* and *Acacia mearnsii* in a Pot trial. Plant Soil 280:267–277. https://doi.org/10.1007/s11104-005-3228-x

Forrester DI, Pares A, O'Hara C et al (2013) Soil organic carbon is increased in mixed-species plantations of eucalyptus and nitrogen-fixing acacia. Ecosystems 16:123–132. https://doi.org/10.1007/s10021-012-9600-9

Fortes JLO (2000) Reabilitação de depósito de rejeito do refino de bauxita com o uso de resíduos industriais e leguminosas arbóreas. Tese, Universidade Federal do Rural do Rio de Janeiro

Franzluebbers AJ (2002) Water infiltration and soil structure related to organic matter and its stratification with depth. Soil Till Res 66:197–205. https://doi.org/10.1016/S0167-1987(02)00027-2

Fraser LH, Pither J, Jentsch A et al (2015) Worldwide evidence of a unimodal relationship between productivity and plant species richness. Science 349:302–305. https://doi.org/10.1126/science.aab3916

Froufe LCM (1999) Decomposição de serapilheira e aporte de nutrientes em plantios puros e consorciados de eucalyptus grandis maiden, pseudosamanea guachapele dugand e acacia mangium willd. Tese, Universidade Federal Rural do Rio de Janeiro

Gartner TB, Cardon ZG (2004) Decomposition dynamics in mixed-species leaf litter. Oikos 104:230–246

Gholz HL, Wedin DA, Smitherman SM et al (2000) Long-term dynamics of pine and hardwood litter in contrasting environments: toward a global model of decomposition. Glob Change Biol 6:751–765. https://doi.org/10.1046/j.1365-2486.2000.00349.x

Giller KE, Cadisch G (1995) Future benefits from biological nitrogen fixation: An ecological approach to agriculture. Plant Soil 174(1–2):255–277

Gonçalves JL d M, Alvares CA, Higa AR et al (2013) Integrating genetic and silvicultural strategies to minimize abiotic and biotic constraints in Brazilian eucalypt plantations. Forest Ecol and Manag 301:6–27. https://doi.org/10.1016/j.foreco.2012.12.030

Güsewell S (2004) N: P ratios in terrestrial plants: variation and functional significance. New Phytol 164:243–266. https://doi.org/10.1111/j.1469-8137.2004.01192.x

Hassink J (1997) The capacity of soils to preserve organic C and N by their association with clay and silt particles. Plant Soil 191:77–87. https://doi.org/10.1023/A:1004213929699

Hättenschwiler S, Coq S, Barantal S, Handa IT (2011) Leaf traits and decomposition in tropical rainforests: revisiting some commonly held views and towards a new hypothesis. New Phytol 189:950–965. https://doi.org/10.1111/j.1469-8137.2010.03483.x

Hoosbeek MR, Scarascia-Mugnozza GE (2009) Increased litter build up and soil organic matter stabilization in a poplar plantation after 6 years of atmospheric CO_2 enrichment (FACE): final results of POP-EuroFACE compared to other forest FACE experiments. Ecosystems 12:220–239. https://doi.org/10.1007/s10021-008-9219-z

van Huysen TL, Perakis SS, Harmon ME (2016) Decomposition drives convergence of forest litter nutrient stoichiometry following phosphorus addition. Plant Soil 406:1–14. https://doi.org/10.1007/s11104-016-2857-6

IBA (2017). Relatório Ibá 2017

Inagaki M, Kamo K, Miyamoto K et al (2011) Nitrogen and phosphorus retranslocation and N:P ratios of litterfall in three tropical plantations: luxurious N and efficient P use by *Acacia mangium*. Plant Soil 341:295–307. https://doi.org/10.1007/s11104-010-0644-3

Inagaki M, Kamo K, Titin J et al (2010) Nutrient dynamics through fine litterfall in three plantations in Sabah, Malaysia, in relation to nutrient supply to surface soil. Nutr Cycl Agroecosyst 88:381–395. https://doi.org/10.1007/s10705-010-9364-6

IPNI (2012) In: Bruulsema TW, Fixen PE, Sulewski GD (eds) 4R Plant nutrition manual: a manual for improving the management of plant nutrition. International Plant Nutrition Institute, Norcross, GA. http://www.ipni.net/article/IPNI-3255 (22 November 2017)

Janzen HH (2004) Carbon cycling in earth systems—a soil science perspective. Agric Ecosyst Environ 104(3):399–417

Jesus GL d, Silva IR, Almeida LFJ et al (2015) Produtividade do eucalipto, atributos físicos do solo e frações da matéria orgânica influenciadas pela intensidade de tráfego e resíduos de colheita. Rev Bras Cienc Solo 39:1190–1203. https://doi.org/10.1590/01000683rbcs20140494

Kaye JP, Resh SC, Kaye MW, Chimner RA (2000) Nutrient and carbon dynamics in a replacement series of eucalyptus and albizia trees. Ecology 81:3267–3273. https://doi.org/10.1890/0012-9658(2000)081[3267:NACDIA]2.0.CO;2

Kindel A, Garay I, do CCAFS, Lima JA de S (2003) Quantificação dos horizontes húmicos e dinâmica da decomposição de material foliar em solos florestais. Comunicado Técnico EMBRAPA 1:1–8

Kirkby CA, Kirkegaard JA, Richardson AE et al (2011) Stable soil organic matter: A comparison of C:N:P:S ratios in Australian and other world soils. Geoderma 163:197–208. https://doi.org/10.1016/j.geoderma.2011.04.010

Kirkby CA, Richardson AE, Wade LJ et al (2014) Nutrient availability limits carbon sequestration in arable soils. Soil Biol Biochem 68:402–409. https://doi.org/10.1016/j.soilbio.2013.09.032

Koutika L-S, Epron D, Bouillet J-P, Mareschal L (2014) Changes in N and C concentrations, soil acidity and P availability in tropical mixed acacia and eucalypt plantations on a nutrient-poor sandy soil. Plant Soil 379:205–216. https://doi.org/10.1007/s11104-014-2047-3

Koutika L-S, Mareschal L (2017) Acacia and eucalypt change P, N and C concentrations in POM of Arenosols in the Congolese coastal plains. Geoderma Reg 11:37–43. https://doi.org/10.1016/j.geodrs.2017.07.009

Koutika L-S, Richardson DM (2019) Acacia mangium Willd.: benefits and threats associated with its increasing use around the world. Forest Ecosyst 6:2. https://doi.org/10.1186/s40663-019-0159-1

Kuzyakov Y, Friedel JK, Stahr K (2000) Review of mechanisms and quantification of priming effects. Soil Biol Biochem 32:1485–1498

Laclau J-P, de MoraesGonçalves JL, Stape JL (2013) Perspectives for the management of eucalypt plantations under biotic and abiotic stresses. For Ecol Manage 301:1–5

Laclau J-P, Ranger J, de MoraesGonçalves JL et al (2010) Biogeochemical cycles of nutrients in tropical Eucalyptus plantations: Main features shown by intensive monitoring in Congo and Brazil. Forest Ecol and Manag 259:1771–1785. https://doi.org/10.1016/j.foreco.2009.06.010

Lal R (2005) Forest soils and carbon sequestration. Forest Ecol and Manag 220:242–258. https://doi.org/10.1016/j.foreco.2005.08.015

Landsberg JJ, Gower ST (1997) Soil organic matter and decomposition (Chapter 6). In: Landsberg JJ, Gower ST (eds) Applications of physiological ecology to forest management. Academic, San Diego, pp 161–184

Lange M, Eisenhauer N, Sierra CA et al (2015) Plant diversity increases soil microbial activity and soil carbon storage. Nat Commun 6:6707. https://doi.org/10.1038/ncomms7707

Leal JR, Velloso ACX (1973) Adsorção de fosfato em latossolos sob vegetação de cerrado. Pesqui Agropecu Bras 8:81–88

Lehmann J, Kleber M (2015) The contentious nature of soil organic matter. Nature 528:60–68. https://doi.org/10.1038/nature16069

Leite FP, Barros NF, Novais RF, Fabres AS (1998) Acúmulo e distribuição de nutrientes em Eucalyptus grandis sob diferentes densidades populacionais. Rev Bras Cienc Solo 22:419–426. https://doi.org/10.1590/S0100-06831998000300007

Lima AMN, Silva IR, Neves JCL et al (2006) Soil organic carbon dynamics following afforestation of degraded pastures with eucalyptus in southeastern Brazil. Forest Ecol and Manag 235:219–231. https://doi.org/10.1016/j.foreco.2006.08.331

Lloyd J, Bird MI, Veenendaal EM, Kruijt B (2001) Should phosphorus availability be constraining moist tropical forest responses to increasing CO_2 concentrations? In: Schulze ED, Heimann M, Harrison S, Holland E, Lloyd J, Prentice IC, Schimel D (eds) Global biogeochemical cycles in the climate system. Academic, San Diego, pp 95–114

Malhi Y, Wood D, Baker TR et al (2006) The regional variation of aboveground live biomass in old-growth Amazonian forests. Glob Change Biol 12:1107–1138. https://doi.org/10.1111/j.1365-2486.2006.01120.x

Maquere V, Laclau JP, Bernoux M et al (2008) Influence of land use (savanna, pasture, Eucalyptus plantations) on soil carbon and nitrogen stocks in Brazil. Eur J Soil Sci 59:863–877. https://doi.org/10.1111/j.1365-2389.2008.01059.x

Marklein AR, Winbourne JB, Enders SK et al (2016) Mineralization ratios of nitrogen and phosphorus from decomposing litter in temperate versus tropical forests. Glob Ecol Biogeogr 25:335–346. https://doi.org/10.1111/geb.12414

Mochiutti S, de Queiroz JAL, Junior NJM (2006) Produção de serapilheira e retorno de nutrientes de um povoamento de taxi-branco e de uma floresta secundária no Amapá. Boletim de Pesquisa Florestal, Colombo, pp 3–20

Motta PEF, Curi N, Siqueira JO, Van Raij B, Neto AEF, Lima JM (2002) Adsorção e formas de fósforo em latossolos: influência da mineralogia e histórico de uso. Revista Brasileira de Ciência do Solo 26(2):349–359

Negi JD, Sharma SC (1996) Mineral nutrition and resource conservation in Eucalyptus plantation and other forest covers in India. In: Attwill PM, Adams MA (eds) Nutrition of eucalyptus. CSIRO, Collingwood, pp 399–416

Nottingham AT, Whitaker J, Turner BL et al (2015) Climate warming and soil carbon in tropical forests: insights from an elevation gradient in the Peruvian Andes. Bio Science 65:906–921. https://doi.org/10.1093/biosci/biv109

Nouvellon Y, Laclau J-P, Epron D et al (2012) Production and carbon allocation in monocultures and mixed-species plantations of *Eucalyptus grandis* and *Acacia mangium* in Brazil. Tree Physiol 32:680–695. https://doi.org/10.1093/treephys/tps041

Novais RF, Barros N (1997) Sustainable agriculture and forestry production systems on acid soils: Phosphorus as a case-study. In: Moniz AC, Furlani AMC, Schaffert RE, Fageria NK, Rosolem CA, Cantarella H (eds) Plant-soil interactions as low pH: Sustainable agriculture and forestry production, Viçosa, pp 39–51

Novotny EH, Rodrigues AF, Balieiro F de C et al (2013) Avaliação das alterações estruturais da serapilheira de florestas plantadas em decomposição por meio da espectroscopia vibracional

(FTIR) aliada à análise de componentes principais (PCA). In: Embrapa Solos-Resumo em anais de congresso (ALICE)

Ometto JC (1981) Bioclimatologia vegetal. Agronomica Ceres, Sao Paulo

Oorts K, Vanlauwe B, Cofie OO et al (2000) Charge characteristics of soil organic matter fractions in a Ferric Lixisol under some multipurpose trees. Agrofor Syst 48:169

Pandey R, Zinta G, AbdElgawad H et al (2015) Physiological and molecular alterations in plants exposed to high [CO_2] under phosphorus stress. Biotechnol Adv 33:303–316. https://doi.org/10.1016/j.biotechadv.2015.03.011

Parrotta JA (1999) Productivity, nutrient cycling, and succession in single- and mixed-species plantations of Casuarina equisetifolia, Eucalyptus robusta, and Leucaena leucocephala in Puerto Rico. For Ecol Manage 124(1):45–77

Payn T, Carnus JM, Freer-Smith P, Kimberley M, Kollert W et al (2015) Changes in planted forests and future global implications. Forest Ecol and Manag 352:57–67

Pereira AP d A, de Andrade PAM, Bini D et al (2017) Shifts in the bacterial community composition along deep soil profiles in monospecific and mixed stands of *Eucalyptus grandis* and *Acacia mangium*. PLoS One 12:e0180371. https://doi.org/10.1371/journal.pone.0180371

Ponge J-F (2013) Plant–soil feedbacks mediated by humus forms: A review. Soil Biol Biochem 57:1048–1060. https://doi.org/10.1016/j.soilbio.2012.07.019

Poorter L, Bongers F (2006) Leaf traits are good predictors of plant performance across 53 rain forest species. Ecology 87:1733–1743

Qiao Y, Miao S, Silva LCR, Horwath WR (2014) Understory species regulate litter decomposition and accumulation of C and N in forest soils: A long-term dual-isotope experiment. Forest Ecol and Manag 329:318–327. https://doi.org/10.1016/j.foreco.2014.04.025

Quesada CA, Lloyd J, Schwarz M et al (2010) Variations in chemical and physical properties of Amazon forest soils in relation to their genesis. Biogeosciences 7:1515–1541. https://doi.org/10.5194/bg-7-1515-2010

Rachid CTCC, Balieiro FC, Fonseca ES et al (2015) Intercropped silviculture systems, a key to achieving soil fungal community management in eucalyptus plantations. PLoS One 10:e0118515. https://doi.org/10.1371/journal.pone.0118515

Rachid CTCC, Balieiro FC, Peixoto RS et al (2013) Mixed plantations can promote microbial integration and soil nitrate increases with changes in the N cycling genes. Soil Biol Biochem 66:146–153. https://doi.org/10.1016/j.soilbio.2013.07.005

Rachid CTCC (2013) Biodisponibilidade de nutrientes e estrutura microbiana do sistema solo-serapilheira em floresta plantada mista de *Eucalyptus urograndis* e *Acacia mangium*. Universidade Federal do Rio de Janeiro, Rio de Janeiro, p 115. (PhD in Science Thesis)

Raven JA, Franco AA, de Jesus EL, Jacob-Neto J (1990) H+ extrusion and organic-acid synthesis in N2-fixing symbioses involving vascular plants. New Phytol 114:369–389. https://doi.org/10.1111/j.1469-8137.1990.tb00405.x

Reis MDGF, Barros NF (1990) Ciclagem de nutrientes em plantio de eucalipto. In: de Barros NF, de Novaes RF (eds) Relação Solo-Eucalipto. Centro de ciências agrárias, Viçosa

Reis MDGF, Kimmins JP, de Rezende GC, de BNF (1985) Acumulo de biomassa em uma sequência de idade de *Eucalyptus grandis* plantado no cerrado em duas áreas com diferentes produtividades. Rev Arvore:149–162

Resh SC, Binkley D, Parrotta JA (2002) Greater soil carbon sequestration under nitrogen-fixing trees compared with eucalyptus species. Ecosystems 5:217–231. https://doi.org/10.1007/s10021-001-0067-3

Rigatto PA, Dedecek RA, Mattos JLM (2005) Influência dos atributos do solo sobre a produtividade de Pinus taeda. Revi Árv 29:701–709. https://doi.org/10.1590/S0100-67622005000500005

Rocha PV, Ataíde DHS, Lima JSS, Santos FM, Chaer GM, Balieiro FC (2019) Preparo do solo e leguminosa arbórea fixadora de N_2 consorciada com eucalipto afetam os estoques de carbono e nitrogênio de solo arenoso. Simpósio Brasileiro de Solos Arenosos, III, Campo Grande

Roscoe R, Machado PLOA (2002) Fracionamento físico do solo em estudos da matéria orgânica. Embrapa Agropecuária Oeste, Dourados, p 86

Sanchez P (1979) Soil fertility and conservations for agroforestry systems in the humid tropics of Latin America. In: O M PAH (ed) Soils research in agroforestry. ICRAF, Nairobi

Sanches PA (1997) Changing tropical soil fertility paradigms: from Brazil to Africa and back. In: Moniz AC, Furlani AMC, Schaffert RE, Fageria NK, Rosolem CA, Cantarella H (eds) Plant-soil interactions at low pH: sustainable agriculture and forestry production. Viçosa, Sociedade Brasileira de Ciência do Solo, pp 19–28

Sands R, Mulligan DR (1990) Water and nutrient dynamics and tree growth. Forest Ecol Manag 30:91–111. https://doi.org/10.1016/0378-1127(90)90129-Y

Santana GS, Knicker H, González-Vila FJ et al (2015) The impact of exotic forest plantations on the chemical composition of soil organic matter in Southern Brazil as assessed by Py–GC/MS and lipid extracts study. Geoderma Reg 4:11–19. https://doi.org/10.1016/j.geodrs.2014.11.004

Santana RC, de Barros NF, Novais RF et al (2008) Alocação de nutrientes em plantios de eucalipto no Brasil. Rev Bras Cien Solo 32:2723–2733. https://doi.org/10.1590/S0100-06832008000700016

Santos FM, Balieiro F de C, Ataíde DH dos S et al (2016) Dynamics of above ground biomass accumulation in monospecific and mixed-species plantations of Eucalyptus and Acacia on a Brazilian Sandy soil. Forest Ecol Manag 363:86–97. https://doi.org/10.1016/j.foreco.2015.12.028

Santos FM, Balieiro F de C, Fontes MA, Chaer GM (2018) Understanding the enhanced litter decomposition of mixed-species plantations of Eucalyptus and *Acacia mangium*. Plant Soil 423:141

Santos FM, Chaer GM, Diniz AR, Balieiro F de C (2017) Nutrient cycling over five years of mixed-species plantations of Eucalyptus and Acacia on a sandy tropical soil. Forest Ecol Manag 384:110–121. https://doi.org/10.1016/j.foreco.2016.10.041

Scheidegger A, Borkovec M, Sticher H (1993) Coating of silica sand with goethite: preparation and analytical identification. Geoderma 58:43–65. https://doi.org/10.1016/0016-7061(93)90084-X

Schlesinger WH, Lichter J (2001) Limited carbon storage in soil and litter of experimental forest plots under increased atmospheric CO_2. Nature 411:466. https://doi.org/10.1038/35078060

Schmidt MWI, Torn MS, Abiven S, Dittmar T, Guggenberger G, Janssens IA, Kleber M, Kögel-Knabner I, Lehmann J, Manning DAC, Nannipieri P, Rasse DP, Weiner S, Trumbore SE (2011) Persistence of soil organic matter as an ecosystem property. Nature 478(7367):49–56

Senesi N, Loffredo E (1999) The chemistry of soil organic matter. In: Sparks DL (ed) Soil physical chemistry, pp 239–370

Silva LCR, Corrêa RS, Doane TA et al (2013) Unprecedented carbon accumulation in mined soils: the synergistic effect of resource input and plant species invasion. Ecol Appl 23:1345–1356. https://doi.org/10.1890/12-1957.1

Silva CF, Carmo ER, Martins MA, Freitas MSM, Pereira MG, Silva EMR (2015) Deposition and nutritional quality of the litter of pure stands of *Eucalyptus camaldulensis* and *Acacia mangium*. Biosci J 31:1081–1091

Silva EV, Goncalves JLM, Coelho SRF, Moreira RM, Mello SLM, Bouillet J-P, Jourdan C, Laclau J-P (2009) Dynamics of fine root distribution after establishment of monospecific and mixed species plantations of *Eucalyptus grandis* and *Acacia mangium*. Plant Soil 325:305–318

Sokol NW, Bradford MA (2019) Microbial formation of stable soil carbon is more efficient from belowground than aboveground input. Nat Geosci 12(1):46–53

Sollins P, Homann P, Caldwell BA (1996) Stabilization and destabilization of soil organic matter: mechanisms and controls. Geoderma 74:65–105. https://doi.org/10.1016/S0016-7061(96)00036-5

de Souza MJH, Ribeiro A, Leite HG et al (2006) Disponibilidade hídrica do solo e produtividade do eucalipto em três regiões da Bacia do Rio Doce. Revista Árvore 30(3):399–410

Stape JL, Binkley D, Ryan MG et al (2010) The Brazil Eucalyptus Potential Productivity Project: Influence of water, nutrients and stand uniformity on wood production. For Ecol Manag 259:1684–1694. https://doi.org/10.1016/j.foreco.2010.01.012

Stape JL, Binkley D, Ryan MG, do Nascimento Gomes A (2004) Water use, water limitation, and water use efficiency in a Eucalyptus plantation. Bosque 25(2)

Tilman D (1999) The ecological consequences of changes in biodiversity: a search for general principles101. Ecology 80:1455–1474. https://doi.org/10.1890/0012-9658(1999)080[1455:TECOCI]2.0.CO;2

Tilman D, Reich PB, Knops J et al (2001) Diversity and productivity in a long-term grassland experiment. Science 294:843–845. https://doi.org/10.1126/science.1060391

Tisdall JM, Oades JM (1982) Organic matter and water-stable aggregates in soils. J Soil Sci 33:141–163. https://doi.org/10.1111/j.1365-2389.1982.tb01755.x

Vadez V, Lim G, Durand P, Diem HG (1995) Comparative growth and symbiotic performance of four Acacia mangium provenances from Papua New Guinea in response to the supply of phosphorus at various concentrations. Biol Fertil Soils 19(1):60–64

Vance CP (2001) Symbiotic nitrogen fixation and phosphorus acquisition. Plant nutrition in a world of declining renewable resources. Plant Physiol 127(2):390–397

Venterink HO (2011) Legumes have a higher root phosphatase activity than other forbs, particularly under low inorganic P and N supply. Plant Soil 347(1–2):137–146

Violle C, Navas M-L, Vile D et al (2007) Let the concept of trait be functional! Oikos 116:882–892. https://doi.org/10.1111/j.0030-1299.2007.15559.x

Voigtlaender M (2012) Produção de biomassa aérea e ciclagem de nitrogênio em consórcio de genótipos de Eucalyptus com *Acacia mangium*. Universidade de São Paulo, São Paulo

Voigtlaender M, Brandani CB, Caldeira DRM et al (2019) Nitrogen cycling in monospecific and mixed-species plantations of *Acacia mangium* and Eucalyptus at 4 sites in Brazil. Forest Ecol Manag 436:56–67. https://doi.org/10.1016/j.foreco.2018.12.055

Whitehead D (1998) Regulation of stomatal conductance and transpiration in forest canopies. Tree Physiol 18:633–644. https://doi.org/10.1093/treephys/18.8-9.633

Wieder WR, Cleveland CC, Townsend AR (2009) Controls over leaf litter decomposition in wet tropical forests. Ecology 90:3333–3341. https://doi.org/10.1890/08-2294.1

Withers PJA, Rodrigues M, Soltangheisi A et al (2018) Transitions to sustainable management of phosphorus in Brazilian agriculture. Sci Rep-UK 8:2537. https://doi.org/10.1038/s41598-018-20887-z

Zhou W-J, Sha L-Q, Schaefer DA et al (2015) Direct effects of litter decomposition on soil dissolved organic carbon and nitrogen in a tropical rainforest. Soil Biol Biochem 81:255–258. https://doi.org/10.1016/j.soilbio.2014.11.019

Zhou Z, Zhang Z, Zha T et al (2013) Predicting soil respiration using carbon stock in roots, litter and soil organic matter in forests of Loess Plateau in China. Soil Biol Biochem 57:135–143. https://doi.org/10.1016/j.soilbio.2012.08.010

Zinn YL, Resck DVS, da Silva JE (2002) Soil organic carbon as affected by afforestation with Eucalyptus and Pinus in the Cerrado region of Brazil. Forest Ecol Manag 166:285–294. https://doi.org/10.1016/S0378-1127(01)00682-X

Chapter 5
Soil Bacterial Structure and Composition in Pure and Mixed Plantations of *Eucalyptus* spp. and Leguminous Trees

Caio Tavora Coelho da Costa Rachid

5.1 Introduction

It is in the first centimeters belowground that the most diverse and rich biodiversity of our planet can be found. In fact, a single soil sample can harbor billions of organisms assembling thousands to millions of species of bacteria, archaea, and fungi (Whitman et al. 1998; Lozupone and Knight 2007). In this environment, they form very complex communities with a highly branched network of ecological interactions (Bonfante and Anca 2009; Fuhrman 2009).

Bacteria populations far exceed fungi and archaea in numbers (Rachid et al. 2013; Siles and Margesin 2016), and they play important roles, such as in nutrient turnover, phytohormone production, and biocontrol that affect soil functioning and plant productivity (Chaparro et al. 2012).

Obviously, such rich and diverse community poses a challenge to be studied and understood. In a scale of an average bacterium, 1 cm can represent a huge distance, with very different environmental conditions in terms of atmosphere, pH, and quantity and quality of organic and inorganic matter, among others. All these factors, along with temperature, water, and biological interactions, will affect the structure of the bacterial community (Fierer 2017).

However, what exactly bacterial community structure means? It means all the abundance and composition of bacteria in a given environment. Still more specifically, how many species it encompasses (bacterial richness, alpha diversity), who they are (bacterial composition), how many they are (bacterial abundance), and how they vary among sites from the same environment (beta-diversity). All these factors together compose the bacterial structure and show all aspects of the bacterial diversity (Konopka 2009).

C. T. C. d. C. Rachid (✉)
Institute of Microbiology, Federal University of Rio de Janeiro, Ilha do Fundão, RJ, Brazil
e-mail: caiorachid@micro.ufrj.br

E. J. Bran Nogueira Cardoso et al. (eds.), *Mixed Plantations of Eucalyptus and Leguminous Trees*, https://doi.org/10.1007/978-3-030-32365-3_5

How bacterial community structure is modulated is still in debate in literature. Many studies have shown that there are some factors with very strong effects over the bacterial community. Among them, the pH, salinity, and organic matter content (Fierer and Jackson 2006; Lozupone et al. 2006; Lozupone and Knight 2007; Ding et al. 2015) are the strongest modulators. Some authors argue that the bacterial community suffers very little influence of the plant coverage (Fierer 2017). However, the impact of the plant on the soil bacterial community, especially in the rhizosphere zone (soil region very close and under influence of the plant roots), has also been very well documented (Berg and Smalla 2009; Mitchell et al. 2010; Rachid et al. 2013). In this context, mixed plantations are one of the best models to demonstrate how plants can influence the soil microbial community, and how it can cause impact on the system's ecology and functioning.

Regarding soil and plant characteristics, previous studies have shown that *Eucalyptus* stands intercropped with leguminous trees affect positively a series of soil attributes increasing soil organic matter and other attributes. Soil nitrate concentration, biomass productivity, and water use efficiency, among others, are also increased, when compared to pure *Eucalyptus* plantations (Forrester et al. 2005, 2006; Laclau et al. 2008; Balieiro et al. 2008; Voigtlaender et al. 2012; Rachid et al. 2013; Koutika et al. 2014). Soil's available P in mixed plantations has increased at the beginning of the rotation but decreased at the end, especially in sandy soils (Rachid et al. 2013; Koutika et al. 2014), leaving to an apparent P (and N, of course) control over the litter decomposition (Santos et al. 2017).

Behind most of these soil changes, there is the essential role of the microorganism associated with the forestry system. The co-cultivation efficacy relies, among other factors, on the ability of the legume trees to establish symbiotic relationships with nitrogen-fixing bacteria, which, in turn, can increase soil N levels and improve nutrient cycling (Rachid et al. 2013; Bernhard-Reversat 1988; Santos et al. 2017; Tchichelle et al. 2017; Voigtlaender et al. 2019). Additionally, the introduction of another plant species can change bacteria, archaea, and fungi communities associated with soil, altering the complex biological network responsible for nutrient turnover and soil health. Changes in the microbial community, therefore, could lead to changes in litter decomposition and soil nutrient balance, which in turn will reflect in plant productivity (van der Heijden et al. 2008; Chaparro et al. 2012).

In this chapter, we present briefly the role of the bacterial community in forestry ecosystems and its response from the resulting interaction of mixing different tree species in a co-cultivation system. Finally, we explore the biotechnological potential of the microorganisms to improve the sustainability of wood production.

5.2 Soil Bacterial Functioning

The soil microbiome (all microbes of a given site) can cause impacts on the plant productivity by many ways. For example, symbiotic nitrogen-fixing bacteria, such as Rhizobiales, *Azospirillum*, and *Frankia*, among others, can improve soil N levels, and bacteria and archaeal nitrifiers, such as *Nitrosomonas*, *Nitrosococcus*, and *Nitrososphaera*, can change the predominant form of inorganic nitrogen. Many microbial groups including the bacteria *Bacillus* and the fungi *Aspergillus* can increase P availability and acquisition by plants by exudation of organic acids. Additionally, microorganisms can interfere in competitive interactions, helping or suppressing the growth of a given plant group by phytohormone production or inhibiting a pathogen by the production of antimicrobial substances (van der Heijden et al. 2008). These are some examples of specific functions played by well-described organisms. In nature, many other microorganisms, most of which are still unknown, play these and many other roles, in highly diverse communities forming a very complex network of interaction.

5.3 Soil Bacterial Communities in Pure and Mixed *Eucalyptus* Plantations

The soil bacterial community of *Eucalyptus* plantations has been described for a number of sites in different locations around the globe. In a broader view of soil bacterial composition, *Eucalyptus* plantations do not differ from other cultivations, with most of the studies showing predominance of the Proteobacteria phylum, followed by Acidobacteria, Actinobacteria, Planctomycetes, and Firmicutes among others, as evaluated in Brazil (Silveira et al. 2006; Pereira et al. 2017; Cuer et al. 2018; Rachid unpublished) and China (Li et al. 2018). However, each study showed special aspects of the community when compared to native forests or other plant species cultivated in the same area.

In a Brazilian traditional forest-producing region, located in Minas Gerais state, Cuer et al. (2018) studied two adjacent *Eucalyptus urograndis* plantations—a recently logged site that harbored new seedlings and an adult plantation—and compared them to a site hosting native vegetation. They showed that plant harvesting and implementation of a new rotation can significantly change the soil bacterial community in a short term, mainly by reducing the relative abundance of Acidobacteria and increasing Actinobacteria. When comparing with an adjacent field with native forest, it was shown that *Eucalyptus* plantations had a distinct bacterial community structure. The authors suggested that the main driver of the observed differences in microbial community structure was land use type, as opposed to management practices (i.e., plantation, logging, and initiation of new rotations) or soil characteristics. Surprisingly, the *Eucalypt* plantation showed higher bacterial diversity compared to the fragment of native rainforest. This result was confirmed in an independent sampling, 9 months after the first (Monteiro et al. unpublished data). Higher bacterial diversity in Eucalypt compared to a secondary forest has also been demonstrated in China (Lan et al. 2017). Harboring a higher diversity is frequently associated with an indication of higher stability in the ecosystem. However, the authors argue that in this case it is probably correlated with higher primary productivity in young forests.

The development of *Eucalypt* plantations was also associated to increases in bacterial community biomass. The proposed mechanisms are diverse. In *Eucalyptus grandis,* it has been associated with increased inputs of C from residues or roots (Zhang et al. 2012), and in *Eucalyptus urophylla* it was suggested that it could be caused by the decrease of the dissolved organic carbon (DOC) of the soil. The reduced DOC would give a competitive advantage to bacteria compared to fungi and would increase the ratios of bacteria:fungi along the chronosequence (Wu et al. 2013). Contrastingly, the opposite was also suggested in which *E. camaldulensis* could negatively impact the microbial community, reducing the microbial biomass and catabolic diversity, mainly due to soil acidification and accumulation of phenolic compounds (Soumare et al. 2016). Evaluating soil microbiological attributes in pure and mixed forest plantations of *Acacia mangium* and *Eucalyptus grandis* at the end of a second rotation, Zagatto et al. (2019) did not observe any change among treatments for Cmic, CO_2-C, dehydrogenase, and Cmic/Ct, but only for qCO_2-C. The qCO_2-C was lower for *Acacia* plantations than in pure and mixed *Eucalyptus*. On the other hand, at the beginning of this rotation (27 and 39 months after planting) Pereira et al. (2018) detected a significant higher microbial C content in *Acacia* and mixed plantations than in monoculture of *Eucalyptus*.

The general pattern shows the prevalence of increased bacterial biomass or activity in this forestry system. This is often considered a good indicator of soil health. An active microbial community is important to keep nutrient cycling and soil fertility, which will direct influence crop productivity and sustainability.

However, these studies were based on indirect measurements, mostly in microbial biomass and phospholipid fatty acid (PFLA) contents. Few studies used more precise techniques to quantify bacterial communities, such as real-time PCR (qPCR). The qPCR can provide a measurement of copies of a given gene per gram of soil. Usually, bacterial communities are quantified through the quantification of the gene coding for the subunit of the 16S rRNA, a molecular marker for bacteria. Studies in three different locations in Brazil showed no change in bacterial abundance comparing *Eucalyptus* to native forests or to *Acacia mangium* plantations in upper layers of soil (Rachid et al. 2013; Pereira et al. 2017; Cuer et al. 2018). Therefore, it is very likely that most of the impacts caused by *Eucalyptus* on the soil bacterial community are qualitative and not quantitative.

In recent years, studies of soil microbiology under Eucalypt plantations shed light on different management regimes, on which we will focus from now on to describe the intercropping systems. The first report on soil bacterial community in mixed Eucalypt plantations arose in 2013 and was based on the association of Eucalypt with *Acacia mangium*. Using indirect measurements of the community, such as microbial respiration, dehydrogenase activity, C_{mic}, and N_{mic}, Bini et al. (2013) showed that these attributes present high variation in the initial stage of development of the mixed stands (from 2 to 20 months). The high variability poses a challenge on data interpretation, without a clear pattern, but considering all attributes with other variables in multivariate analysis, they showed that mixed stands are different from monocultures starting at month two. According to the authors, despite the high variability of the data within treatments, the co-cultivation established a new microbial status, with synergistic effects on the two plant species in maintaining and stimulating biogeochemical cycling with beneficial effects over the system.

Later, this analysis was repeated with older stands (27 and 39 months) (Pereira et al. 2018). The study confirmed that nutrient cycling in pure *Eucalypt* systems is different from when *Acacia* is present, with changes in enzymatic and metabolic activity. While some enzymes, such as urease, amidase, and dehydrogenase, tend to present higher activity in pure Eucalypt plantations, the intercropped system tends to present higher microbial biomass C and increased C and N concentrations in soil organic matter, which was interpreted as a gain in soil health by the authors.

Direct measurements were also performed to assess how mixed plantations interfere in the bacterial structure. Using DGGE and qPCR for bacteria (16S rRNA), and genes involved in nitrogen cycling (nirK, amoA, nifH), Rachid et al. (2013) showed that mixed plantation resulted in the integration of the bacterial community present in the monocultures. Additionally, they showed that acacia stands presented higher amounts of nitrifiers and lower amounts of denitrifiers in the soil, and this could be directly linked to the higher levels of nitrate found in acacia monocultures and mixed plantations, which is an indication of better nutrient status of the soil.

Using high-throughput DNA sequencing, the bacterial community of *Eucalyptus* and Acacia monoculture and mixed plantations was studied with taxonomic detail in the superficial soil layer. The results showed that the soil bacterial community integration occurred, but with higher influence of Acacia on the structuring of the soil bacteria, compared to *Eucalyptus*, especially in the first 3 m belowground (Pereira et al. 2017).

Very few studies addressed the soil bacterial community in mixed plantations of *Eucalypt* with legume trees, other than Acacia. Among the exceptions, there is one study of *Eucalypt* with *Sesbania*, in which the plant consortium was evaluated for its capacity for reforestation of degraded lands. The study measured the microbial respiration and the hydrolysis of fluorescein diacetate (FDA) in monospecific and mixed systems, and showed that mixed systems present less seasonal variation and higher microbial activity (de Oliveira Paulucio et al. 2017).

We are still far from really understanding completely how plants modulate the soil microbial community. However, altogether, these studies leave no doubt about the influence of the plant on the soil microbiome. While there are some authors who argue that Eucalypt plantations can cause negative impact on the soil microbial community, recent studies have shown no basis for this affirmation. Still, they have also shown that the metabolic potential of the soil microbial community can be improved when *Eucalypt* is associated with a leguminous tree. In fact, the relationship between these plants is so intimate that the presence of one plant can interfere in the endophyte colonization of the other plant, as previously demonstrated with *Eucalypt* and *Acacia* (Fonseca et al. 2018).

Obviously, the decision whether *Eucalypt* should be or not be mixed with other plants cannot be taken only under the soil microbial perspective. Productivity, logistics, and economy are factors to be analyzed, and other factors could be incorporated in this decision, depending on the ecosystem services expected by the mixed planted forest (see Chap. 11). However, the soil microbial community should always be considered, since the microbial community is the foundation of soil fertility, health, and sustainability and, in an ultimate analysis, these will turn back into productivity and economy.

Despite the numerous questions, there is a strong indication that the consortium of Eucalypt with a legume tree can integrate the soil bacterial community, increasing microbial activity and system stability with direct benefits to soil biogeochemistry, as discussed in other chapters of this book.

5.4 Plant Growth-Promoting Endophytic Bacteria and the Potential for *Eucalyptus*

It has been proposed that each one of 300,000 species of superior plants harbor endophytic bacteria inside their tissues (Strobel et al. 2004). In fact, studies have shown that inside plant leaves, stem, and roots it is possible to find a rich bacterial community (Gottel et al. 2011; Bodenhausen et al. 2013; Akinsanya et al. 2015). During the evolution, plants and endophytic bacteria developed a symbiotic relationship, with mutual benefits (van der Heijden et al. 2008; Rout 2014). Plants provide shelter and organic compounds for bacterial nutrition, while bacteria can play many positive roles for plants.

The benefits of endophytic bacteria for plants include nutritional improvement through biological nitrogen fixation, phosphorus bioavailability, and iron uptake (Hallmann et al. 1997); growth promotion through phytohormone production, which can promote plant rooting and stem development (Bent et al. 2001); and also fitness improvement through defense against pathogens and hydric stress among others (Sala et al. 2007; Ferreira et al. 2008).

Eucalypt endophytic bacteria have been studied by culture-dependent and -independent techniques. Ferreira et al. (2008) showed that seeds from many *Eucalypt* species harbor endophytic bacteria. Among the bacterial genera found in seeds and seedlings were *Bacillus*, *Paenibacillus*, *Enterococcus*, and *Methylobacterium*. The vertical transmission (from one generation to another) of a given bacterium is suggested as an indicator that symbiotic relationships of the plant with the transmitted microorganisms are of great importance for the species fitness (Zilber-Rosenberg and Rosenberg 2008).

Miguel et al. (2016) showed using DGGE—a molecular typing technique—that *Eucalyptus* leaves harbor a complex endophytic microbial community, however, with a very similar profile among different plants and over different stages of development. They also showed the occurrence of diazotrophic bacteria inside the leaves. Among the cultured microbial community, they reported the isolation of *Pantoea, Stenotrophomonas, Massilia, Paenibacillus, Terrabacter, Rhizobium, Agrobacterium, Novosphingobium, Micrococcus, Streptomyces, Pseudoxanthomonas, Caulobacter*, and *Ochrobactrum*. More recently, the first study using high-throughput DNA sequencing technology to understand the endophytic microbial community associated with *Eucalyptus* roots was published (Fonseca et al. 2018). The authors reported an unprecedented biodiversity living inside *Eucalyptus* roots, with the occurrence of approximately 360 different bacterial genera with the most abundant ones being *Mycobacterium, Bradyrhizobium, Streptomyces, Bacillus, Actinospica*, and *Burkholderia*. They showed that many of these genera were associated with nitrogen-fixing bacteria and also that environmental factors can change the community structure inside the roots.

The occurrence of large amounts of diazotrophic bacteria associated with *Eucalyptus* leaves and roots is surprising. This is not because these bacteria do not colonize trees. For instance, *Acacia* has promiscuous nodulation, being capable to symbiotically associate with *Rhizobium* and *Bradyrhizobium* among other nitrogen-fixing bacteria (Galiana et al. 1990, 1994; Le Roux et al. 2009; see Chapter 6), and these associations are of great importance in plant nutrition and nutrient cycling. However, *Acacia* is a legume tree, and it is a common knowledge that this family of trees developed symbiotic association with nitrogen-fixing bacteria. However, *Eucalyptus* has never been considered a plant which benefits from this association. Only recently we learned that nitrogen-fixing bacteria are highly abundant also in *Eucalyptus* (Miguel et al. 2016; Fonseca et al. 2018), and despite the fact that they do not form nodules, we cannot disregard their potential to improve *Eucalyptus* nutrition.

Due to the ability to improve plant development, there is a great interest in the selection of microorganisms as plant inoculum. In general, these studies include the cultivation of endophytic bacteria in culture medium, followed by in vitro tests to evaluate the capacity of the bacteria to perform specific functions, such as phosphorus solubilization and indole acetic acid (IAA) production, to grow in the absence of nitrogen (indication of biological nitrogen fixation), to produce siderophores, or to inhibit a given phytopathogen. The best bacterial lineages usually are identified and tested in vivo, with inoculation of seed or micro-propagated plants.

Paz et al. (2012) tested seven selected endophytic *Bacillus* sp. lineages regarding their ability to improve plant rooting and growth. When tested in vivo they found that only one of them significantly increased the growth of the root and aerial parts of *Eucalyptus* plantlets. Mafia et al. (2009) evaluated effectiveness of ten plant growth-promoting rhizobacteria (including the genera *Pseudomonas*, *Bacillus*, *Stenotrophomonas*, and *Frateuria*) for the control of mini-cutting rot of *Eucalyptus* caused by *Cylindrocladium candelabrum* and *Rhizoctonia solani*. They showed that one lineage of *Pseudomonas fulva* reduced the incidence of mini-cutting rot, under nursery conditions, by 33% compared to the control, and by 27% compared to a fungicide treatment. Teixeira et al. (2007) evaluated the rooting effect of 107 lineages of bacteria isolated from *Eucalyptus* rhizosphere on seedlings propagated by mini-cuttings. They found ten isolates with promising results, based on rooting efficiency and root biomass improvements. The best lineage was affiliated to *Bacillus subtilis* and induced a 219% increase in rooting frequency and a 223% increase in root biomass compared to the control.

It is very common, during the development of a bacterial inoculum, to get frustrating results when testing the effects in vivo. The main reason is that the in vitro conditions are quite different from the in vivo ones. It is impossible, with the actual methods, to cultivate most of the microorganisms living associated with the plants. In general, the ones selected are those fast-growing microorganisms with high affinity to the culture conditions. However, very frequently they are not part of the dominant microbial community, and because they cannot thrive in a highly competitive natural environment, they fail to colonize the plant and, therefore, fail to develop growth-promoting activity. For this reason, it was proposed that a bacterial consortium for plant inoculation should be developed on some naturally abundant bacterial genera (Fonseca et al. 2018).

In recent research (unpublished data), Fonseca and colleagues got a deeper evaluation of plant growth-promoting capacity of some bacterial lineages isolated from *Eucalyptus* roots. They contrasted the list of the most abundant genera found in *Eucalyptus* with the list of bacterial lineages with the best capacity in inorganic phosphorus solubilization, phytate mineralization, IAA production, and nitrogen fixation capacity. From these, they selected four lineages, belonging to genera *Paraburkholderia, Methylobacterium, Paenibacillus*, and *Mesorhizobium*, to formulate a bacterial consortium. The results showed that the inoculation of mini-cutting propagated seedlings can significantly improve seedling survival, below- and aboveground biomass, and average plant height.

5.5 Final Comments

The use of plant growth-promoting endophytic bacteria represents a great tool for improving *Eucalyptus* and other forestry species cultivation. Still, it is a highly unexplored world for forestry companies with many gain opportunities. Despite the costs and time associated with the selection and tests of bacterial lineages for this application, the use of biotechnology has very low costs, with the capacity to be applied to any size production and with minimal changes in the seedling production workflow. The development of this strategy should be encouraged and understood as a low-cost green technology, which beneficial effects include nutrient optimization, higher survival rates, and better plant fitness.

References

Akinsanya MA, Goh JK, Lim SP et al (2015) Metagenomics study of endophytic bacteria in Aloe vera using next-generation technology. Genomics Data 6:159–163

Balieiro FDC, Pereira MG, Alves BJR, de Resende AS, Franco AA (2008) Soil carbon and nitrogen in pasture soil reforested with *Eucalyptus* and *Guachapele*. Rev Bras Ciênc Solo 32(3):1253–1260

Bent E, Tuzun S, Chanway CP et al (2001) Alterations in plant growth and in root hormone levels of lodgepole pines inoculated with rhizobacteria. Can J Microbiol 47:793–800

Berg G, Smalla K (2009) Plant species and soil type cooperatively shape the structure and function of microbial communities in the rhizosphere. FEMS Microb Ecol:1–13

Bernhard-Reversat F (1988) Soil nitrogen mineralization under a Eucalyptus plantation and a natural Acacia forest in Senegal. For Ecol Manag 23(4):233–244

Bini D, Santos CA d, Bouillet JP et al (2013) *Eucalyptus grandis* and *Acacia mangium* in monoculture and intercropped plantations: Evolution of soil and litter microbial and chemical attributes during early stages of plant development. Appl Soil Ecol 63:57–66

Bodenhausen N, Horton MW, Bergelson J (2013) Bacterial communities associated with the leaves and the roots of Arabidopsis thaliana. PLoS One 8:e56329

Bonfante P, Anca I-A (2009) Plants, mycorrhizal fungi, and bacteria: a network of interactions. Annu Rev Microbiol 63:363–383

Chaparro JM, Sheflin AM, Manter DK et al (2012) Manipulating the soil microbiome to increase soil health and plant fertility. Biol Fertil Soils 48:489–499

Cuer CA, Rodrigues R de AR, Balieiro FC et al (2018) Short-term effect of Eucalyptus plantations on soil microbial communities and soil-atmosphere methane and nitrous oxide exchange. Sci Rep 8:15133

de Oliveira Paulucio V, da Silva CF, Martins MA et al. (2017) Reforestation of a degraded area with Eucalyptus and Sesbania: microbial activity and chemical soil properties. Rev Bras Cienc do Solo 41:1–14.

Ding J, Zhang Y, Wang M et al (2015) Soil organic matter quantity and quality shape microbial community compositions of subtropical broadleaved forests. Mol Ecol 24:5175–5185

Ferreira A, Quecine MC, Lacava PT et al (2008) Diversity of endophytic bacteria from Eucalyptus species seeds and colonization of seedlings by *Pantoea agglomerans*. FEMS Microbiol Lett 287:8–14

Fierer N (2017) Embracing the unknown: disentangling the complexities of the soil microbiome. Nat Rev Microbiol 15:579–590

Fierer N, Jackson RB (2006) The diversity and biogeography of soil bacterial communities. Proc Natl Acad Sci U S A A103:626–631

Fonseca E d S, Peixoto RS, Rosado AS et al (2018) The microbiome of eucalyptus roots under different management conditions and its potential for biological nitrogen fixation. Microb Ecol 75:183–191

Forrester D, Bauhus J, Cowie A (2005) On the success and failure of mixed-species tree plantations: lessons learned from a model system of and. For Ecol Manag 209:147–155

Forrester DI, Bauhus J, Cowie AL et al (2006) Mixed-species plantations of Eucalyptus with nitrogen-fixing trees: A review. For Ecol Manag 233:211–230

Fuhrman JA (2009) Microbial community structure and its functional implications. Nature 459:193–199

Galiana A, Chaumont J, Diem HG et al (1990) Nitrogen-fixing potential of *Acacia mangium* and *Acacia auriculiformis* seedlings inoculated with *Bradyrhizobium* and *Rhizobium* spp. Biol Fertil Soils 9:261–267

Galiana A, Prin Y, Mallet B et al (1994) Inoculation of *Acacia mangium* with alginate beads containing selected *Bradyrhizobium* strains under field conditions: long-term effect on plant growth and persistence of the introduced strains in soil. Appl Environ Microbiol 60:3974–3980

Gottel NR, Castro HF, Kerley M et al (2011) Distinct microbial communities within the endosphere and rhizosphere of *Populus deltoides* roots across contrasting soil types. Appl Environ Microbiol 77:5934–5944

Hallmann J, Quadt-Hallmann A, Mahaffee WF et al (1997) Bacterial endophytes in agricultural crops. Can J Microbiol 43:895–914

Konopka A (2009) What is microbial community ecology. ISME J 3:1223–1230

Koutika LS, Epron D, Bouillet JP, Mareschal L (2014) Changes in N and C concentrations, soil acidity and P availability in tropical mixed acacia and eucalypt plantations on a nutrient-poor sandy soil. Plant Soil 379:1–12. https://doi.org/10.1007/s11104-014-2047-3

Laclau JP, Bouillet JP, Gonçalves JLM et al (2008) Mixed-species plantations of *Acacia mangium* and *Eucalyptus grandis* in Brazil. 1. Growth dynamics and aboveground net primary production. For Ecol Manag 255:3905–3917

Lan G, Li Y, Wu Z et al (2017) Soil bacterial diversity impacted by conversion of secondary forest to rubber or eucalyptus plantations: a case study of Hainan Island, South China. For Sci 63:87–93
Le Roux C, Tentchev D, Prin Y et al (2009) Bradyrhizobia nodulating the *Acacia mangium* × A. *auriculiformis* interspecific hybrid are specific and differ from those associated with both parental species. Appl Environ Microbiol 75:7752–7759
Li J, Lin J, Pei C et al (2018) Variation of soil bacterial communities along a chronosequence of Eucalyptus plantation. Peer J 6:e5648
Lozupone CA, Knight R (2007) Global patterns in bacterial diversity. Proc Natl Acad Sci 104:11436–11440
Lozupone C, Hamady M, Knight R (2006) UniFrac—an online tool for comparing microbial community diversity in a phylogenetic context. BMC Bioinformatics 7:371
Mafia RG, Alfenas AC, Maffia LA et al (2009) Plant growth promoting rhizobacteria as agents in the biocontrol of eucalyptus mini-cutting rot. Trop Plant Pathol 34:10–17
Miguel PSB, de Oliveira MNV, Delvaux JC et al (2016) Diversity and distribution of the endophytic bacterial community at different stages of Eucalyptus growth. Antonie van Leeuwenhoek. Int J Gen Mol Microbiol 109:755–771
Mitchell RJ, Hester AJ, Campbell CD et al (2010) Is vegetation composition or soil chemistry the best predictor of the soil microbial community? Plant Soil 333:417–430
Paz ICP, Santin RCM, Guimarães AM et al (2012) Eucalyptus growth promotion by endophytic *Bacillus* spp. Genet Mol Res 11:3711–3720
Pereira AP de A, Andrade PAM de, Bini D et al (2017) Shifts in the bacterial community composition along deep soil profiles in monospecific and mixed stands of *Eucalyptus grandis* and *Acacia mangium*. Kuramae EE (ed.). PLoS One 12:e0180371
Pereira APA, Zagatto MRG, Brandani CB et al (2018) Acacia changes microbial indicators and increases C and N in soil organic fractions in intercropped eucalyptus plantations. Front Microbiol 9:1–13
Rachid CTCC, Balieiro FC, Peixoto RS et al (2013) Mixed plantations can promote microbial integration and soil nitrate increases with changes in the N cycling genes. Soil Biol Biochem 66:146–153
Rout ME (2014) The plant microbiome, 1st edn. Elsevier, Amsterdam
Sala VMR, Silveira APD, Cardoso EJBN (2007) Bactérias diazotróficas associadas a plantas não-leguminosas. In: Siveira APD, Freitas SS (eds) Micirobiota Do Solo e Qualidade Ambiental. Capinas, p 312
Santos FM, Chaer GM, Diniz AR, Balieiro FC (2017) Nutrient cycling over five years of mixed-species plantations of Eucalyptus and Acacia on a sandy tropical soil. For Ecol Manag 384:110–121
Santos FM, Balieiro FC, Fontes MA, Chaer GM (2017) Understanding the enhanced litter decomposition of mixed-species plantations of *Eucalyptus* and *Acacia mangium*. Plant Soil 141–155
Siles JA, Margesin R (2016) Abundance and diversity of bacterial, archaeal, and fungal communities along an altitudinal gradient in alpine forest soils: what are the driving factors? Microb Ecol 72:207–220
Silveira ÉLD, Pereira RM, Scaquitto DC et al (2006) Bacterial diversity of soil under eucalyptus assessed by 16S rDNA sequencing analysis. Pesqui Agropecuária Bras 41:1507–1516
Soumare A, Sall SN, Sanon A et al (2016) Changes in soil pH, polyphenol content and microbial community mediated by *Eucalyptus camaldulensis*. Appl Ecol Environ Res 14:1–19
Strobel G, Daisy B, Castillo U et al (2004) Natural products from endophytic microorganisms. J Nat Prod 67:257–268
Tchichelle SV, Mareschal L, Koutika LS, Epron D (2017) Biomass production, nitrogen accumulation and symbiotic nitrogen fixation in a mixed-species plantation of eucalypt and acacia on a nutrient-poor tropical soil. For Ecol Manag 403:103–111
Teixeira DA, Alfenas AC, Mafia RG et al (2007) Rhizobacterial promotion of eucalypt rooting and growth. Brazilian J Microbiol 38:118–123
van der Heijden MG, Bardgett RD, van Straalen NM (2008) The unseen majority: soil microbes as drivers of plant diversity and productivity in terrestrial ecosystems. Ecol Lett 11:296–310

Voigtlaender M, Brandani CB, Caldeira DRM, Tardy F, Bouillet J-P, Gonçalves JLM, Moreira MZ, Leite FP, Brunet D, Paula RR, Laclau J-P (2019) Nitrogen cycling in monospecific and mixed-species plantations of *Acacia mangium* and Eucalyptus at 4 sites in Brazil. For Ecol Manag 436:56–67

Voigtlaender, M, Laclau JP, Gonçalves JLM, Piccolo MC, Moreira MZ, Nouvellon Y, Ranger J, Bouillet JP (2012) Introducing *Acacia mangium* trees in *Eucalyptus grandis* plantations: consequences for soil organic matter stocks and nitrogen mineralization. Plant Soil 352:99–111

Whitman WB, Coleman DC, Wiebe WJ (1998) Prokaryotes: The unseen majority. Proc Natl Acad Sci 5:6578–6583

Wu JP, Liu ZF, Sun YX et al (2013) Introduced *Eucalyptus urophylla* plantations change the composition of the soil microbial community in subtropical china. L Degrad Dev 24:400–406

Zagatto MRG et al (2019) Interactions between mesofauna, microbiological and chemical soil attributes in pure and intercropped *Eucalyptus grandis* and *Acacia mangium* plantations. For Ecol Manag 433:240–247

Zhang D, Zhang J, Yang W et al (2012) Effects of afforestation with Eucalyptus grandis on soil physicochemical and microbiological properties. Soil Res 50:167

Zilber-Rosenberg I, Rosenberg E (2008) Role of microorganisms in the evolution of animals and plants: the hologenome theory of evolution. FEMS Microbiol Rev 32(5):723–735

Chapter 6
Biological Nitrogen Fixation (BNF) in Mixed-Forest Plantations

Sergio Miana de Faria, Fabiano de Carvalho Balieiro, Ranieri Ribeiro Paula, Felipe Martini Santos, and Jerri Edson Zilli

6.1 Introduction

The emergence of symbiosis between leguminous plants and bacteria of the Rhizobia group (bacteria able to induce nodules and fix nitrogen) remains clouded in mystery and speculation (Sprent 1994; Brockwell et al. 2005; Doyle 2016; Parniske 2008). However, because the legume plants also possess the ability to take up N from soil, it is assumed that they were associated with diazotrophic bacteria (nitrogen-fixing bacteria) primitively in a parasitic mode, later evolving to infection, nodulation, and fixation patterns (Faria et al. 1987; Sprent 1994, 2007). It is important to realize that symbiosis is not obligatory for the plant or bacteria, but when associated both symbionts have ecological advantages of survival and competition (Sprent 2007).

The reason why some legume species do not nodulate, even though different leguminous species are capable of similarly accumulating N in the tissue, is not fully understood. For example, the genus *Cassia* comprises about 30 tree species, in which nodulation and nitrogen fixation remain unconfirmed. On the other hand, the genus *Chamaecrista* (closely related to *Cassia* genus) with more than 250 species that are herbs, shrubs, and arboreal types is an exclusive nodulant (de Faria et al.

S. M. de Faria · J. E. Zilli (✉)
EMBRAPA Agrobiology, Brazilian Agricultural Research Corporation, Seropédica, RJ, Brazil
e-mail: jerri.zilli@embrapa.br

F. de Carvalho Balieiro
EMBRAPA Soils, Brazilian Agricultural Research Corporation, Rio de Janeiro, RJ, Brazil

R. R. Paula
Center for Agricultural Sciences and Engineering, Federal University of Espirito Santo, Alegre, ES, Brazil

F. M. Santos
Federal Rural University of Rio de Janeiro, Seropédica, RJ, Brazil

E. J. Bran Nogueira Cardoso et al. (eds.), *Mixed Plantations of Eucalyptus and Leguminous Trees*, https://doi.org/10.1007/978-3-030-32365-3_6

2010; Sprent 2007, 2009; de Faria et al. 1999; Moreira et al. 1992). It is intriguing, however, that these two leguminous genera with similar capability to accumulate N in their tissues vary in their nodulation features.

Is it possible that the ability of some legumes to accumulate N may have exerted a selection pressure driving towards symbioses, being a way of legumes adapting to the increasing demand for N in a limiting environment? If so, and given that the biological nitrogen fixation (BNF) event in prokaryotes is so old, it is questionable why plants would not have acquired the ability to form a structure ("rhizoplast"), resembling an organelle. Moreover, why most plants do not nodulate, including some legumes, is unclear. It is probable that the selection pressure was not strong enough for the plants to acquire *nif* genes (genes found in all diazotrophs required for structure, biosynthesis, and regulation of nitrogenase, the enzyme responsible for fixing N) and the plant-bacteria relationship is still evolving (Postgate 1992; Coba de la Peña et al. 2018). Probably, some groups of plants (especially the Papilionoideae subfamily and Mimosoideae clade) shared a close relationship with bacteria that had this enzymatic apparatus and took advantage of this association (Polhill et al. 1981; Sprent 2007).

BNF is the primary N intake form in agroecosystems, promoting equilibrium between atmospheric N_2, being the reactive forms incorporated in soil and organisms. For a number of reasons (such as the cost of production and environmental impact of synthetic fertilizers), BNF has become indispensable for sustainable agriculture on the planet (Crews and Peoples 2004), with Brazil being an excellent example of leveraging this process in annual crops, such as soybeans, beans, and others (Hungria and Mendes 2015).

Besides, BNF has also been useful in the restoration of degraded areas, part of the recovery technologies, which are based on the introduction of pioneering, fast-growing N_2 -fixing legumes (Franco et al. 1995; Chaer et al. 2011; Balieiro et al. 2018). In this case, the inoculation of legume seeds with rhizobia supplies plants with nitrogen often scarce on severely degraded lands, improves soil quality, and supports plant growth and ecological succession (Parrota et al. 1997; Franco and Faria 1997; Batterman et al. 2013).

Tropical soils, like most in Brazil, are highly weathered and poor in nutrients and organic matter, which require the supply of nutrients for adequate plant growth. Forest plantations, whether pure or mixed, depend on external inputs for adequate growth, although water availability exerts substantial control over growth and biomass accumulation (Stape et al. 2010; Moraes Gonçalves et al. 2013). This is especially important when it comes to fast-growing species, such as eucalyptus, which can accumulate about 155 kg ha^{-1} of N in biomass during the first year after planting (Laclau et al. 2010). It has been shown, however, that the introduction of N_2-fixing legumes in a consortium with fast-growing non-N_2-fixing species may be a strategy to promote gains in biomass production (wood), decrease the dependence on chemical nitrogen fertilizer by companion species (Chap. 1), and contribute to several soil processes, especially those dependent on the soil organic matter (see Chaps. 4 and 10).

This chapter addresses the biological fixation of N_2 as a critical ecological facilitation strategy in mixed-forest plantations, allowing N fixation to act positively on the development of non-N_2 -fixing species. The mixture of fast-growing N_2-fixing species is capable of improving the quality of soil organic matter, improving N status in the system and crop productivity (Forrester et al. 2005; Voigtlaender et al. 2012; Rachid et al. 2013; Santos et al. 2018). A compilation of information, without the commitment to exhaust the literature, was made by trying to give numbers to the benefits of nodulating bacteria. The potential use of the plant microbiome and growth-promoting microorganisms in forestry is still incipient, but very promising. This subject is discussed in Chap. 5.

6.2 Nodulating Bacteria and Symbiosis Establishment

Into the plant kingdom, members of the family Leguminosae have the ability to interact with diazotrophic bacteria and form nodules (Fig. 6.1). The nodules can be located in the roots of major legume species and in the stem of a few, as seen in the genera *Aeschynomene*, *Sesbania*, and *Neptunia* that grow in the flooded regions (Fig. 6.1).

The term "rhizobium" has always represented a group of gram-negative, obligate aerobic, non-endospore-forming alpha-proteobacteria that induce plants to form nodules through highly complex molecular signaling (Parniske 2008; Doyle 2011, 2016; Clúa et al. 2018). The taxonomy of nitrogen-fixing bacteria that associate with legumes has been frequently reviewed, especially as they have been identified as nodulating (Chen et al. 2006; Peix et al. 2015; Andrews and Andrews 2017).

Nowadays, rhizobia represent several lineages within the alpha-proteobacteria, and hence the term "rhizobia" does not represent a single taxon but refers to a polyphyletic cluster of bacterial lineages having similar functions. Most known rhizobia

Fig. 6.1 Spherical stem nodules of *Aeschynomene* sp. (approx. 3 mm in diameter); spherical radicular nodule of *Dalbergia nigra* (approx. 2 mm in diameter) and branched root nodules of *Andira nitida* (approx. 5–10 mm in length)

still belong to the family Rhizobiaceae [*Rhizobium*, *Ensifer* (*Sinorhizobium*), *Allorhizobium*, *Parahizobium*, *Neorhizobium*, *Shinella*], Phyllobacteriaceae, *Mesorhizobium*, *Aminobacter*, *Phyllobacterium*, Brucellaceae (*Ochrobactrum*, *Methylobacterium*, *Microvirga*), Bradyrhizobiaceae (*Bradyrhizobium*), Xanthobacteraceae (*Azorhizobium*), and Hyphomicrobiaceae (*Devosia*), but some others belong to beta-proteobacterial genera in the family Burkholderiaceae (*Paraburkholderia*, *Cupravidus*, and *Trinickia*) (Andrews and Andrews 2017; Peix et al. 2015; Sprent et al. 2017).

For the establishment of a mutual symbiosis, as is usual in the symbiosis between rhizobia and nodulating legumes, it is necessary that a series of physical, biochemical, physiological, and environmental factors complement each other. Several authors (Moreira and Siqueira 2006; Parniske 2008; Doyle 2011, 2016) describe that the fundamental stages for the establishment of symbioses are (1) preinfection, in which symbionts are recognized and interactions occur between surface bacteria and plant; (2) plant infection by the bacteria and formation of nodules; and (3) functioning of nodules, i.e., nitrogen fixation. According to the authors, several dozen genes are involved in the process of N_2 fixation in nodule-fixing bacteria, which influence everything from the recognition of the host plant by the bacteria to the transport of carbon from the plant to the bacteroid (the active form of nitrogen-fixing bacteria).

The pink color inside the nodule indicates the effectiveness of nodulation and the efficiency of nitrogen fixation. It shows the presence of active leghemoglobin, which is needed to supply oxygen at low tension for the nodules to function. The oxygen tension inside the nodules, which is usually low, is necessary because nitrogenase (the enzyme responsible for nitrogen fixation) is irreversibly inhibited in the presence of high O_2 concentration (Raymond et al. 2004). The efficiency of nodulation can be measured by the benefits of symbiosis to the host and the system as a whole, such as higher production of plant biomass, significant accumulation of nutrients (including N), and even transfer, directly or indirectly, of N to non-N_2- fixing plants.

Among the principal genera of nodulating bacteria of native forest legumes studied from the Amazon, Cerrado, Caatinga, and Brazilian Atlantic Forest *Bradyrhizobium*, *Rhizobium*, *Ensifer*, *Mesorhizobium*, and *Paraburkholderia* are the most common (Moreira and Siqueira 2006; Bournaud et al. 2013; da Silva et al. 2014; Zilli et al. 2014; Reis Jr et al. 2010).

The induction of nodules of forest legumes occurs through several bacterial genera; however, there is a certain specificity of response in terms of efficiency in nitrogen fixation (de Faria et al. 1999). This specificity appears to increase as symbionts coevolve in the same geographic region. For example, the species *Mimosa pudica* and even other members of Mimosa tribes are efficiently and almost exclusively nodulated by bacteria of the genus *Rhizobium* in Central America, whereas, in the Brazilian Cerrado, nodulation is almost exclusively by *Paraburkholderia* (Bontemps et al. 2016). Local climate conditions and specifically edaphic (pH) factors contribute to the emergence of this specificity (Bontemps et al. 2016; Pires et al. 2018; Reis Jr et al. 2010). Patterns relating to strains of *Rhizobium* nodulating species originated in Central America and *Paraburkholderia* in South America have also been observed in the *Calliandra* genus (tribe Ingae) (Silva et al. 2018).

Species of the *Acacia* genus, commonly used in mixed plantations, are nodulated by both *Rhizobium* and *Bradyrhizobium* and less frequently by *Ensifer*, *Mesorhizobium*, and *Paraburkholderia* (Lawrie 1981; Barberi et al. 1998, Sakrouhi et al., 2016). However, there are important differences in the nodulation efficiency and N_2 fixation even within a genus and species to which the strains belong (Galiana et al. 2002).

6.3 The Ability of the Forest Legumes to Nodulate and Fix Nitrogen

The family Leguminosae was recently reviewed and classified into six subfamilies: Duparquetioideae, Cercidoideae, Dialioideae, Detarioideae, Papilionoideae, and Caesalpinioideae, with the latter also encompassing the traditional subfamily Mimosoideae, which became a clade of Caesalpinioideae (LPWG 2017). The nodulation is mostly concentrated in the subfamily Papilionoideae, in which about 97% of the investigated species can nodulate. In this subfamily, only some tribes and genera do not nodulate, such as Dipteryxeae, part of the Dalbergieae (*Vaitarea* and *Vataereopsis*), Swartzieae (only some species of the genus *Swartzia* nodulate), and some genera of Sophoreae. Similarly, in the subfamily Papilionoideae, the Mimosoid clade within the subfamily Caesalpinioideae, 95% of the species fix nitrogen associated with rhizobia. For the other members of the old subfamily Caesalpinioideae, until recently, only 25% were associated with rhizobia-producing nodules and these are concentrated in some genera of the Cassieae (*Chamaecrista*), Caesalpinieae (*Melanoxylon*, *Moldenhawrea*, *Tachigali*, *Dimorphandra*, and other genera in this tribe) (Allen and Allen 1981, de Faria et al. 1989).

Nodulation and consequently the BNF benefits usually occur when nitrogen is scarce in the environment, and therefore in mature forests it is rare to find nodules in the species capable of associating with rhizobia (de Faria et al. 1984; Winbourne et al. 2018; Piotto et al. 2009), with reduced BNF contributions (Nardoto et al. 2014). Likewise, nitrogen fixation will occur at a higher intensity when the species requires more nitrogen, that is, during its exponential growth phase. Although the majority of soils have native bacteria capable of nodulating tree species typically used in mixed plantations, significant gains in establishment, growth, and productivity can be obtained when seedlings are inoculated with selected strains (Franco and de Faria 1999; Galiana et al. 2002).

6.4 Brazilian Rhizobia Selection Program for Leguminous Trees

Embrapa Agrobiologia is a pioneer and internationally recognized for its work in the selection and maintenance of diazotrophic bacterial germplasm associated with native and introduced leguminous trees. Since 1960, this research center isolated over 5000 strains of rhizobia from different regions of Brazil. More than 2600 botanical specimens were investigated for nodulation capacity, which includes more than 80 genera and 400 forest species reported for the first time as nodulants or non-nodulants. It is worth remembering that the pioneering work developed with tree species was performed by Dr. Döbereiner and her group in the 1960s, including the preliminary studies on host specificity of the sabiá (*Mimosa caesalpiniifolia*) (Campelo and Döbereiner 1969). Native of the Caatinga biome, this legume originated from Caatinga (Brazilian Northeast), and is widely distributed throughout the country as it has several uses such as live fences, charcoal, firewood, erosion control, forage, honey flowers, and others.

A program to obtain and select strains of rhizobia for legumes initially requires the confirmation of the ability of the isolate to induce nodulation in a host, but in Brazil, given the large diversity of legume species and the limited knowledge of the flora it is often necessary first to evaluate nodulation capacity.

Roots of individual plants can be examined directly in the field for the presence of nodules. If present, the nodules can be collected for bacteria in the laboratory. Subsequent purification and selection of the most efficient isolates for nitrogen fixation in the target plant species can be performed. The presence of nodulation can also be confirmed in the greenhouse by inoculating seed collected in the field where a target legume grows with a set of several bacteria from different groups. The inoculation with a mixture of rhizobia strains of different origin along with the soil from the native location where the legume grows is another strategy.

After confirmation of nodulation in specific species, the selection of most efficient strains for biological N_2 fixation is the next step. It is important to note that some specific responses exist in terms of nitrogen fixation efficiency by a particular group and/or several bacterial strains, and that due to such preferences it may be necessary to select the most efficient strain for each forest species.

Embrapa develops trials that are divided into three phases and follow the official rules of the Ministry of Agriculture, Livestock and Food Supply—MAPA (Brazil 2011). In the first phase, each legume species is tested aseptically in "Leonard jars" containing a mixture of sand and vermiculite (Vincent 1970), with strains of several different origins. In this phase, the nature of the isolates (if they are rhizobia) is confirmed, besides the N_2 fixation potential. The best strains are tested in soil, and unsterilized conditions (second-phase test). This phase evaluates the competitiveness and efficiency of bacteria in comparison to those in the native soil. Nursery and field conditions are part of the third phase.

Studies developed over the last decades by Embrapa Agrobiologia have led to the selection of several rhizobia strains for different legume species. These strains were

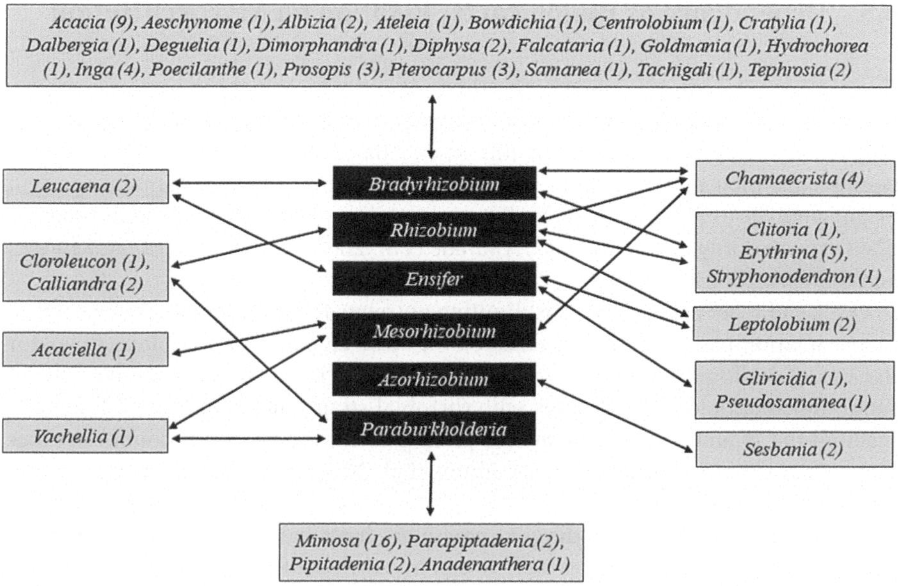

Fig. 6.2 Legume tree genera and respective rhizobia genera with strain efficient in the BNF symbiosis. The numbers in parentheses indicate the number of legume species within each genus for which strains of rhizobium have already been selected

selected for approximately 90 forest species belonging to 38 genera (Fig. 6.2). For most legume genera studied (60%), among 57 species, strains of the genus *Bradyrhizobium* were selected (Fig. 6.2), followed by the genera *Rhizobium* and *Paraburkholderia*. It is clear, therefore, that the *Bradyrhizobium* genus not only is the most common symbiont of native and introduced legumes growing in Brazil, but also tends to be the most efficient for N_2 fixation among most genera (Fig. 6.2). However, this cannot be generalized as plant-microbe specificity involving another bacteria genus that can be important. For example, within *Mimosa*, 16 species that were identified and selected strains all are members of the genus *Paraburkholderia* (Fig. 6.2). The same has also been observed for the plant genera *Piptadenia*, *Parapiptadenia*, and *Anadenathera* that probably only are nodulated by *Paraburkholderia* strains (Fig. 6.2).

For 24 of these legumes, there is at least one strain authorized by MAPA for the production of inoculants (Brazil 2011). Other forest legumes, for which although currently there is at least one strain already selected, are not yet included in the official MAPA list, which still requires significant efforts, including support from the industry to validate the efficiency for registration (Fig. 6.2). This is the case of the introduced forest legume such as some *Acacia* species, and most of the native species already tested or with potential for use in mixed planting, such as the species within the genera *Enterolobium*, *Inga*, *Erythrina*, *Mimosa*, *Dalbergia*, *Tachigali*, and others.

6.5 Dependence of Biological N_2 Fixation on Mycorrhization

Several microorganisms colonize the rhizosphere, which include bacteria, actinomycetes, and fungi. These microbes perform activities that are related to the physiology and nutrition of plants. In this sense, the decomposition of soil organic compounds, their mineralization, BNF, release of substances that stimulate growth or antagonism to pathogens, as well as availability of nutrients are important for plant growth (Grayston et al. 1997; Andrade et al. 2000; Kuiper et al. 2004; Moreira and Siqueira 2006).

As N and P are usually the most limiting nutrients for plant growth in the tropics, more attention has been paid to research on these elements and to alternatives for the use of biological inputs, such as inoculation with rhizobia and mycorrhizal fungi. Inoculation of tree legumes with rhizobial strains and mycorrhizal fungi can meet all the N and P requirements for plant growth, taking into account the other factors that are not limiting (Oliveira Júnior et al. 2016; Patreze et al. 2004; Moreira and Siqueira 2006).

Several tree species, including *A. mangium*, have the ability to associate with arbuscular mycorrhizal fungi as well as ectomycorrhizal fungi, in addition to establishing efficient nodulation with rhizobia. Mycorrhizal fungi can help increase the biological nitrogen fixation, by enhancing P availability that is in demand for the BNF process, as well as the nitrogen-fixing bacteria tend to influence mycorrhizal colonization. This pattern of synergistic response is commonly observed in the Mimosoid clade that responds to both types of symbioses (Oliveira Júnior et al. 2016; Bournaud et al. 2017).

In fact, there are complex interactions between legumes with their symbiotic partners, which is the result of an old coevolution leading plants and microsymbionts to respond more or less effectively to this interaction (Parniske 2008). Variations in responses by both mycorrhizal fungi and rhizobial inoculation are typical, because they are also associated with plant genetics and microsymbiotic performance (Monteiro 1990a; Patreze and Cordeiro 2004). Monteiro (1990a) studied the interaction between *Mimosa caesalpiniifolia* and *M. scabrella* with rhizobia and arbuscular mycorrhizal fungi and concluded that the microsymbionts acted synergistically for the production of biomass and nutrient accumulation in plants, with biomass production exceeding 400% compared to the treatment without the microorganisms even with nutrient addition. In the same way Founoune et al. (2002) evaluated the influence of two isolates of the ectomycorrhizal fungi, *Pisolithus* sp. (COI 007, COI 024), and one isolate of *Scleroderma dictyosporum* (Sd 109) on the growth of *A. mangium* and the synergy with rhizobium inoculation. Compared to the control treatment that lacked inoculation with both the symbionts, *A. mangium* plants treated with COI 007 and Sd 109, respectively, had significantly higher biomass of roots and leaves. In addition, treatment with COI 007 resulted in a higher number of nodules per plant. In the case of *Acacia holosericea*, however, the highest number of nodules was present when inoculated with the COI 024 isolate.

These results show that, although symbiosis with mycorrhizal fungi is a rule among higher plants, there may be additional benefits from certain isolates (Moreira and Siqueira 2006; Shiavo and Martins 2002; Diagne et al. 2013).

The vast majority of the tree legumes can benefit from the association with mycorrhizal fungi and through this association the BNF is improved as well. However, certain groups of legumes not only benefit from this tripartite association, but are also highly dependent on the mycorrhization to establish efficient nodulation, even when supplied with phosphorus (Jesus et al. 2005). For example, recent studies have shown that *Piptadenia gonoacantha* is only capable of inducing the formation of inefficient nodules in the absence of mycorrhiza, and in this case, the color and shape of the nodules formed are different (Bournaud et al. 2017; Oliveira Júnior et al. 2016).

The importance of field-level and nursery studies with mycorrhizal fungi (mycorrhizal fungi and diazotrophic bacteria) is essential to evaluate their efficiency. Laboratory and greenhouse conditions often do not represent the tougher field conditions, even though nursery conditions for seedling production can be similar to those in the greenhouse. For practical purposes inoculation with rhizosphere soil of plants growing in the field will provide well-adapted AMF.

6.6 The Contribution of BNF in Mixed-Forest Plantations

N_2-fixing trees, mainly species from the Leguminosae family, have been widely used to improve N status of non-N_2-fixing species in agroforestry systems (Mafongoya et al. 1998) and mixed-forest planting for timber production (Binkley and Giardina 1997; Richards et al. 2010) and for recovery of degraded lands (Franco and Faria 1997; Chaer et al. 2011). However, the contribution of BNF (percentage of N derived from the atmospheric fixation—% Ndfa) to tree and shrub species under field conditions is not easy to evaluate, mainly due to the difficulties in estimating the amount of N accumulated in the above- and belowground plant components (Khanna 1998; Boddey et al. 2000). Interactions with the abiotic (climate and soil in particular) and biotic factors (inter- and intraspecific interaction of mixed plantations) also complicate these estimations, since they affect competition and facilitation between plants and species, especially in mixed plantations (see Chap. 2).

6.6.1 Measuring the Biological Nitrogen Fixation (BNF) in Woody Perennial Species

Determining the BNF contribution in trees and shrubs, both in planted forests and agroecosystems or in the native forests, has been the subject of several studies and reviews (Boddey et al. 2000; Galiana et al. 2004; Gehring and Vlek 2004; Gehring et al. 2005; Bouillet et al. 2008; Chalk 2016; Paula et al. 2018). Among the methodologies developed for the quantification of BNF and its applicability to woody perennial species either under greenhouse (pots) or field conditions, Peoples et al. (1989) cited acetylene reduction analysis (ARA), determination of relative abundance of ureides in plant sap, and use of ^{15}N (isotopic enrichment and natural abundance) isotope dilution (ID) techniques. The N balance and nitrogen accretion method can also be used to estimate the total N input (kg ha^{-1}) via BNF (Peoples et al. 1989; Forrester et al. 2007; Voigtlaender et al. 2018) in different ecosystems. It should be noted, however, that each of these methodologies mentioned has specifications and limitations (see Boddey et al. 2000; Unkovich et al. 2008; Chalk 2016).

The ARA method uses the activity of the nitrogenase enzyme because under high acetylene concentration, it can be used as a substrate to be reduced to ethylene. This analysis represents a qualitative evaluation of BNF as a point analysis of the nitrogenase activity in the nodules from the plant. The evaluation of the abundance of ureides (allantoins and allantoic acid) in the xylem relies on the ability of the species to transport these compounds preferentially, to the detriment of nitrate and other amino compounds, such as asparagine and glutamine (Peoples et al. 1996). For example, plants from the genus *Acacia*, the most transported BNF products, are asparagine and glutamine (Brockwell et al. 2005).

Isotopic dilution (ID) using the natural abundance of ^{15}N is currently a good option to determine the proportion of N derived from the BNF from the air (% Ndfa) under field conditions (Boddey et al. 2000). This method relies on the fact that under the same natural condition plants that fix some or all of their nitrogen will have lower ^{15}N signal than plants that obtain their entire N from the soils, which are ^{15}N enriched.

For the method based on ^{15}N enrichment (E), the soil is enriched with a labeled fertilizer and paired plots—one containing the legume and the other an N_2-fixing reference plot—are used for the application. However, the ID (NA or E) technique presents some limitations as plant selection, tissue sampling, unpredictability in the levels of ^{15}N, N available to plants from organic matter decomposition, quality and quantity of organic matter, and selective absorption of N sources by ecto- and endomycorrhizal fungi (Högberg 1997; Natelhoffer and Fry 1988; Boddey et al. 2000; Gehring and Vlek 2004).

For the N balance and N accretion method, the BNF rate is estimated as the difference in accumulated N in the plant biomass (aerial and root), in the litter deposited in the soil, or in the soil between plots of the non-N_2-fixing species and plots containing the N_2-fixing species (monocultures or mixed species). Thus, it is assumed that the differences (in kg ha^{-1}) in N in the treatment are mainly due to the biological process. The amounts of N_2 fixation could be underestimated if the N in the belowground area is not considered, especially in planted forests (Forrester et al. 2007).

In addition, Chalk (2016) suggested that the uncertainty in BNF rate could be attributed to the B-value (the relative isotopic abundance of legumes growing in N-free medium). The reasons are as follows: (1) it is not usually determined similarly in the field or pot experiments; (2) it is dependent on the rhizobial strain used as the inoculant; and (3) it depends on the part of the plant tissue sampled. Differences in the root density among the plants in each treatment (monocultures and mixed plantation) of the topsoil, and variations in nitrate and ammonium availability, were reported by Bouillet et al. (2008) as factors contributing to uncertainty in N_2 fixation estimates. These can lead to differences in ™^{15}N of mineral N uptake (NH_4^+ is less depleted in ^{15}N than NO_3^-) by both plant species from soil or from fertilizer application. The costs of the enriched fertilizer and the isotope analyses also restrict the use of these two techniques (NA or E).

6.6.2 Higher Nitrogen Fixation in Mixed Plantations

The rate of the BNF estimates (% Ndfa) in tree species under mixed planting conditions is scarce in Brazil, especially when dealing with native flora. Increased attention was paid to *Acacia mangium* because of the increased growth seen in degraded lands and low-fertility soils in the 1980–1990s of the last century (Franco and Faria 1999; Coelho et al. 2007), additionally to the interest of forestry companies in studies of silvicultural performance and interaction of the species with eucalyptus at the end of the twentieth century (Harwood and Nambiar 2014; Parrota and Knowles 1999), and the growing demands in Southeast Asia (Harwood and Nambiar 2014; Balieiro et al. 2018).

A few reports estimating the %Ndfa in pure and mixed plantations with legumes, using the NA or E techniques, across the Brazilian states of São Paulo and Rio de Janeiro, showed a significant contribution of BNF to mixed plantations (Balieiro et al. 2004, Paula et al. 2018), thus corroborating the work performed in other locations, such as Puerto Rico (Parrotta et al. 1996) and Ivory Coast (Tchichele et al. 2016). However, Forrester et al. (2007) studied mixed plantations of *Acacia mearnsii* and *E. globulus*, but observed opposite results. The higher BNF in mixed plantations was attributed to the elevated N requirement of eucalyptus under mixed plantation regimes. This requirement led to a strong competition for soil N by the plants and a consequent elevation of the N demand in the system (Balieiro et al. 2004; Paula et al. 2018), by the high litter decomposition rate (Santos et al. 2016) and soil N mineralization (Voigtlaender et al. 2019). Table 6.1 contains the %Ndfa estimates for tree legumes in mixed and pure plantations in Brazil and other countries, using natural abundance (NA) and enrichment of ^{15}N (E) techniques.

The results in Table 6.1 corroborate the work of Brockwell et al. (2005), who reported that the biological nitrogen fixation rates observed under the field conditions for *Acacia*, shrubs, and tree species occur in the 2–90% range. The authors further describe that this range denotes the genetic variability within the genus *Acacia*, the efficiency of strains and different species of rhizobia, and the different estimation techniques of BNF.

In general, BNF contributions are higher when the planting is younger (Parrotta et al. 1996; Paula et al. 2018; Balieiro et al. unpublished date; Balieiro et al. 2002 —Tables 6.1 and 6.2) and with local infertile soil (Bernhard- Reversat et al. 1996;

Table 6.1 Percentage of percentage of nitrogen derived from atmosphere (Ndfa, %) for native and exotic woody periannias l species estimated by natural abundance (NA) and enrichment ^{15}N (E) techniques, under pure and mixed plantation

N_2-fixing species	Reference species[(a)]	Country	Aboveground component or litter sampled	Age months	Ndfa %	Method	Reference
Peltophorum dubium (mixed, 50:50)	*Eucalyptus grandis*	Brazil	Le	24		NA	Coelho et al. (2007)
Inga sp. (mixed, 50:50)	*E. grandis*	Brazil	Le	24	74	NA	Coelho et al. (2007)
Mimosa scabrella (mixed, 50:50)	*E. grandis*	Brazil	Le	24	92	NA	Coelho et al. (2007)
Mimosa caesalpiniaefolia (mixed, 50:50)	*E. grandis*	Brazil	Le	24	74	NA	Coelho et al. (2007)
Pseudosamanea guachapele (G100)	*E. grandis* (E100)	Brazil	Le+Ba+Br+st[#]	84	17–36	NA	Balieiro et al. (2004)
P. guachapele (E50:G50)	*E. grandis* (E100)	Brazil	Le+Ba+Br+st[#]	84	34–84	NA	Balieiro et al. (2004)
Acacia mangium (A100)	*E. grandis* (E100)	Brazil	Le	39	0–14	E	Paula et al. (2018)
A. mangium (E50:A50)	*E. grandis* (E100)	Brazil	Le	60	30–52	E	Paula et al. (2018)
A. mangium (A100)	*E. urograndis* (E100)		Le+Ba+Br+St+Cr+Mr+Fr+Li[#]	24	59	E	Tchichelle et al. (2017)
A. mangium (E50:A50)	*E. urograndis* (E100)		Le+Ba+Br+St+Cr+Mr+Fr+Li[#]	24	64	E	Tchichelle et al. (2017)
A. mangium (A100)	*E. grandis* (E100)	Brazil	Le+Ba+Br+St+Cr+Mr+Fr+Li[#]	30	20	NA	Bouillet et al. (2008)
A. mangium (E50:A50)	*E. grandis* (E100)	Brazil	Le+Ba+Br+St+Cr+Mr+Fr+Li[#]	30	10	NA	Bouillet et al. (2008)
A. mangium (E50:A50)	*E. grandis* (E100)	Brazil	Le+Ba+Br+St+Cr+Mr+Fr+Li[#]	30	59	E	Bouillet et al. (2008)
A. mangium (A100) (plots low fertility)	*E. urograndis* (E100)	Ivory Coast	Le	19	27	NA	Galiana et al. (2002)
A. mangium (A100) (plots medium fertility)	*E. urograndis* (E100)	Ivory Coast	Le	19	64	NA	Galiana et al. (2002)
A. mangium (A100) (plots medium fertility)	*E. urograndis* (E100)	Ivory Coast	Le	19	67	NA	Galiana et al. (2002)

Leucaena leucocephala (E50:L50)	*E. robusta*	Puerto Rico	Le	12	98	E	Parrota et al. (1996)
Leucaena leucocephala (E50:L50)	*E. robusta*	Puerto Rico	Le	42	98	E	Parrota et al. (1996)
A. mearnsii (A100)	*E. globulus* (E100)	Australia	Le	120	52–107	NA	Forrester et al. (2007)
A. mearnsii (E50:A:50)	*E. globulus* (E100)	Australia	Le	120	8–17	NA	Forrester et al. (2007)
A. mangium (A100)	*E. urograndis* (E100)	Brazil	Le	12	65	NA	Balieiro et al. (data not published)
A. mangium (A100)	*E. urograndis* (E100)	Brazil	Le	24	26	NA	Balieiro et al. (data not published)
A. mangium (A100)	*E. urograndis* (E100)	Brazil	Le	60	0	NA	Balieiro et al. (data not published)
A. mangium (E50:A50)	*E. grandis* (E100)	Brazil	Le	12	70	NA	Balieiro et al. (data not published)
A. mangium (E50:A50)	*E. grandis* (E100)	Brazil	Le	24	50	NA	Balieiro et al. (data not published)
A. mangium (E50:A50)	*E. grandis* (E100)	Brazil	Le	60	0	NA	Balieiro et al. (data not published)

(a) non-N_2-fixing species used to calculate the %Ndfa, (b) componet used by authors to estimate the %Nfda, *Le* leaves, *Ba* barck, *Br* branches, *St* stem; *Cr* coarse roots, *Mr* medium roots, *Fr* fine roots, *Li* litter

Table 6.2 N_2 fixed (kg ha^{-1}) by woody perennial legumes in pure and mixed plantations based on difference in aboveground N accumulation using the accretion method

N_2-fixing species (A)	Reference species (B)	Country	Age (months)	N acumulated (kg ha^{-1})	N accretion (A−B or (A+B)-B	Reference
Acacia mangium (A100)	*Eucalyptus grandis* (E100)	Brazil	39	204.6	−19.6	Paula et al. (2018)
A. mangium (E50:A50)	*E. grandis* (E100)	Brazil	39	444.7	+220.5	Paula et al. (2018)
	E. grandis (E100)	Brazil		224.2		Paula et al. (2018)
Acacia mangium (A100)	*E. urograndis* (E100)	Brazil	60	186.3	+17.6	Santos et al. (2018)
A. mangium (E50:A50)	*E. urograndis* (E100)	Brazil	60	231.8	+63.1	Santos et al. (2018)
A. mangium (E100:A100)	*E. urograndis* (E100)	Brazil	60	285.5	+116.8	Santos et al. (2018)
	E. urograndis (E100)	Brazil	60	168.7		Santos et al. (2018)
Mimosa scabrella (mixed, 50:50)	*E. grandis*	Brazil	24	317.7	+170.9	Coelho et al. (2007)
Mimosa caesalpiniaefolia (mixed, 50:50)	*E. grandis*	Brazil	24	149.6	+2.8	Coelho et al. (2007)
A. mangium	*E. grandis*	Brazil	24	188.4	+41.6	Coelho et al. (2007)
	E. grandis	Brazil	24	146.8		Coelho et al. (2007)
A. mangium (A100)		Brazil				Balieiro et al. (2004)
	E. grandis (E100)	Brazil	60	410.5		Balieiro et al. (2002)
Pseudosamanea guachapele (G100)	*E. grandis* (E100)	Brazil	60	756.8	−57.0	Balieiro et al. (2002)
P. guachapele (E50:G50)	*E. grandis* (E50:G50)	Brazil	60	467.5	+289.3	Balieiro et al. (2002)
	E. grandis (E100)	Brazil	60	410.5		Balieiro et al. (2002)
Acacia mangium (A100)	*Eucalyptus grandis* (E100)	Ivory Coast	24	30.1	+23.8	Tchichele et al. (2017)
A. mangium (E50:A50)	*E. grandis* (E100)	Ivory Coast	24	23.5	+17.2	Tchichele et al. (2017)
	E. grandis (E100)	Ivory Coast	24	6.3		Tchichele et al. (2017)

(continued)

Table 6.2 (continued)

N_2-fixing species (A)	Reference species (B)	Country	Age (months)	N acumulated (kg ha^{-1})	N accretion (A−B or (A+B)-B	Reference
A. mearnsii (A100)	*E. globulus* (E100)	Brazil	120	496	+358	Forrester et al. (2007)
A. mearnsii (A50:E50)	*E. globulus* (E100)	Brazil	120	480	+342	Forrester et al. (2007)
	E. globulus (E100)	Brazil	120	138		Forrester et al. (2007)
A. mangium (A100)—Itatinga	*E. grandis* (E100)	Brazil	72	2050	+413	Voigtlander et al. (2018)
A. mangium (E50:A50)—Itatinga	*E. grandis* (E100)	Brazil	72	1836	+199	Voigtlander et al. (2018)
	E. grandis (E100)	Brazil	72	1637		Voigtlander et al. (2018)
A. mangium (A100)—Bofete	*E. grandis* (E100)	Brazil	72	2242	+348	Voigtlander et al. (2018)
A. mangium (E50:A50)—Bofete	*E. grandis* (E100)	Brazil	72	2111	+217	Voigtlander et al. (2018)
	E. grandis (E100)	Brazil	72	1895		Voigtlander et al. (2018)
A. mangium (A100)—Luiz Antônio	*E. urograndis* (E100)	Brazil	72	1622	–	Voigtlander et al. (2018)
A. mangium (E50:A50)—Luiz Antônio	*E. urograndis* (E100)	Brazil	72	1848	+84	Voigtlander et al. (2018)
	E. urograndis (E100)	Brazil	72	1764		Voigtlander et al. (2018)
A. mangium (A100)—Santana do Paraíso	*E. urograndis* (E100)	Brazil	72	3258	+367	Voigtlander et al. (2018)
A. mangium (E50:A50)—Santana do Paraíso	*E. urograndis* (E100)	Brazil	72	3168	+277	Voigtlander et al. (2018)
	E. urograndis (E100)	Brazil	72	2891		Voigtlander et al. (2018)

Galiana et al. 2002, Balieiro et al. 2004). Both factors are related to the upregulation of N_2 fixation depending on the soil N status (Vitousek et al. 2002; Galiana et al. 2002). Galiana et al. (2002) observed the spatial variability in %Ndfa for *A. mangium* as a result of soil fertility variation between plots. The %Ndfa reached 64 and 67% in blocks II and III, respectively, versus 27% in block I, following a parallel increase in N and P soil content.

The total N accumulated in the biomass, litterfall, and soil-derived BNF is calculated as the difference in total N found in N_2-fixing species and reference species (non-fixing) (Parrotta et al. 1996; Forrester et al. 2007). From some previous work, the additional amount of N introduced by tree legumes in mixed and pure planting conditions is estimated, in order to measure the benefits of introducing legumes in these systems. As shown in Table 6.2, the total N accumulated in the aerial parts of the plants at a given stage is underestimated, as much of the N_2 fixed may be related to the roots (coarse and fine) and the litter (deposited and on the ground). Although the N_2-fixed N by the legume in the plantation is proportional to its capacity to compete in a specific local, the BNF contribution is always higher in the mixed plantation than in the eucalyptus monocultures. It is expected to contribute up to 60 kg ha^{-1} $year^{-1}$ under mixed-forest plantations. These values corroborate with previous reports on plantations under field conditions, with *Acacia* spp. in Africa and Australia, with up to 50 kg ha^{-1} $year^{-1}$ of fixed N (Sprent 1993; Sutherland and Sprent 1993). As much as BNF contribution of leguminous species depends on its adaptability and growth in local edaphoclimatic conditions, it is imperative that breeding and selection of these species be carried out for different Brazilian conditions, as it has been done in Southeast Asia (Griffin et al. 2015).

Likewise, the new mixed planting arrangement using Brazilian native species may offer some benefits, especially for the biological conservation and the associated ecosystem services (see Chaps. 10 and 12). Silvicultural management of these plantations needs to be better understood as the pruning, thinning, or proper cleaning and maintenance of the plants are activities that could disturb the system and affect BNF in the legumes.

6.6.3 Nitrogen Transfer between Plants in Mixed Plantations

Although N transfer between plants occurs in both directions, i.e., from N_2-fixing tree to non-N_2-fixing and vice versa, the magnitude of the transfer is greater from N_2-fixing tree to non-N_2-fixing species (see review by Chalk et al. 2014). Several studies show that 0–50% of N contained in plants associated with N_2-fixing trees could be derived from such transfers. Due to the transfer, non-N_2-fixing trees growing in a consortium of N_2-fixing trees sometimes accumulate more N in their biomass than the individually growing trees. This additional N is assumed to derive from the transfer (Chalk et al. 2014).

The N present in the N_2-fixing trees can be transferred directly or indirectly to non-N_2-fixing species growing within a consortium (Munroe and Isaac 2014) (Fig. 6.3). Low-molecular-weight nitrogen compounds, such as nitrate, ammonium, and amino acids, are transferred directly between the plants without transforming, from the root exudates or by the action of mycorrhiza. The decomposition of the vegetative tissue, above- and belowground, and its mineralization by soil microorganisms lead to the indirect transfer of N between plants. These N transfer rates vary in speed and significance, and are poorly understood, as well as the limiting or

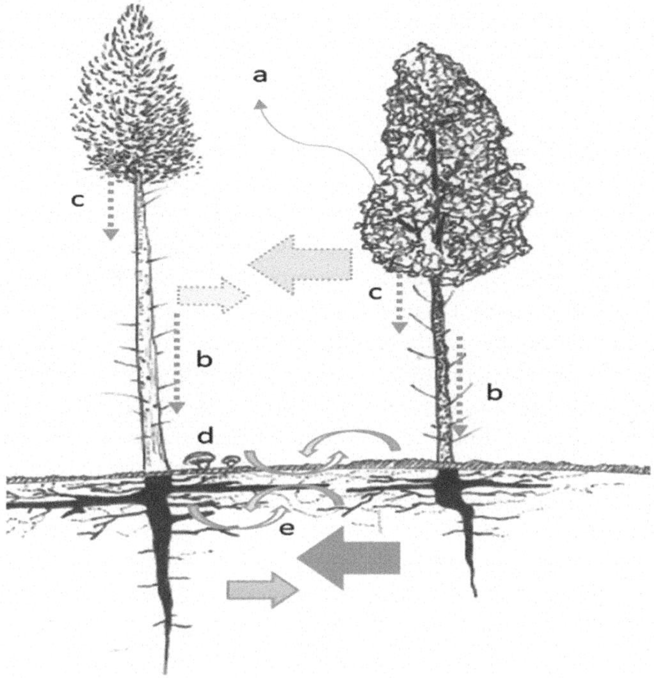

Fig. 6.3 Schematic showing pathways of N transfer between trees in mixed-species plantation with non-N_2-fixing and N_2-fixing trees. Direct transfer of N begins with an N compound with low molecular weight without undergoing transformation by (**a**) ammonium volatilization, (**b**) stem-flow, (**c**) throughfall, (**d**) soluble N in litterfall and pruning residues, and (**e**) root exudates, soluble N in nodule and roots, and mycorrhiza network. Indirect transfer of N occurs after transformation of N substance by (**d**) decomposition and mineralization of litterfall and pruning residues, (**e**) root and nodule decomposition and mineralization, and *by mycorrhizal network (both ectomycorrhizal and arbuscular fungi)*

facilitating factors of this transfer (Mafongoya et al. 1998; Munroe and Isaac 2014; Chalk et al. 2014; Peoples et al. 2015).

Among the main source of N potentially transferable in the mixed plantation, the branches and green leaves of litterfall and the fine roots and nodules represent the major sources of N in quantitative terms. It is important to notice however that the amount of N accumulated in the fine roots is still poorly understood (Mafongoya et al. 1998; Munroe and Isaac 2014; Peoples et al. 2015). The biomass of these roots (diameter <2 mm) at a depth of 17 m was estimated for pure and mixed plantations of acacia and eucalyptus at 4 years of age in Itatinga, Brazil (Germon et al. 2018). Acacia trees produced approximately 4.2 tons ha^{-1} of fine roots in the mixed plantation up to a depth of 12 m. Using an average of 2.3% N content in fine acacia roots (Paula 2015), the N content in the fine roots was 98.3 kg ha^{-1}.

Living tissue derived from pruning or harvesting is an important source of N since the total and soluble N content is higher than that in the senescent tissues. Paula (2015) used crop residues (i.e., leaves, branches, and bark) of ^{15}N-enriched *A. mangium* and *E. grandis* in eucalyptus seedlings to trace the path of the residue-derived N in the soil-plant system. Three months after the application of these residues, young leaves of eucalyptus seedlings were significantly enriched with ^{15}N when they received *A. mangium* residues, which was not the case when the plants received *E. grandis* residues or unlabeled residues. This result shows rapid transferability of a large fraction of soluble N from the legume to other plants.

Direct N transfer from legumes through root exudation and mycorrhizal represents a substantial source of N during tree growth (Munroe and Isaac 2014). This transfer is important because it occurs in the short term (e.g., hours, days), and even trees distant from the source can benefit from the transferred N (Paula et al. 2015). Paula et al. (2015) applied potassium nitrate enriched with ^{15}N to the stem of *A. mangium* trees and observed values of ^{15}N above natural abundance in eucalyptus tissues located within a radius of up to 6.2 m around the acacia plants marked for 60 days after the application. N transfer between plants is facilitated by the presence of both arbuscular mycorrhizae and ectomycorrhizas associated with the vast majority of plant species and can be modulated by source and drain relationship (He et al. 2003). These organisms are able to absorb mineral and organic forms of N derived from N_2-fixing species and assimilate N as needed before transferring to plants growing in a consortium (He et al. 2003; Munroe and Isaac 2014). Ectomycorrhizae can also act on the transfer of N between plants, as they can break down complex organic compounds present in the soil and transform them into forms that are assimilated by plants (He et al. 2003). Estimates of N transfer through mycorrhizae from legumes to non-legumes vary between 20 and 50% of accumulated N (He et al. 2003).

Direct approaches to estimate N transfer involve the application of a nitrogen source enriched with ^{15}N to the nitrogen-fixing tree, and subsequent isotope tracing in the tissues of the reference species (Chalk et al. 2014). The N_2-fixing tree can be labled with ^{15}N via foliar absorption, via injection in the branches and stem, and by root absorption, each one with its particularities (Yasmin et al. 2006; Chalk et al. 2014). Paula et al. (2015) used the ^{15}N values observed in the fine roots of *E. grandis* and *A. mangium* to calculate the N ratio of *E. grandis* derived from *A. mangium* and concluded that the average N transfer reached values of approximately 43%. Based on mass balance, the authors calculated the proportion of ^{15}N injected into the stem of *A. mangium* that was transferred to *E. grandis* trees within a radius of 6.2 m around the acacia, which reached an estimate of N transfer of approximately 3%.

Other potential high-throughput N transfer routes involve leaf leaching, foliar ammonia gas release, root leaching, and herbivory of nodules (Peoples et al. 2015). The first two routes were studied by Paula (2015). At different dates after ^{15}N application in *A. mangium* trees, ^{15}N was determined in the stemflow and throughfall samples collected below the labled acacia, as well as samples derived from collectors installed above the acacia to capture ammonia. The researcher observed that there was no enrichment of these sources of N with ^{15}N above the natural one, and that both sources of N had negative values of $\delta\ ^{15}N$.

Although many advances are being made to understand the facilitation and the ecological relationships involving the transfer of N between plants (mainly N_2-fixing to non-N_2-fixing species), it is also urgent that studies with key nutrients for the process of decomposition, growth, and biological stabilization of N_2 and stabilization of soil organic matter, such as P, are initiated.

6.7 Brazilian Native Legume Tree Species with Potential for Mixed Plantations

In this section, we present some Brazilian legume tree species with potential for mixed planting with *Eucalyptus* spp. The N_2-fixing legume tree species were categorized into two groups: "fertilizer" and "timber" species. The former one included fast-growing species that have high rates of N_2 fixation, which can be used to increase the N and other nutrient levels through aboveground biomass pruning, litter deposition, and/or root exudates and decomposition. In addition, they are species that generally produce light to moderately heavy wood with lower commercial-value timber. The second group included the species which present longer rotation than *Eucalyptus* spp. and produce wood for multiple uses with high commercial value.

6.7.1 Fertilizing Legume Trees

The genera *Enterolobium*, *Erythrina*, *Inga*, and *Mimosa* include tree species with high levels of nodulation in natural environments or under controlled conditions (de Faria et al. 2006; Canosa et al. 2012; Lorenzi 1992). They can be suggested as species of the Brazilian flora with the potential to be introduced in mixed plantations with *Eucalyptus* spp. Table 6.3 lists some of these species.

In the genus *Enterolobium*, popularly known as "tamboril," it is possible to distinguish *E. maximum* and *E. contortisiliquum* as potential species to be introduced in the mixed plantations with *Eucalyptus* spp. *E. maximum*, an Amazonian species, has wood with easy workability and good finishing, for use in boats, toys, household utensils, and plates (Souza et al. 2002). *E. contortisiliquum* is seen along the Brazilian east-coast, including the Atlantic rainforest and Caatinga biomes. It has lightwood (density of 0.54 g cm^{-3} at 12% of moisture content), which can be used in the manufacture of boats and crates. The flowers are mellifluous, and the fruits contain saponin, a substance used in the manufacture of soap.

Inga and *Erythrina* are commonly used as arboreal components of the agroforestry systems with banana, cocoa, and rubber trees in the state of Bahia and the Amazon. Frequently, both genera have also been planted under different arrangements of agroforestry systems in Latin America (Bolivia, Peru, and Colombia).

Table 6.3 Species of "fertilizer" tree legumes of Brazilian flora suggested for mixed plantations with *Eucalyptus* spp.

Genus	Species	Geographic distribution (Brazilian states[a])	Phytogeographical domains (Brazilian biomes)
Erythrina	*E. verna* Vell.	North (AC); Southeast (ES, MG, RJ, SP)	Amazon, Atlantic rainforest
	E. poeppigiana (Walp.) O. F. Cook	North (AC, AM, PA, RO)	Amazon
	E. fusca Lour.	North (AC, AM, AP, PA, RO) Midwest (MT)	Amazon, Cerrado
Inga	*I. edulis* Mart.	North (AC, AM, AP, PA, RO, RR); Northeast (BA, PB, PE); Midwest (MT); Southeast (ES, MG, RJ, SP) South (PR, SC)	Amazon, Caatinga, Cerrado, Atlantic rainforest
	I. laurina (Sw.) Willd.	North (AC, AM, PA); Northeast (BA, CE, MA, PB, PE); Midwest (DF, GO, MS, MT); Southeast (ES, MG, RJ, SP) South (PR)	Amazon, Caatinga, Cerrado, Atlantic rainforest
	I. cinnamomea Spruce ex Benth.	North (AC, AM, AP, PA, RO)	Amazon
Enterolobium	*E. maximum* Ducke	North (AC, AM, PA, RO, RR); Midwest (MT)	Amazon
	E. contortisiliquum (Vell.) Morong	Northeast (BA, CE, PB, PE, PI, RN); Midwest (DF, GO, MS, MT); Southeast (MG, RJ, SP); South (PR, RS, SC)	Caatinga, Cerrado, Atlantic rainforest
Anadenanthera	*A. colubrina* var. *cebil* (Griseb.) Altschul	Northeast (BA, CE, PB, PE, PI, RN, SE); Midwest (DF, GO, MS, MT); Southeast (MG)	Caatinga, Cerrado, Atlantic rainforest
	A. colubrina (Vell.) Brenan var. *colubrina*	Northeast (BA); Southeast (MG, RJ, SP); South (PR)	Caatinga, Cerrado, Atlantic rainforest
	A. peregrina (L.) Speg. var. *peregrina*	North (AM, PA, RR); Midwest (DF, GO, MS); Southeast (MG)	Amazon, Cerrado
	A. peregrina var. *falcata* (Benth.) Altschul	Northeast (BA, PB); Midwest (MS, MT); Southeast (MG, RJ, SP); South (PR)	Caatinga, Cerrado, Atlantic rainforest
Mimosa	*M. scabrella* Benth.	Southeast (MG, RJ, SP); South (PR, RS, SC)	Atlantic rainforest
	M. caesalpiniaefolia Benth.	North (AM, PA, RO); Northeast (AL, BA, CE, MA, PB, PE, PI, RN); Midwest (DF, GO, MS); Southeast (ES, MG, RJ, SP); South (PR, SC)	Amazon, Caatinga, Cerrado, Atlantic rainforest

[a]Brazilian states abbreviations: *AC* Acre, *AL* Alagoas, *AM* Amazonas, *AP* Amapá, *BA* Bahia, *CE* Ceará, *DF* Distrito Federal, *ES* Espírito Santo, *GO* Goiás, *MA* Maranhão, *MG* Minas Gerais, *MS* Mato Grosso do Sul, *MT* Mato Grosso, *PA* Pará, *PB* Paraíba, *PE* Pernambuco, *PI* Piauí, *PR* Paraná, *RJ* Rio de Janeiro, *RN* Rio Grande do Norte, *RO* Rondônia, *RR* Roraima, *RS* Rio Grande do Sul, *SC* Santa Catarina, *SP* São Paulo, *SE* Sergipe

There are several species of *Inga* in the Brazilian flora that are generally adapted to hot and humid climates. Many of them are responsive to pruning in the aerial part, where all green biomass generated can be enclosed along the planting lines of *Eucalyptus* spp. The trees of *Inga* spp. produce lightwood that can be used for coinage and energy. The fruits are edible and can be commercially exploited for the regional markets. Similarly, the genus *Erythrina* also has trees with light- and softwood, used mainly for crates, furniture linings, shoes, and toys.

Mimosa has smaller trees (varying from 5 to 15 m) distributed in several biomes. *M. scabrella*, for example, is a species found in cold places or altitudes in the south and southeast of Brazil. The wood from this genus has an average density of 0.67 g cm^{-3}, and is widely used in interior finishes, in the manufacture of plywood and packaging and for energy. *M. caesalpiniaefolia* however adapts well to warm, dry, or humid climates. In the Brazilian Northeast, it is commonly cultivated. It is a spiny tree, even though with non-spiny variants, that produces multiple stems, requiring significant maintenance. Some of the tree characteristics are helpful, as living fences. The wood is suitable for firewood, charcoal, cable tools, and external uses such as wood posts. Its wood is long lasting under external conditions, even without chemical treatment.

6.7.2 Timber Species

Among the trees of this group, we highlight the following genera: *Anadenanthera, Bowdichia*, *Centrolobium*, *Dalbergia*, *Hymenolobium*, *Plathymenia*, and *Tachigali* (Table 6.4).

Some species of *Anadenanthera* are known as "angicos" in Brazil. These are fast-growing plants and can be found in several Brazilian regions. *A. colubrina* has heavy woods (density ranging from 0.80 to 1.00 g cm^{-3}) suitable for building (indoor), planks, packaging, firewood, and charcoal. *A. peregrina* has a dense wood (0.70 to 0.97 g cm^{-3}) and is suitable for the manufacture of pieces of rafts, frames, roof slats, rural constructions, and outdoor construction materials such as sleepers, stakes, fence posts, and posts. In addition to the wood uses, the mixed plantations of *Anadenanthera* with *Eucalyptus* offer honey from the flowers.

Species of the genus *Bowdichia* are popularly known as "sucupira" in Brazil. *B. nitida* is an Amazonian species with the potential to reach heights of up to 35 m in natural conditions. The most common uses include the timber for making furniture, decorative laminates, bridges, and civil and naval constructions (Souza et al. 1997). The wood is very dense (exceeding 0.96 g cm^{-3}) and dark brown in color. *B. virgilioides* is distributed in the different Brazilian biomes and generally reaches heights of about 20 m. The wood is of high density (0.91 g cm^{-3}), is long-lasting, and is used in construction (outdoor areas) and furniture.

The genus *Centrolobium* has some of the important timber species such as *C. tomentosum*, *C. robustum*, and *C. paraense*, known in Brazil as "putumuju" or "araribá." The woods of these species are heavy, dense (over 0.75 g cm^{-3}), and easy

Table 6.4 Timber species legumes with late rotations of Brazilian flora suggested for mixed plantations with *Eucalyptus* spp.

Genus	Species	Geographic distribution (Brazilian states[a])	Phytogeographical domains (Brazilian biomes)
Plathymenia	*P. reticulata* Benth.	North (PA); Northeast (BA, CE, MA, PI); Midwest (DF, GO, MS, MT); Southeast (ES, MG, RJ, SP); South (PR)	Amazon, Caatinga, Cerrado, Atlantic rainforest
Dalbergia	*D. nigra* (Vell.) Allemãoex Benth.	Northeast (AL, BA, PB, PE, SE); Southeast (ES, MG, RJ, SP); South (PR)	Atlantic rainforest
Tachigali	*T. vulgaris* L. G. Silva & H. C. Lima	North (AM, PA, TO); Northeast (BA, CE, MA, PI); Midwest (DF, GO, MS, MT); Southeast (MG, SP)	Amazon, Caatinga, Cerrado
Centrolobium	*C. robustum* (Vell.) Mart. ex Benth.	Northeast (BA); Southeast (ES, MG, RJ, SP)	Atlantic rainforest
	C. tomentosum Guillem. ex Benth.	Northeast (BA); Midwest (DF, GO); Southeast (ES, MG, RJ, SP); South (PR)	Caatinga, Cerrado, Atlantic rainforest
	C. paraense Tul.	North (RR)	Amazon
Hymenolobium	*H. modestum* Ducke	North (AM, PA)	Amazon
	H. petraeum Ducke	North (AM, AP, PA); Northeast (MA)	Amazon
	H. excelsum Ducke	North (AM, PA)	Amazon
Bowdichia	*B. virgilioides* Kunth	North (AM, AP, PA, RO, RR, TO); Northeast (AL, BA, CE, MA, PB, PE, PI, RN, SE); Midwest (DF, GO, MS, MT); Southeast (ES, MG, SP)South (PR)	Amazon, Caatinga, Cerrado, Atlantic rainforest, Pantanal
	B. nitida Spruce ex Benth.	North (AC, AM, AP, PA, RO, RR)	Amazon

[a]Brazilian states abbreviations: *AC* Acre, *AL* Alagoas, *AM* Amazonas, *AP* Amapá, *BA* Bahia, *CE* Ceará, *DF* Distrito Federal, *ES* Espírito Santo, *GO* Goiás, *MA* Maranhão, *MG* Minas Gerais, *MS* Mato Grosso do Sul, *MT* Mato Grosso, *PA* Pará, *PB* Paraíba, *PE* Pernambuco, *PI* Piauí, *PR* Paraná, *RJ* Rio de Janeiro, *RN* Rio Grande do Norte, *RO* Rondônia, *RR* Roraima, *RS* Rio Grande do Sul, *SC* Santa Catarina, *SP* São Paulo, *SE* Sergipe, *TO* Tocantins

to work. The wood colors range from brown to yellow, with veins or orange spots. The wood is employed typically in luxury carpentry and fine furniture, but it is also used in civil and naval constructions, and in hydraulic work.

Dalbergia nigra, known as "jacarandá-da-Bahia," produces one of the most beautiful and premium woods of the Atlantic rainforest. Consequently, it happens to be one of the threatened species facing extinction in the Brazilian forests. In Brazil, there are other timber species of the same genus, for example, *D. spruceana*, which is found in the Amazon. These species produce heavy (ranging from 0.80 to 1.00 g cm^{-3}), smooth, fine-textured, natural-toned, dark-colored (sometimes blasted) wood that offers an excellent finish to luxury furniture and interior decoration. It is one of the well-known Brazilian woods and used in the manufacture of musical instruments (piano, violin, and others).

Species of the genus *Hymenolobium* are known commercially as "angelim", although this vernacular name has also been attributed to other Amazonian legume tree species, such as *Dinizia excelsa* Ducke ("angelim-vermelho, angelim pedra"), *Vatairea paraensis* Ducke, *Vatairea sericea* (Ducke) Ducke, *Vataireopsis speciosa* Ducke ("angelim-amargoso"), and *Pithecellobium racemosum* (Ducke) Killip (angelim-rajado). However, only *Hymenolobium* and *Pithecellobium* species are N_2 fixing. Some *Hymenolobium* species reach heights up to 40 to 50 m and 80 to 100 cm of diameter at breast height (DBH) in their natural habitats, with rectilinear and cylindrical shafts up to 25 m in length. The wood has a reddish-brown core, with darker brown spots due to oil-resin exudation and pale-brown sapwood. The wood is of medium to high density (0.71 g cm^{-3}), with easy workability, and offers a good finish. It is currently one of the largest woods used in the Brazilian domestic market, and commonly used in the manufacture of furniture and civil construction (beams, rafters, frames, linings, and others).

Plathymenia foliolosa, popularly known as "vinhático," is widely distributed in Brazil. The vinhático trees reach a height between 15–30 m and 40–70 cm of DBH under natural conditions. A striking feature of the adult trees is that the bark emerges from the trunk as large plaques. The wood is light (density of 0.50 g cm^{-3}) and has easy workability and longer durability. The color of the wood ranges from yellow-gold to yellow-brown, and is therefore commonly used in luxury articles, furniture and civil construction (decorative interior panels), and internal ship finishing.

Tachigali vulgaris, known as "tachi-branco or tachi-dos-campos," has been widely cultivated for over two decades in monoculture stands in northern Brazil and has shown good silvicultural potential. Castro et al. (1990) found annual mean increments (AMI) in height, DBH, and volume of 2.2 m year $^{-1}$, 2.9 cm year^{-1}, and 9.2 m^3 year^{-1}, respectively, when evaluating 3.5-year-old monocultures, established with a spacing of 3 m × 3 m. Narducci (2014) found AMI of 2.53 m year^{-1} and height of 2.05 cm year^{-1} at DBH in 7.5-year-old monocultures planted with a spacing of 4 m × 4 m. *T. vulgaris* presents a medium to high wood density (0.60 g cm^{-3} to 0.74 g cm^{-3}) which can be suitable for the production of sawwood and roundwood, especially posts, beams, and civil construction, and for energy purposes. It is considered to be moderately dense wood (0.65 g cm^{-3} to 0.81 g cm^{-3}) (Carvalho 2005).

6.7.3 Other Introduced-Potential Species (Trees and Shrubs)

In Brazil, some legume trees were introduced and have been cultivated/domesticated by farmers and foresters. Such species could also be tested in different arrangements of mixed-species plantations with *Eucalyptus* and/or other native species, including the cultivation of shrubby legumes for green manure production. All these species are listed in Table 6.5.

6.8 Final Considerations

In recent years, studies on nitrogen-fixing tree species have been intensified, including in mixed-forest plantations. The ability to fix nitrogen and accumulate large amounts of N as part of their biomass confers adaptive characteristics to the legumes that excel over other species. There is a large diversity of legume trees, such as the early and fast growers, and some which are slow growers but produce better quality timber.

Most studies on mixed-species plantations seem to be focused on *Acacia* and *Eucalyptus*. From these studies, many technical recommendations are readily available for the productive sector, although its large-scale use is challenging. *A. mangium* and *A. mearnsii* seem to be the main N_2-fixing species studied (Forrester et al. 2005; Bouillet et al. 2013). These species have shown great adaptation to South American edaphoclimatic conditions and other tropical and subtropical countries. Furthermore, both species have been widely cultivated in their native regions (i.e., Southeast Asia and Oceania). This is due to the multiple wood uses, which can be applied to the production of cellulosic pulp, firewood, and charcoal. In Indonesia and Vietnam, the branches and dead leaves are used as fuel and the leaves as fodder for cattle due to their high protein content (Krisnawati et al. 2011). Some non-timber uses still include the production of honey (due to apiculture flowers and the presence of extrafloral nectaries) (Tonini et al. 2010), glue, and tannin extraction from the bark (mainly for *A. mearnsii*). The barks of *A. mangium* additionally provide a good substrate for edible mushrooms (Lim et al. 2011).

However, in Brazil, *A. mangium* behaves as a aggressive and invasive species, which in part seems to be due to its broad N_2-fixing capability even in marginal soils with low nutrients (Souza et al. 2018; Le Maitre et al. (2011) Delnatte and Meyer 2012, Aguiar et al. 2014, Morais and Montagner 2015, see Chap. 11). Its capacity to absorb P and the highly efficient nutrient recycling within the plants make it a strong competitor for this element. Therefore, silvicultural programs must consider the phosphate fertilization management in successive rotations that include these acacia species.

In this context, we stimulate the test with native legumes from Brazil and in neotropical region, in general, in order to generate information on optimal growth conditions of these legumes. In Brazil, despite the high diversity of woody species,

Table 6.5 Other introduced-potential species (trees and shrubs) with potential to use in forest planting

Genera	Species	Origin	Habit	Potential invasion risk reported in Brazil[a]	Uses
Acacia	*A. auriculiformis* Ex Benth.	Southeastern Asia, Australia, Papua New Guinea	Woody	YES+	Fertilizer/ timber
	A. mangium Willd.	Indonesia, Australia, Papua New Guinea	Woody	Yes+	Fertilizer/ timber
	A. mearnsii Willd.	Australia, Papua New Guinea, Tasmania	Woody	Yes	Fertilizer/ timber
Acaciella	*A. angustissima* (Mill.) Kuntze	Central America, Colombia	Woody	Yes	Fertilizer
Albizia	*A. lebbeck* (L.) Benth.	Southern Asia, South Africa, Australia	Woody	Yes+	Fertilizer/ timber
Cajanus	*C. cajan* (L.) Millsp.	Probably from India	Shrubby	No	Fertilizer
Crotalaria	*C. grahamiana* Wight & Arn.	Probably from India	Shrubby	No	Fertilizer
	C. juncea L.	Asia	Shrubby	No	Fertilizer
	C. spectabilis Roth	Asia	Shrubby	No	Fertilizer
Falcataria	*F. moluccana* (Miq.) Barneby and Grimes	Southeastern Asia, Papua New Guinea	Woody	No	Fertilizer
Gliricidia	*G. sepium* (Jacq.) Steud.	Central America	Shrubby/ woody	No	Fertilizer
Leucaena	*L. leucocephala* (Roxb.) Benth.	Central America	Woody	Yes+	Fertilizer
Pithecellobium	*P. dulce* (Roxb.) Benth.	Mexico, Central America and northern of South America	Woody	Yes	Fertilizer
Pseudosamanea	*P. guachapele* (Kunth) Harms	Central America, Colombia, Ecuador, Peru, Venezuela	Woody	Yes+	Fertilizer

(continued)

Table 6.5 (continued)

Genera	Species	Origin	Habit	Potential invasion risk reported in Brazil[a]	Uses
Sesbania	*S. grandiflora* (L.) Pers.	From southeastern Asia to northern Australia	Woody	No	Fertilizer
	S. sesban (L.) Merr.	Africa, Asia	Woody	No	Fertilizer
Tephrosia	*T. vogelii* Hook. f.	Tropical Africa	Shrubby	No	Fertilizer

[a]"+" means high

there are few long-term experiments on silviculture and management of native species under mixed plantations. One of the pioneering works was conducted by Dr. Renato de Jesus and his collaborators at the Reserva Natural Vale (Linhares, ES, Brazil) since the 1970s (Rolim and Piotto 2018). In general, results have shown that many native species (including N_2-fixing trees) present great potential for reforestation and agroforestry systems, increasing the supply of high-quality timber and reducing the pressure on remnants of the Atlantic Forest. Clearly, several barriers must be overcome because many of the native species mentioned above require further characterization and knowledge for their domestication, especially for people relying on timber for their livelihood. Other barriers include the identification of species adapted to the different Brazilian edaphoclimatic conditions, responses to silvicultural treatments (i.e., thinning or pruning regimes), and need to ensure adequate seed availability.

As for the benefits of BNF, it can be a major factor to enhance the productivity and sustainability of a forest plantation. What we still need is the increased adoption of inoculation of legume seeds during the seedling production and transplanting stages. Selecting the appropriate strain and inoculating during these two stages have been improving the symbiosis.

It is recommended that the species discussed in this chapter be prioritized in future studies of *Eucalyptus* plantations mixed with N_2-fixing legumes. At the time of species selection, the farmer should opt for the availability of seeds and seedlings in the region, as well as check for compatibility to the local climate.

This is an activity involving multilocation field trials and selection of genetically superior material with the desired phenotype suitable for forest stands and for timber yield; particularly, the shape and size of the trunk may be key considerations. Besides, additional studies are needed to understand the behavior and interaction of these species in mixed plantations with *Eucalyptus* spp. and others non-N_2-fixing species in diverse Brazilian soil and climate conditions. A series of experiments, collaborations by different institutions, and participation of forestry experts from different regions would be immensely valuable.

References

Aguiar JA, Barbosa RI, Barbosa JB, Mourão M Jr (2014) Invasion of *Acacia mangium* in Amazonian savannas following planting for forestry. Plant Ecol Divers 7(1–2):359–369

Allen ON, Allen EK (1981) The leguminosae: a source book of characteristics use and nodulation. University of Wisconsin Press, Wisconsin, p 812

Andrade AB, Costa GS, Faria SM (2000) Deposição e decomposição da serapilheira em povoamentos de *Mimosa caesalniifolia*, *Acacia mangium* e *Acacia holosericea* com quatro anos de idade em Planossolo. Rev Bras Ciênc Solo 24:777–785

Andrews M, Andrews ME (2017) Specificity in legume-rhizobia symbioses. Int J Mol Sci 18(4):705

Balieiro FC, Dias FC, Franco AA, Campello EFC, Faria SM (2004) Acúmulo de nutrientes na parte aérea, na serapilhiera acumulada sobre o solo e decomposição de filódios de *Acacia mangium* Willd. Ciência Florestal 14:59–65

Barberi A, Carneiro MAC, Moreira FMS, Siqueira JO (1998) Nodulação em leguminosas florestais em viveiros no Sul de Minas Gerais. Cerne 4:145–153

Batterman SA, Hedin LO, van Breugel M, Ransijn J, Craven DJ, Hall JS (2013) Key role of symbiotic dinitrogen fixation in tropical forest secondary succession. Nature 502:224–227

Binkley D, Giardina C (1997) Nitrogen fixation in tropical forest plantations. In: EKS N, Brown AG (eds) Management of Soil, nutrients and water in tropical plantation forests. Australian Centre for International Agricultural Research, Canberra, pp 297–337

Boddey RM, Peoples M, Palmer B, Dart P (2000) Use of the ^{15}N natural abundance technique to quantify biological nitrogen fixation by woody perennials. Nutr Cycl Agroecosystems 57:235–270

Bontemps C, Rogel MA, Wiechmann A, Mussabekova A, Moody S, Simon MF, Moulin L, Elliott GN, Lacercat-Didier L, Dasilva C et al (2016) Endemic Mimosa species from Mexico prefer alpha-proteobacterial rhizobial symbionts. New Phytol 209:319–333

Bouillet JP, Laclau JP, Gonçalves JDM, Moreira MZ, Trivelin PCO et al (2008) Mixed-species plantations of *Acacia mangium* and *Eucalyptus grandis* in Brazil: 2: nitrogen accumulation in the stands and biological N_2 fixation. For Ecol Manag 255(12):3918–3930

Bouillet J-P, Laclau J-P, Gonçalves JLM, Voigtlaender M, Gava JL, Leite FP, Hakamada R, Mareschal L, Mabiala A, Tardy F, Levillain J, Deleporte P, Epron D, Nouvellon Y (2013) Eucalyptus and Acacia tree growth over entire rotation in single- and mixed-species plantations across five sites in Brazil and Congo. Forest Ecology and Management 301:89–101

Bournaud C, de Faria SM, dos Santos JMF, Tisseyre P, Silva M, Chaintreuil C, Gross E, James EK, Prin Y, Moulin L (2013) Burkholderia species are the most common and preferred nodulating symbionts of the Piptadenia group (tribe Mimoseae). PLoS One 8:e63478

Bournaud C, James EK, de Faria SM, Lebrun M, Melkonian R, Duponnois R, Tisseyre P, Moulin L, Prin Y (2017) Interdependency of efficient nodulation and arbuscular mycorrhization in a Brazilian legume tree . Plant, Cell & Environment.

BRAZIL. Ministério da Agricultura, Pecuária e do Abastecimento (2011) INSTRUÇÃO NORMATIVA SDA N° 13, DE 24 DE MARÇO DE 2011. http://www.agricultura.gov.br/assuntos/insumos-agropecuarios/insumos-agricolas/fertilizantes/legislacao/in-sda-13-de-24-03-2011-inoculantes.pdf

Brockwell J, Searle SD, Jeavons AL, Waayers M (2005) Nitrogen fixation in acacias: an untapped resource for sustainable plantations, farm, forestry and land reclamation. ACIAR Monogr 115:132

Balieiro FC, Tonini H, Lima RA (2018) Produção Científica Brasileira (2007-2016) sobre Acacia mangium Willd.: estado da arte e reflexões. Cad Ciên Tecnol 35(1):37–52

Campelo AB, Dobereiner J (1969) Estudo sobre a inoculação cruzada de algumas leguminosas florestais. Pesq Agrop Brasileira 4:67–72

Canosa GA, de Faria SM, de Moraes LFD (2012) Leguminosas florestais da Mata Atlântica brasileira fixadoras de nitrogênio atmosférico Comunicado técnico 144, EMBRAPA Seropédica RJ p 1–12

Carvalho PER (2005) Taxi-branco. Embrapa Florestas, Colombo, p 11. Embrapa Florestas. Circular técnica, 111

Carvalho WD, Mustin K (2017) The highly threatened and little known Amazonian savannahs. Nat Ecol Evol 1:0100

Castro AWV, Yared JAG, Alves RNB, Silva LS, Meirelles SMLB (1990) Comportamento silvicultural de *Sclerolobium paniculatum* (taxi-branco) no Cerrado amapaense. EMBRAPA-UEPAE Macapá, Macapá, p 4. (EMBRAPA-UEPAE Macapá. Comunicado técnico, 7)

Chaer GM, Resende AS, Campello EFC, Boddey RM (2011) Nitrogen-fixing legume tree species for the reclamation of severely degraded lands in Brazil. Tree Physiol 31:139–149

Chalk PM (2016) The strategic role of ^{15}N in quantifying the contribution of endophytic N_2 fixation to the N nutrition of non-legumes. Symbiosis 69:63–80

Chalk PM, Peoples MB, McNeill AM, Boddey RM, Unkovich MJ, Gardener MJ et al (2014) Methodologies for estimating nitrogen transfer between legumes and companion species in agro-ecosystems: a review of ^{15}N-enriched techniques. Soil Biol Biochem 73:10–21

Chen W, James EK, Coenye T, Chou J, Barrios E, de Faria SM, Elliott GN et al (2006) *Burkholderia mimosarum* sp. nov., isolated from root nodules of *Mimosa* spp. from Taiwan and South America. Int J Syst Evol Microbiol 56:1847–1851

Clúa J, Roda C, Zanetti M, Blanco F (2018) Compatibility between legumes and rhizobia for the establishment of a successful nitrogen-fixing symbiosis. Genes 9(3):125

Coba de la Peña T, Fedorova E, Pueyo JJ, Lucas MM (2018) The symbiosome: legume and rhizobia co-evolution toward a nitrogen-fixing organelle? Front Plant Sci 8:2229

Coelho SRF, Gonçalves JLM, Mello SLM, Moreira RM, Silva EV, Laclau JP (2007) Crescimento, nutrição e fixação biológica de nitrogênio em plantios mistos de eucalipto e leguminosas arbóreas. Pesq Agrop Brasileira 42(6):59–768

Crews TE, Peoples MB (2004) Legume versus fertilizer sources of nitrogen: ecological tradeoffs and human needs. Agri Ecosys Environ Amsterdam 102:279–297

de Faria SM, Lewis GP, Sprent JI, Sutherland JM (1989) Occurrence of nodulation in the leguminosae. New Phytol 111:607–619

de Faria SM, de Lima HC, Olivares FL, Melo RB, Xavier RP (1999) Nodulação em espécies florestais, especificidade hospedeira e implicações na sistemática de leguminosae. Inter-relação fertilidade, biologia do solo e nutrição de plantas. In: Siqueira JO, Moreira FMS, Lopes AS, Guilherme LRG, Faquin V, Neto AEF, Carvalho JG (eds) . Sociedade Brasileira de Ciência do Solo Universidade federal de Lavras, Departamento de Ciência do Solo, Lavras, pp 667–686

Delnatte C, Meyer J-Y (2012) Plant introduction, naturalization, and invasion in French Guiana (South America). Biol Invasions 14:915–927

Diagne N, Thioulouse J, Sanguin H, Prin Y, Krasova-Wade T, Sylla S, Galiana A, Baudoin E, Neyra M, Svistoonoff S, Lebrun M, Duponnois R (2013) Ectomycorrhizal diversity enhances growth and nitrogen fixation of *Acacia mangium* seedlings. Soil Biol Biochem 57:468–476

Doyle JJ (2011) Phylogenetic perspectives on the origin of nodulation. Mol Plant Microbe Interact J 24:1289–1295

Doyle JJ (2016) Chasing unicorns: nodulation origins and the paradox of novelty. Am J Bot 103(11):1865–1868

de Faria SM, Diedhiou AG, de Lima HC, Ribeiro RD, Galiana A, Castilho AF, Henriques JC (2010) Evaluating the nodulation status of leguminous species from the Amazonian forest of Brazil. Journal of Experimental Botany 61(11):3119–3127

Faria SM, Franco AA, Jesus RM, Menandro MS, Baitello JB, Mucci ESF, Dobereiner J, Sprent JI (1984) New Nodulating Legume Trees From South-East Brazil. New Phytologist 98(2):317–328

Faria SMD, McInroy SG, Sprent JI (1987) The occurrence of infected cells, with persistent infection threads, in legume root nodules. Can J Bot 65(3):553–558

Faria SM, Lewis GP, Sprent JI, Sutherland JM (1989) Occurrence Of Nodulation In The Leguminosae. New Phytologist 111(4):607–619

Forrester DI, Bauhus J, Cowie AL (2005) On the success and failure of mixed-species tree plantations: lessons learned from a model system of *Eucalyptus globulus* and *Acacia mearnsii*. For Ecol Manag 209:147–155

Forrester DI, Schortemeyer M, Stock WD, Bauhus J, Khanna PK, Cowie AL (2007) Assessing nitrogen fixation in mixed-and single-species plantations of *Eucalyptus globulus* and *Acacia mearnsii*. Tree Physiol 27(9):1319–1328

Founoune H, Duponnois R, BÂ AM (2002) Ectomycorrhization of *Acacia mangium* Willd. and *Acacia holosericea* A. Cunn. ex G. Don in Senegal. Impact on plant growth, populations of indigenous symbiotic microorganisms and plant parasitic nematodes. J Arid Environ 50:325–332

Franco AA, Faria SM (1997) The contribution of N_2-fixing tree legumes to land reclamation and sustainability in the tropics. Soil Biol Biochem 29:897–903

Franco AA, Campello EFC, Dias LE, Faria SM (1995) Use of nodulated and mycorrhizal legume trees of revegetation of residues from bauxite mining. In: international symposium on sustainable agriculture for the tropics—the role of biological nitrogen fixation. Embrapa Agrobiologia/Universidade Federal Rural do Rio de Janeiro, Angra dos Reis. Anais. Rio de Janeiro, pp 80–81

Galiana A, Balle P, N'guessan kang A, Domenach AM (2002) Nitrogen fixation estimated by the ^{15}N natural abundance method in Acacia mangium Willd. inoculated with *Bradyrhizobium* sp. and grown in silvicultural conditions. Soil Biol Biochem 34:251–262

Galiana A, Bouillet JP, Ganry F (2004) The importance of the biological nitrogen fixation by trees in agroforestry. In: Carsky RJ, Sanginga N, Schulz S, Douthwaite B (eds) Symbiotic nitrogen fixation: prospects for enhanced application in tropical agriculture. Baba Barkha Nath Printers, New Delhi, pp 185–199

Gehring C, Vlek PLG (2004) Limitations of the ^{15}N natural abundance method for estimating biological nitrogen fixation in Amazonian forest legumes. Basic Appl Ecol 5:567–580

Gehring C, Vlek PLG, de Souza LAG, Denich M (2005) Biological nitrogen fixation in secondary regrowth and mature rainforest of Central Amazonia. Agric Ecosyst Environ 111:237–2452

Grayston SJ, Vaughan D, Jones D (1997) Rhizosphere carbon flow in trees, in comparison with annual plants: the importance of root exudation and its impact on microbial activity and nutrient availability. Appl Soil Ecol 5(1):29–56

Griffin AR, Chi NQ, Harbard JL, Son DH, Harwood CE et al (2015) Breeding polyploid varieties of tropical acacias: progress and prospects. South Forests 77(1):41–50

Germon A, Guerrini IA, Bordron B, Bouillet J-P, Nouvellon Y, de Moraes Gonçalves JL, Jourdan C, Paula RR, Laclau J-P, (2018) Consequences of mixing Acacia mangium and Eucalyptus grandis trees on soil exploration by fine-roots down to a depth of 17 m. Plant and Soil 424 (1-2):203–220

Harwood CE, Nambiar EKS (2014) Productivity of acacia and eucalypt plantations in Southeast Asia. 2. Trends and variations. Int For Rev 16(2):249–260

He XH, Critchley C, Bledsoe CS (2003) Nitrogen transfer within and between plants through common mycorrhizal networks (CMNs). Crit Review Plant Sci 22:531–567

Högberg P (1997) ^{15}N natural abundance in soil-plant systems. New Phytol 137:179–203

Hungria M, Mendes IC (2015) Nitrogen fixation with soybean: the perfect symbiosis? In: de Bruijn FJ (ed) Biological nitrogen fixation. Wiley, Hoboken, pp 1009–1024. https://doi.org/10.1002/9781119053095.ch99

Jesus EC, Schiavo JA, Faria SM (2005) Dependencia de micorrizaspara a nodulaçao de leguminosas arboreas tropicais. Revista Arvore 29:545–552

Khanna PK (1998) Nutrient cycling under mixed-species tree systems in Southeast Asia. Agrofor Syst 38:99–120

Krisnawati H, Kallio M, Kanninen M (2011) Acacia mangiumWilld.: ecology,silviculture, and productivity. CIFOR, Bogor, Indonesia.

Kuiper I, Lagendijk EL, Bloemberg GV, Lugtenberg BJJ (2004) Rhizoremediation: a beneficial plant-microbe interaction. Mol Plant-Microbe Interact 17:6–15

Laclau JP, Ranger J, de Moraes Gonçalves JL, Maquère V, Krusche AV et al (2010) Biogeochemical cycles of nutrients in tropical Eucalyptus plantations: main features shown by intensive monitoring in Congo and Brazil. For Ecol Manag 259(9):1771–1785

Lawrie AC (1981) Nitrogen Fixation by Native Australian Legumes. Australian Journal of Botany 29(2):143

Le Maitre DC, Gaertner M, Marchante E, Ens EJ, Holmes PM, Pauchard et al (2011) Impacts of invasive Australian acacias: implications for management and restoration. Divers Distrib 17(5):1015–1029

Legume Phylogeny Working Group (2017) A new subfamily classification of the Leguminosae based on a taxonomically comprehensive phylogeny. Taxon 66(1):44–77. https://doi.org/10.12705/661

Lim S, Gan K, Tan Y (2011) Properties of Acacia mangium planted in Peninsular Malaysia. In: ITTO project on improving utilization and value adding of plantation timbers from sustainable sources in Malaysia. Selangor: Forest Research Institute of Malaysia. p. 1–6.

Lorenzi H (1992) Árvores brasileiras: manual de identificação e cultivo de plantas arbóreas nativas no Brasil. Editora Plantarum, Nova Odessa, p 368

Mafongoya PL, Giller KE, Palm CA (1998) Decomposition and nitrogen release patterns of tree prunings and litter. Agrofor Syst 38:77–97

Monteiro SEM (1990a) Resposta de leguminosas arbóreas à inoculação com rizóbio e fungos micorrízicos em solo ácido (Tese de Doutorado). Universidade Federal Rural do Rio de Janeiro, Itaguaí, p 221

Monteiro SEM (1990b) Resposta de leguminosas arbóreas à inoculação com rizóbio e fungos micorrízicos em solo ácido (Tese de Doutorado). Universidade Federal Rural do Rio de Janeiro, Itaguaí, p 221

Moraes Gonçalves JL, Alvares CA, Higa AR, Silva LD, Alfenas AC et al (2013) Integrating genetic and silvicultural strategies to minimize abiotic and biotic constraints in Brazilian eucalypt plantations. For Ecol Manag 301:6–27

Morais TMO, Montagner AEAD (2015) Infestação por Acacia mangium wild em Sistema Silvipastoril, após fogo no Cerrado Amapaense. I Jornada Cientifíca da Embrapa Amapá

Moreira FMS, Siqueira JO (2006) Fixação biológica de nitrogênio atmosférico. In: Moreira FMS, Siqueira JO (eds) Microbiologia e bioquímica do solo. Editora Universidade Federal de Lavras, Lavras, pp 449–542

Moreira FM, da Silva MF, Faria SM (1992) Occurrence of nodulation in legume species in the Amazon region of Brazil. New Phytol 121:563–570

Munroe JW, Isaac ME (2014) N_2-fixing trees and the transfer of fixed-N for sustainable agroforestry: a review. Agron Sustain Dev 34:417–427

Nardoto GB, Quesada CA, Patiño S, Saiz G, Baker TR, Schwarz M, Schrodt F, Feldpausch TR, Domingues TF, Marimon BS, Marimon B-H, Vieira ICG, Silveira M, Bird MI, Phillips OL, Lloyd J, Martinelli LA (2014) Basin-wide variations in Amazon forest nitrogen-cycling characteristics as inferred from plant and soil N: N measurements. Plant Ecology & Diversity 7(1-2):173–187

Narducci TS (2014) Recuperação de áreas de reserva legal: influência da densidade nos indicadores ambientais do plantio de *Sclerolobium paniculatum* Vogel. Dissertação (Mestrado em Ciências Ambientais). Instituto de Geociências, Universidade Federal do Pará, Belém-PA, p 77. Programa de Pós-Graduação em Ciências Ambientais

Natelhoffer KJ, Fry B (1988) Controls on natural nitrogen-15 and carbon-13 abundance in forest soil organic matter. Soil Sci. Soc. Am. J., 52:1633–1640

Oliveira Júnior JQ, Jesus EC, Lisboa FJ, Berbarac RLL, Faria SM (2016) Nitrogen-fixing bacteria and arbuscular mycorrhizal fungi in Piptadenia gonoacantha (Mart.) Macbr. Braz J Microbiol 48:95–100

Paula RR, Bouillet J-P, Ocheuze Trivelin PC, Zeller B, Gonçalves JLM, Nouvellon Y, Bouvet J-M, Plassard C, Laclau J-P (2015) Evidence of short-term belowground transfer of nitrogen from

Acacia mangium to Eucalyptus grandis trees in a tropical planted forest. Soil Biology and Biochemistry 91:99–108
Paula RR, Bouillet J-P, de Moraes Gonçalves JL, Ocheuze Trivelin PC, de C. Balieiro F, Nouvellon Y, de C. Oliveira J, de Deus Júnior JC, Bordron B, Laclau J-P (2018) Nitrogen fixation rate of Acacia mangium Wild at mid rotation in Brazil is higher in mixed plantations with Eucalyptus grandis Hill ex Maiden than in monocultures. Annals of Forest Science 75(1)
Parniske M (2008) Arbuscular mycorrhiza: the mother of plant root endosymbioses. Nat Rev Microbiol 6:763–775
Parrota JA, Knowles OH (1999) Restoring of tropical moist forest on bauxite-mined lands in Brazilian Amazon. Restor Ecol 7(2):103–116
Parrota JA, Knowles OH, Wunderle JM Jr (1997) Development of floristic diversity in 10-year-old restoration forests on a bauxite mined site in Amazonia. For Ecol Manag 99:21–42
Parrotta JA, Baker DD, Fried M (1996) Changes in dinitrogen fixation in maturing stands of Casuarina equisetifolia and Leucaena leucocephala. Can J For Res 26:1684–1691
Patreze CM, Cordeiro L (2004) Nitrogen-fixing and vesicular–arbuscular mycorrhizal symbioses in some tropical legume trees of tribe Mimoseae. For Ecol Manag 196:275–285
Paula, RR (2015) Processos de transferência de N em curto e longo prazo em plantios mistos de Eucalyptus grandis e Acacia mangium. PhD thesis. São Paulo University
Peix A, Ramírez-Bahena MH, Velázquez E, Bedmar EJ (2015) Bacterial associations with legumes. Crit Rev Plant Sci 34:17–42
Peoples MB, Faizah AW, Rerkasem B, Herridge DF (1989) Methods for evaluating nitrogen fixation by nodulated legumes in the field. ACIAR monograph, N° 11. ACIAR 1989:76
Peoples MB, Palmer B, Lilley DM, Duc LM, Herridge DF (1996) Application of ^{15}N and xylem ureide methods for assessing N_2 fixation of three shrub legumes periodically pruned for forage. Plant Soil 182:125–137
Peoples MB, Chalk PM, Unkovich MJ, Boddey RM (2015) Can differences in 15N natural abundance be used to quantify the transfer of nitrogen from legumes to neighbouring non-legume plant species? Soil Biol Biochem 87:97–109
Pires R, Junior FBR, Zilli JE, Fischer D, Hofmann A, James EK, Simon MF (2018) Soil characteristics determine the rhizobia in association with different species of Mimosa in Central Brazil. Plant Soil 423(1–2):411
Piotto D, Montagnini F, Thomas W, Ashton M, Oliver C (2009) Forest recovery after swidden cultivation across a 40-year chronosequence in the Atlantic forest of southern Bahia, Brazil. Plant Ecology 205(2):261–272
Polhill RM, Raven PH, Stirton CH (1981) Evolution and systematics of the Leguminosae. In: Polhill RM, Raven PH (eds) Advances in legume systematics. Royal Botanic Gardens, Kew, London, pp 1–26
Postgate J (1992) The Leeuwenhoek lecture 1992. Bacterial evolution and the nitrogen fixing plant. Phil Trans R Soc Lond B 338:409–416
Rachid CTCC, Balieiro FC, Peixoto RS, Pinheiro YAS, Piccolo MC, Chaer GM, Rosado AS (2013) Mixed plantations can promote microbial integration and soil nitrate increases with changes in the N cycling genes. Soil Biology and Biochemistry 66:146–153
Raymond J, Siefert JL, Staples CR, Blankship RE (2004) The natural history of nitrogen fixation. Mol Biol Evol 21:541–554
Reis FB Jr, Simon MF, Gross E, Boddey RM, Elliott GN, Neto NE et al (2010) Nodulation and nitrogen fixation by *Mimosa* spp. in the Cerrado and Caatinga biomes of Brazil. New Phytol 186(4):934–946
Richards AE, Forrester DI, Bauhus J, Scherer-Lorenzen M (2010) The influence of mixed tree plantations on the nutrition of individual species: a review. Tree Physiol 30:1992–1208
Rolim SG, Piotto D eds. (2018) Silvicultura e tecnologia de espécies da Mata Atlântica. Belo Horizonte, MG: Editora Rona.
Sakrouhi I, Belfquih M, Sbabou L, Moulin P, Bena G, Filali-Maltouf A, Le Quéré A (2016) Recovery of symbiotic nitrogen fixing acacia rhizobia from Merzouga Desert sand dunes in

South East Morocco – Identification of a probable new species of Ensifer adapted to stressed environments. Systematic and Applied Microbiology 39(2):122–131

Santos FM, Balieiro FC, Ataíde DHS, Diniz AR, Chaer GM (2016) Dynamics of aboveground biomass accumulation in monospecific and mixed-species plantations of *Eucalyptus* and *Acacia* on a Brazilian sandy soil. Forest Ecol Manag 363:86–97

Santos FM, Balieiro F de C, Fontes MA, Chaer GM, (2018) Understanding the enhanced litter decomposition of mixed-species plantations of *Eucalyptus* and *Acacia mangium*. Plant and Soil 423 (1-2):141–155

Schiavo JÁ, Martins MA (2002) Produção de mudas de Acácia colonizadas com micorriza e rizóbio em diferentes recipientes. Pesq Agrop Brasileira 38:173–178

Silva K, Meyer S, Rouws LF, Farias EM et al (2014) *Bradyrhizobium ingae* sp. nov., isolated from effective nodules of *Inga laurina* grown in Cerrado soil. Int J Syst Evol Microbiol 64(10):3395–3401

Silva VC, Alves PAC, Rhem MFK, dos Santos JMF, James EK, Gross E (2018) Brazilian species of Calliandra Benth. (tribe Ingeae) are nodulated by diverse strains of Paraburkholderia. Syst Appl Microbiol 41(3):241–250

Stape JL, Binkley D, Ryan MG, Fonseca S, Loos RA, Takahashi EN, Silva CR, Silva SR, Hakamada RE, Ferreira JMA, Lima AMN, Gava JL, Leite FP, Andrade HB, Alves JM, Silva GGC, Azevedo MR (2010) The Brazil Eucalyptus Potential Productivity Project: Influence of water, nutrients and stand uniformity on wood production. Forest Ecology and Management 259(9):1684–1694

Souza M, Magliano M, Camargos J (1997) Madeiras tropicais brasileiras. IBAMA. Laboratório de Produtos Florestais, Brasília, DF, p 152

Souza MH, Magliano MM, Camargos JAA, Souza MR (2002) Madeira tropicais brasileiras, 2nd edn. LPF/IBAMA, Brasília, p 152

Souza AO, Chaves MPSR, Barbosa RI, Clement CR (2018) Local ecological knowledge concerning the invasion of Amerindian lands in the northern Brazilian Amazon by Acacia mangium (Willd.). J Ethnobiol Ethnomed 14:33

Sprent JI (1993) The role of the nitrogen fixation in primary succession on land. In: Miles J, Walton DWH (eds) Primary succession on land. Blackwell Scientific, Oxford, pp 209–219

Sprent JI (1994) Evolution and diversity in the legume-rhizobium symbiosis: chaos theory? Plant Soil 161:1–10

Sprent JI (2007) Evolving ideas of legume evolution and diversity: a taxonomic perspective on the occurrence of nodulation. New Phytol 174:11–25. https://doi.org/10.1111/j.1469-8137.2007.02015.x

Sprent JI (2009) Legume nodulation: a global perspective. Wiley-Blackwell, West Sussex. https://doi.org/10.1002/9781444316384

Sprent JI, Ardley J, James EK (2017) Biogeography of nodulated legumes and their nitrogen-fixing symbionts. New Phytol 215:40–56

Sutherland JM, Sprent LI (1993) Nitrogen fixation by legume trees. In: Subba Rao NS, Rodriguez-Barrueco C (eds) Symbioses in nitrogen-fixing trees. Oxford/IBH, New Delhi, pp 32–63

Tchichelle SV, Epron D, Mialoundama F, Koutika LS, Harmand J-M, Bouillet J-P, Mareschal L, (2016) Differences in nitrogen cycling and soil mineralisation between a eucalypt plantation and a mixed eucalypt and plantation on a sandy tropical soil. Southern Forests: a Journal of Forest Science 79(1):1–8

Tchichelle SV, Mareschal L, Koutika LS, Epron D (2017) Biomass production, nitrogen accumulation and symbiotic nitrogen fixation in a mixed-species plantation of eucalyptus and acacia on a poor tropical soil. Forest Ecol Management 403:103–111

Tonini H, Angelo DH, Conceicao JS, Herzog FA (2010) Silvicultura da Acacia mangium em Roraima. In: Tonini H, HalfelD-VIeira BA, SJR S (eds) Acacia mangium: características e seu cultivo em Roraima. Embrapa Informação Tecnológica e Embrapa Roraima, Brasília e Boa Vista, pp 76–9

Unkovich MJ, Herridge D, Peoples MB, Cadisch G, Boddey B, Giller K, Alves B, Chalk P (2008) Measuring plant-associated nitrogen fixation in agricultural systems (ACIAR Monograph, 136). Australian Centre for International Agricultural Research, Canberra, p 258

Vincent JM (1970) A manual for the practical study of root-nodule bacteria. Blackwell Scientific, Oxford, p 164

Vitousek PM et al (2002) Towards an ecological understanding of biological nitrogen fixation. Biogeochemistry 57:1–45

Voigtlaender M, Laclau J-P, Gonçalves JLM, Piccolo MC, Moreira MZ, Nouvellon Y, Ranger J, Bouillet J-P (2012) Introducing Acacia mangium trees in Eucalyptus grandis plantations: consequences for soil organic matter stocks and nitrogen mineralization. Plant and Soil 352(1-2):99–111

Voigtlaender M, Brandani CB, Caldeira DRM, Tardy F, Bouillet J-P, Gonçalves JLM, Moreira MZ, Leite FP, Brunet D, Paula RR, Laclau J-P (2018) Nitrogen cycling in monospecific and mixed-species plantations of Acacia mangium and Eucalyptus at 4 sites in Brazil. Ann For Sci 75:14. https://doi.org/10.1007/s13595-018-0695-9

Voigtlaender M, Brandani CB, Caldeira DRM, Tardy F, Bouillet J-P et al (2019) Nitrogen cycling in monospecific and mixed-species plantations of Acacia mangium and Eucalyptus at 4 sites in Brazil. For Ecol Manag 436:56–67

Winbourne JB, Feng A, Reynolds L, Piotto D, Hastings MG, Porder S (2018) Nitrogen cycling during secondary succession in Atlantic Forest of Bahia, Brazil. Scientific Reports 8(1)

Yasmin K, Cadisch G, Baggs EM (2006) Comparing 15N-labelling techniques for enriching above- and below-ground components of the plant-soil system. Soil Biol Biochem 38:397–400

Zilli JE, Baraúna AC, da Silva K, de Meyer SE, Farias ENC, Kaminski PE, da Costa IB, Ardley JK, Willems A, Camacho NN et al (2014) Bradyrhizobium neotropical sp. nov., isolate from effective nodules of Centrolobium paraense. Int J Syst Evol Microbiol 64:3950–3957

Chapter 7
Mycorrhiza in Mixed Plantations

Maiele Cintra Santana, Arthur Prudêncio de Araujo Pereira, Bruna Andréia de Bacco Lopes, Agnès Robin, Antonio Marcos Miranda Silva, and Elke Jurandy Bran Nogueira Cardoso

7.1 Introduction

The mutualistic association between plant roots and soil fungi, which results in the enhancement of plant health, is denominated mycorrhiza. The term mycorrhiza originates from Greek and was proposed by the German botanist Albert Bernhard Frank, in 1885, in which "myco" means fungus and "rhiza" means root (Frank and Trappe 2005). In this interaction, the plants, through photosynthesis, provide energy and carbon for the survival and multiplication of symbiotic fungi (Smith and Read 2008; van der Heijden et al. 2015). In addition, the mycorrhizal hyphal system increases the area of root exploration in the soil, being important mainly for the efficient absorption of nutrients and water by plants (Cardoso et al. 2010; Smith and Smith 2011).

M. C. Santana (✉) · B. A. de Bacco Lopes · A. M. M. Silva · E. J. Bran Nogueira Cardoso
Department of Soil Science, University of São Paulo, "Luiz de Queiroz"
College of Agriculture, Piracicaba, SP, Brazil
e-mail: mcsantana@usp.br

A. P. A. Pereira
Department of Soil Science, University of São Paulo, "Luiz de Queiroz"
College of Agriculture, Piracicaba, SP, Brazil

Soil Science Department (Pici Campus), Federal University of Ceará, Fortaleza, Ceará, Brazil

A. Robin
Department of Soil Science, University of São Paulo, "Luiz de Queiroz"
College of Agriculture, Piracicaba, SP, Brazil

Eco & Sols, Univ. Montpellier, CIRAD, INRA, IRD, Montpellier SupAgro, Montpellier, France

E. J. Bran Nogueira Cardoso et al. (eds.), *Mixed Plantations of Eucalyptus and Leguminous Trees*, https://doi.org/10.1007/978-3-030-32365-3_7

Mycorrhizae are classified into seven types: arbuscular mycorrhiza (AM), ectomycorrhiza (ECM), ericoid mycorrhiza and orchidoid mycorrhiza, ectendomycorrhiza, arbutoid mycorrhiza, and monotropoid mycorrhiza (Harley and Smith 1983; Brundrett 2004). Most groups of vascular plants are able to form mycorrhizae, while only a few families, such as Brassicaceae, Cyperaceae, and Proteaceae, do not develop this association (Souza et al. 2006). Specifically, AM and ECM are the most studied in forests due to their important role in the maintenance of biodiversity and ecosystem productivity in agriculture and forestry (Berbara et al. 2006). Normally, most plants associate only with one mycorrhizal type; however studies regarding the occurrence and diversity of mycorrhizas in Brazil have shown that plants from the genera eucalyptus and acacia are able to form, either alone or simultaneously, both AM and ECM (Pereira 2015; Santana et al. 2016; Santana 2017). It has been assumed that mycorrhizal colonization is restricted to the topsoil, considered the arable part of the soil, and the mycorrhizal association in deeper regions of the soil profile is poorly understood. However, recent research has shown that this type of colonization can occur in deep soil layers, down to 8 m, as will be detailed below (Pereira 2015; Pereira et al. 2018; Robin, A. personal data).

Currently, in addition to conventional techniques, such as spore morphology and root colonization rates, molecular technology has been widely used for the taxonomic classification and characterization of mycorrhizal communities (Anderson and Cairney 2004; Gasparotto et al. 2010; Clasen et al. 2018). Starting with the extraction of nucleic acids from the fungi, it is possible to get information regarding the structure of the fungal community, which can be accessed through analysis and construction of cloned libraries, sequencing, and fingerprinting techniques (Lambais et al. 2005; Gasparotto et al. 2010). These techniques have helped to understand the diversity and ecology of mycorrhizal fungi and their impact on plant development.

Thus, our main objective of this chapter is to describe succinctly the most important structures of arbuscular mycorrhiza and ectomycorrhiza, and to detail these associations in pure and mixed eucalyptus and acacia plantations, highlighting the main results obtained in Brazil. We hope that this information contributes to the understanding of the interactions that occur in these planting systems, in order to assist forest production and environmental sustainability.

7.2 Arbuscular Mycorrhizal Fungi (AMF)

The arbuscular mycorrhizal symbioses result from the association of arbuscular mycorrhizal fungi (AMF), belonging to the phylum *Glomeromycota* with the root system of more than 90% of the vascular plants (Wang and Qiu 2006; Willis et al. 2013). These fungi present an obligate biotrophic behavior because they depend on a living host to fulfill their biological cycle. In this association, the host plant supplies the fungus with more than 10% of the photosynthates produced and, in return, the plants are benefited by improving their nutritional status, especially in soils where the nutrient supply is low or unbalanced (Smith and Read 2008; Cardoso et al. 2010).

An interesting point is that AMF are not very selective with regard to their host plants; that is, they show almost no host specificity (Santos et al. 2006; Smith and Read 2008). Nevertheless, there is a great variability in the outcome or in the effectiveness of different host plant-AMF combinations, influenced by both the host plant and the fungus, and modulated by edaphoclimatic conditions, especially P (Cardoso et al. 2017). Thus, the growth-promoting effect of an AMF depends on the level of available P in the soil. The more P, the less will be the positive effect of AMF inoculation. This is doubtless one of the causes for AMF being extremely effective in P-depleted tropical soils, much more than in soils of greater fertility. On the other hand, even in the presence of mycorrhizas, a minimum of available phosphate in soils is required to result in satisfactory plant growth, and this minimum varies according to the plant, endophyte, and soil characteristics. Mycorrhizal endophytes colonize the roots much more extensively under low-P conditions. As soil P concentrations increase, the root colonization decreases strongly (Nogueira and Cardoso 2006; Cardoso et al. 2017).

The formation of AMF mycorrhiza starts with the exchange of signals between the host plant and the AMF propagules, with the exudation by the roots of compounds that stimulate the branching of the fungal hyphae. These hyphae, when they enter in contact with the surface of the roots, differentiate into appressoria, and penetrate through the epidermis (Lambais 2010). Within the roots, the hyphae may grow inter- and intracellularly in the cortical tissue, not ever invading the meristematic region and vascular tissues. These hyphae differ in arbuscules or hyphal coils, structures that morphologically describe the two types of AM, Arum type and Paris type (Gallaud 1905). The Arum type is the most common, identified by the growth of inter- and intracellular hyphae and the production of intracellular arbuscules in the cells of the root cortex. The Paris type is defined by cell-to-cell growth of intracellular hyphal coils (Smith and Smith 2011).

A prominent feature of the Arum-type morphology is the intercellular growth of hyphae in a longitudinal manner through the root. Arbuscules arise on short side branches from these intercellular hyphae, typically at right angles to the main root axis (Smith and Smith 2011). Coils of the Paris type of mycorrhiza often, but not invariably, become arbusculate; that is, they develop arbuscule branches from one or more loci on the coil (Gallaud 1905; Smith and Smith 2011; Dickson 2004; Requeña et al. 2007). Arbuscules and hyphal coils are extremely important structures in the process of exchanging metabolites and nutrients between the symbionts, but the mechanisms controlling their development and functioning are little known (Lambais 2010).

Other characteristic components of some AM are vesicles, lipid-rich structures whose function is a nutrient reserve (Moreira and Siqueira 2006), but they do not occur in all AMF species. In AMF genera as *Gigaspora* and *Scutellospora*, that do not present such structures, normally one can find extraradical organelles with an equivalent function, called secondary cells. Besides, AMF produce extraradical fungal spores present in the soil and, in some cases, in the roots (Smith and Read 2008). AMF spores vary in size from 22 to 1050 μm (Stürmer 2012), and it is through their descriptions that the identification of AMF species traditionally is carried out (Schenck and Perez 1990; INVAM (http://invam.caf.wvu.edu)). The MA does not produce macroscopic morphological alterations in the host roots, so that it is necessary to treat them by means of clarification and dyeing processes and to examine them under the microscope (Siqueira et al. 2002).

Another feature is the presence of a minimum number of spores in the substrate to result in root colonization, the infection potential. This potential likewise depends on the biotic and abiotic conditions, which modulate the colonization intensity (Santana 2017; Steffen et al. 2010; Moreira and Siqueira 2006; Smith and Read 2008).

7.3 Ectomycorrhizal Fungi (ECMF)

Ectomycorrhiza is a symbiotic relationship that occurs between fungi belonging to the phyla *Basidiomycota* and *Ascomycota* (Clasen et al. 2018), with roots of various species of Gymnosperms and Angiosperms (Brundrett 1991). We estimate that there are more than 5000 fungal species that form ECM, predominantly with tree species of temperate climatic regions (Futai et al. 2008). In tropical regions, most tree species form AM; however, the study of ECM in these regions is driven by the extensive use of exotic species in reforestation programs such as Pinus, Eucalyptus, and Acacia (Kasuya et al. 2010).

In ECM symbioses, fungi also use energetic organic compounds, and in turn promote the mineralization of organic forms of the nutrients and solubilize minerals through the production of organic acids, making these elements available for absorption (Shah et al. 2016). The formation of ECM changes the environment of the rhizosphere, which becomes unfavorable for most pathogens of the root system (Garbaye 1991). ECM symbiosis is characterized by presenting a fungal mantle, a hyphal layer formed externally to the root epidermis, and a Hartig net, with growth of the hyphae in the intercellular spaces of the root cortex (Agerer 1995; Vesk et al. 2000; Brundrett 2002). Extending from the mantle there are fungal rhizomorphs, similar to plant roots, essential for the connection of the fungus with the soil to produce fruiting bodies (Smith and Read 2010).

The fungal mantle helps in the transfer of nutrients to the plants and acts as a mechanism of protection against pathogens (Peterson and Bonfante 1994). In the mantle, there also occurs the synthesis of reserve compounds such as glycogen, polyphosphates, and proteins (Peterson and Bonfante 1994). Starting at the rhizomorphs the hyphae extend outwardly in all directions through the soil, forming a dense network connecting different plants, conducive to the exchange of organic and inorganic nutrients. It also helps the roots to assess an increased soil volume for water and nutrients (Brundrett 2002; Shah et al. 2016). The Hartig net growing in between the cortex cells has a great contact interface with the plant and is the site of nutrient exchange between the symbionts. In angiosperms, the Hartig net is generally limited to only a few more external plant cell layers in the cortex, whereas in gymnosperms it may comprise the intercellular spaces of all cell layers, though never surpassing the outer limit of the cortex imposed by the root endodermis (Kasuya et al. 2010).

Unlike AMF, ECMF are not obligate biotrophs and can be cultivated on some culture media. ECM modifies the morphology of the plant roots and may inhibit the formation of root hairs, which are replaced by fungal hyphae (Kasuya et al. 2010). The ECM roots can take different forms, as nodular, pyramidal, bifurcated, and coralloid, among others, with a highly variable color spectrum of the fungal mycelium (black, red, yellow, brown, white, etc.), which will equally define the external coloration of the root mantle (Agerer 2001; Tedersoo and Brundrett 2017). Variations in mantle structure, surface ornamentations, staining, and presence of rhizomorphs are used, along with chemical and immunological tests, to characterize the ECM and for identification of the associated fungus (Clasen et al. 2018), although nowadays there is opening up of the opportunity of molecular technology for the taxonomic classification of these fungi (Suz et al. 2008).

7.4 AM and ECM Symbiosis in Pure and Mixed Plantation of *Eucalyptus* sp. and *Acacia* sp.

The occurrence of mycorrhiza in eucalyptus species was mentioned for the first time in 1917 by van der Bijl (Barros et al. 1978), but the interest in this association began to gain prominence with the first attempts to classify and describe the structure of mycorrhiza in eucalyptus by Chilvers and Pryor (1965). The work of these researchers complemented by others, such as Levisohn (1958), shows that apparently all species of eucalyptus are able to form mycorrhiza. From this period on, the study of the effect of mycorrhizal fungi on eucalyptus growth and survival has been addressed in the literature. Researchers have shown that symbiosis provides a great success of these plants in forest plantations (Chen et al. 2007, 2014; Futai et al. 2008; Souza et al. 2008; Chiquete 2011; Bini et al. 2018).

Similar to eucalyptus, acacia plants also associate symbiotically with mycorrhizal fungi (Pereira 2015; Santana 2017). Besides that, acacia species with few exceptions nodulate and fix nitrogen with root nodule bacteria in the range of 20 to 300 kg ha^{-1} year^{-1} (Dommergues 1987). Mycorrhiza enhances nutrient absorption, particularly P, and water uptake by acacia species and improves their nitrogen fixation, which enables them to establish in marginalized lands in the tropics (Requeña et al. 2001). These associations contribute to their tolerance to drought, and induce resistance against soil pathogens (Smith and Read 2008). These associations, in general, enable many of the acacia species to perform well in degraded soils with high acidity, high salinity, high aluminum saturation, and low soil fertility (Craig et al. 1991).

As we have seen, the roots of *Eucalyptus* and *Acacia* can be colonized by mycorrhizal fungi and there have been reports that these plants can be colonized by both types of mycorrhiza, AM or ECM either alone or together. Pereira (2015) and Santana (2017), through morphological characterization, reported colonization of AM and ECM in both eucalyptus and acacia roots in pure and mixed plantations in Brazil. While eucalyptus and acacia roots can establish symbioses with both AMF and ECM fungi, several studies show a succession in the establishment of the two fungal communities. Symbiosis with AMF fungi is predominantly present when trees are young, followed by symbiosis with ECM fungal communities (Bellei et al. 1992; Oliveira et al. 1997). The two symbioses can, however, be observed at the same time. This ability to form both types of symbiosis is very interesting from a nutrient acquisition point of view, with the AMF symbiosis being especially recognized for their assimilation of P, and the ECM symbiosis may play a more important role in the assimilation of N (Read 1991).

Bini et al. (2018) studied the root colonization rates by AMF in *Eucalyptus grandis* and *Acacia mangium*, in monocultures and consortia, at 7, 14, and 20 months after planting, and found that there was a medium-to-high AMF colonization rate (above 34%), especially at 14 months. In monocultures of *A. mangium* (A) there were the highest colonization rates in all sampling times, although at 14 and 20 months there was no significant difference between monocultures and consortia (*A. mangium* + *E. grandis* (A + E) and *E. grandis* + *A. mangium* (E + A)). In pure *E. grandis* (E) plantations, there was a lower AMF colonization rate. When evaluating the same treatments, however, at 48 months, Pereira et al. (2018) found lower AMF colonization rates, with mean percentages varying between close to 0 and 10%. In *A. mangium*, at this time, there was a still greater decline in AMF root colonization although high percentages of ECM were observed. At 48 months, however, the root colonization of *Acacia* by AMF regressed strongly, while ECM showed high colonization rates (Table 7.1).

In a different field experiment, similar to that of Bini (2012) and Pereira (2015), Santana (2017) showed that, at 24 months, the ECMF colonization rate is higher (49% in the 0–10 cm layer of soil and 12% in the 20–50 cm) when compared to the colonization rate by AMF (30% in the 0–10 cm layer and 10% in the 50–100 cm layer) (Fig. 7.1). These data are similar to those reported by Pereira (2015), who found higher ECM values in relation to AMF in all evaluated plants. This difference in symbiotic fungal communities according to the age of eucalyptus and acacia plantations could have consequences also in the carbon cycle. Indeed, recent studies have shown a difference in rate of decomposition between tree species associated with AMF fungi or ECM fungi, with the litter of the AMF trees decomposing faster (Midgley et al. 2015; Taylor et al. 2016). Variations of exudation rates have been detected, and trees associated with ECM exude more carbon than AMF trees (Yin

Table 7.1 Percent root colonization by AMF or ECMF in pure or mixed plantations of *Eucalyptus grandis* (E) and *Acacia mangium* (A) at different times after transplanting of seedlings

	Months after transplantation				
	AMF				ECM
Treatments	7	14	20	48	48
A. mangium (A)	51 aA	53 aA	45 aB	7.1 bC	33.3 bB
E. grandis (E)	35 cB	45 bA	38 bB	9.3 aC	52.3 aA
A. mangium in consortium (A + E)	42 bB	52 aA	44 aB	2.0 cC	34.6 bB
E. grandis in consortium (E + A)	43 bB	50 abA	42 abB	0.7 dC	33.0 bB

Here we summarize the data obtained by Bini (2012) and Pereira (2015)
Adapted from Bini (2012) and Pereira (2015), cited by Cardoso et al. (2017)

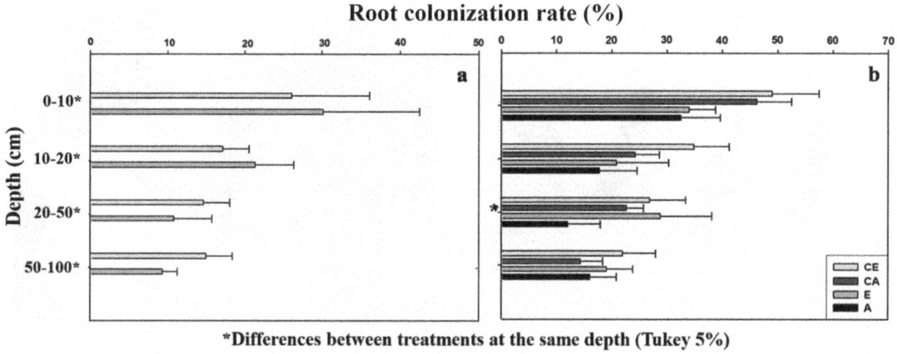

Fig. 7.1 Colonization by AMF in a mixed plantation (E+A) (**a**) and ECMF (**b**) in Eucalyptus growing in consortium (CE), Acacia in consortium (CA), Eucalyptus (E) and Acacia (A) in monoculture along a soil depth gradient (cm) (Santana 2017)

et al. 2014; Liese et al. 2018). The differences in traits between AMF and ECM fungi (such as the amount of rhizomorphs and the presence of pigments) (Aguilar-Trigueros et al. 2014; Churchland and Grayston 2014) can have important impacts on soil carbon storage and a better understanding of the impact of associations between the two mycorrhizal communities is an important issue.

Despite the fact that AMF colonization was found in eucalyptus roots planted in a consortium with acacia, Santana (2017) did not find arbuscular mycorrhiza in *Acacia mangium* plants in this forest system (neither in pure nor in mixed stands with eucalyptus) (Figs. 7.1 and 7.4c). *A. mangium*, besides forming symbiosis with AMF, also associates symbiotically with nitrogen-fixing bacteria (NFB), forming a tripartite symbiosis: AMF–plant–NFB (Carvalho and Moreira 2010). As both microorganisms of this symbiosis depend on the C sources offered by the plant for their survival, the costs of maintaining the tripartite symbiosis for the plant are considerable (Mortimer et al. 2008). In this way, the plant needs mechanisms that allow it to control both symbioses according to its needs. In a study with plants inoculated with AMF or NFB, in which inoculation with the second symbiont was done 20 days after inoculation with the first symbiont, it was possible to observe that the pre-establishment of NFB suppressed subsequent formation of AMF and vice versa (Bethlenfalvay et al. 1985).

In the experiment performed by Santana (2017), we reported the formation of NFB nodules formed, in addition to ECM in roots of *A. mangium* (Fig. 7.2). Furthermore, it was found, through chemical and multivariate analyses, that N and P concentrations in the root correlated negatively with AMF root colonization, and acacia plants presented high mineral nutrient concentrations when compared with Eucalyptus roots. These data may be the explanation for the absence of mycorrhiza, probably indicating that there was no need for AMF colonization at this stage of acacia development. However, when in optimal conditions the tripartite leguminous symbiosis corresponds to a maximal return in growth and productivity of the plant, as occurred in an experiment with soybeans as test plants (Cardoso 1985), which

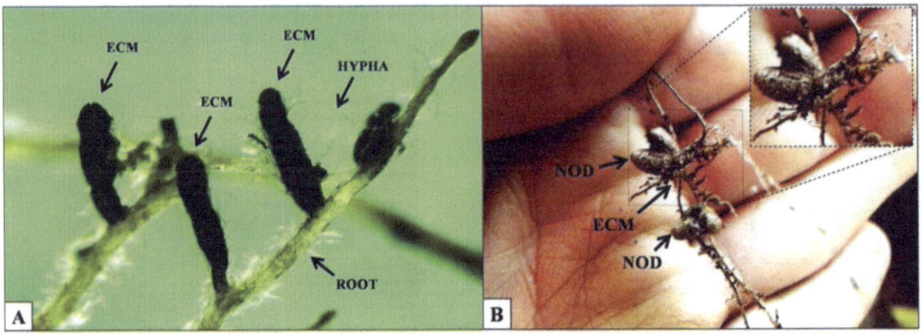

Fig. 7.2 Root colonization by ectomycorrhizal fungi (A) ECM associated with roots of *Acacia mangium* involving the fungus *Cenococcum* sp., (B) ECM associated with *Eucalyptus grandis* probably involving *Pisolithus* sp. fungi. Photo A taken by photo camera and photo B taken by binocular stereoscopic microscope with 40-fold magnification. *NOD* Nitrogen-fixing bacterial nodule (Santana, M.C., personal data)

corroborated the results of a previous study with a similar outline of Bethlenfalvay et al. (1985).

Analyzing the total number of AMF spores and colonization of *A. mangium* roots, Santana (2017) observed that the presence of spores in the soil did not correlate with the intensity of root colonization. This result is in agreement with the data of Steffen et al. (2010), who observed that the number of spores in the substrate did not correspond to the degree of colonization of *E. grandis* roots. According to Moreira and Siqueira (2006) and Smith and Read (2008), the presence of mycorrhizal spores in the soil or substrate will result in nutritional and adaptive benefits to the host plant, if the biotic and abiotic conditions of the site allow colonization. Through the morphological analysis of AMF spores, Santana (2017) found the fungi *Glomus macrocarpum*, *Acaulospora mellea*, *Racocetra* sp., and *Gigaspora* sp. (Fig. 7.3). These fungi were considered of common occurrence in areas with eucalyptus plantations (Gomes and Trufem 1998) and *A. mangium* (Caproni et al. 2005). Cavagnaro et al. (2007) found, in an experiment with plants of *Lycopersicon esculentum*, that different species of the genus *Glomus* can form either Arum- or Paris-type mycorrhiza, and reported that the genera *Gigaspora* and *Scutellospora* formed mycorrhiza of the Paris type in this experiment.

Santana (2017) demonstrated the formation of Paris-type mycorrhizae in roots of *E. grandis* in monoculture or in a consortium with *A. mangium* (Fig. 7.4). Complete absence of arbuscules or of coils in eucalyptus roots was also reported by Campos et al. (2011) in another survey, while Pereira et al. (2018) reported the presence of Paris-type mycorrhizae in eucalyptus roots. However, Malajczuk et al. (1981) observed the presence of both types of mycorrhizae, with hyphal coils and arbuscules in another plantation, in *Eucalyptus diversicolor* and *Eucalyptus marginata*. Since there is still a great knowledge gap regarding the functional aspects involved in both types, we suggest that, in future studies, the morphotype of the fungus and eventual successional stages should also be reported and not just the presence or absence of the symbiosis (Berbara et al. 2006).

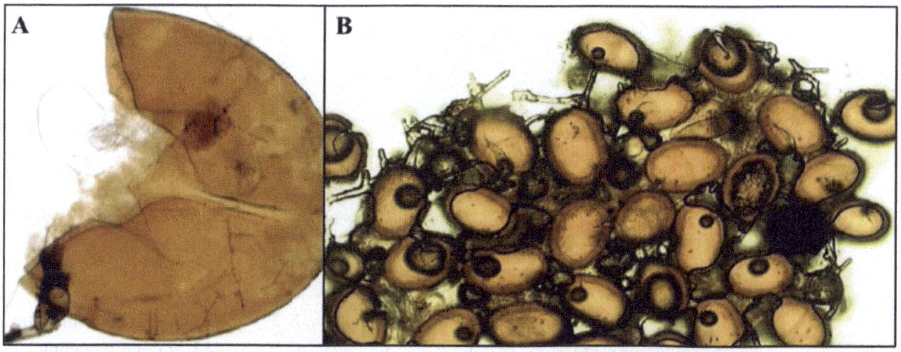

Fig. 7.3 Spores of arbuscular mycorrhizal fungi. (**a**) *Gigaspora* sp. spore found in a consortium of Eucalyptus and Acacia, in the 0–10 cm soil layer (20×). (**b**) *Glomus macrocarpum* spores, found in the same plantation, in the 10–20 cm soil layer (10×) (Photos: Denise de Lourdes C. Mescolotti (USP-ESALQ) Santana (2017))

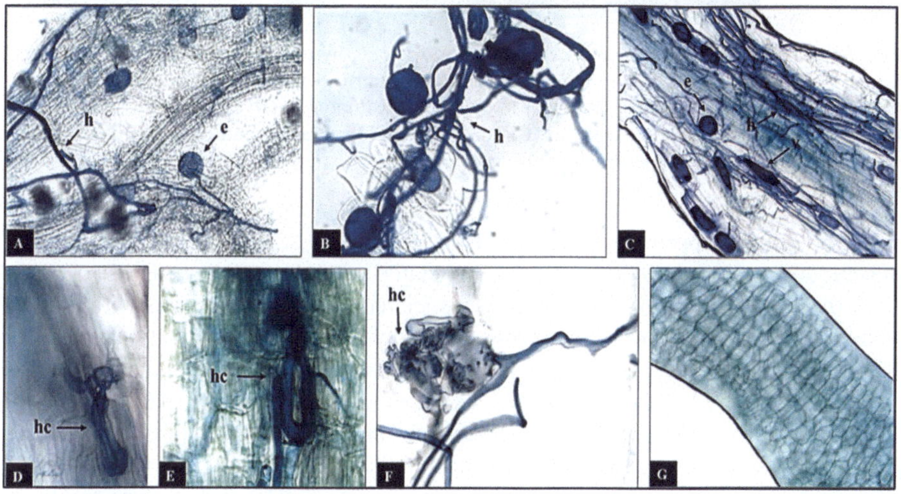

Fig. 7.4 Arbuscular mycorrhizal colonization. In **a–c** there are hyphae (h), vesicles (v), and spores (e) although no arbuscules. In **d–f** the presence of hyphal coils (hc), **g** shows a root without colonization. A and G with 10-fold magnification; **b–f** with 40-fold magnification (optical microscope) (Santana 2017)

Many ECM fungi have a broad host range while others are more specific and colonize certain hosts or host genera (Molina et al. 1992). For AM fungi, so far no convincing evidence was presented demonstrating that these are host specific, although host preferences and host selectivity have been widely reported (Helgason et al. 1998; Vandenkoornhuyse et al. 2003; Torrecillas et al. 2012). A high-throughput sequencing study (using the technology 454) carried out on pure eucalyptus or acacia plantations in the Congo (Pointe-Noire) on the AMF fungi community revealed the predominant presence of the genus *Rhizophagus* associated with the

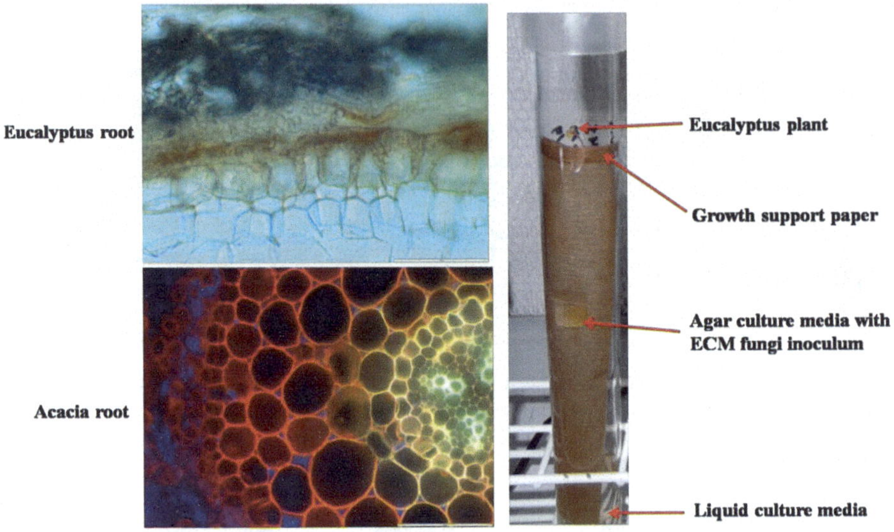

Fig. 7.5 Colonization of *E. grandis* and *A. mangium* roots by the strain *Pisolithus* sp. 45, after 90 days of inoculation. Microscopic images from the transversal root cuttings show the Hartig nets. For acacia roots, fungal cells appear in blue (Uvitex fluorescent dye to stain the intercellular cortex cell spaces). Photos: Yves Prin (Cirad, UMR LSTM, France) and Ranieri Ribeiro Paula (USP-ESALQ) (Robin A., personal data)

roots of acacia (approximately 80% of *Rhizophagus* sequences associated with acacia roots against 18% of sequences associated with eucalyptus roots). In opposition, the genus *Gigaspora* was mainly associated with eucalyptus roots (approximately 60% of *Gigaspora* sequences associated with the roots of eucalyptus against 8% of these sequences associated with acacia roots) (Robin A., personal data).

When the same ECM fungus colonizes the roots of both species, a different colonization of root tissues was observed between the roots of Eucalyptus and Acacia. A study made in controlled conditions showed that the same strain of Pisolithus sp. 45 could colonize both Eucalyptus and Acacia roots. However, interestingly, a microscopic cross section clearly showed a difference in colonization. Pisolithus penetrates only the first few layers of the epidermis of the eucalyptus roots whereas, for the acacia, the fungus reaches the endodermis (Fig. 7.5).

7.5 Mycorrhiza in Deep Soil Layers

Few studies were conducted on mycorrhizal fungi associated with roots in deep soil layers, despite the growing interest in studies of microbial communities in deep soils (Li et al. 2014; Gocke et al. 2017; Pereira et al. 2017; Zheng et al. 2017), with studies in mixed plantations being even rarer. A recent survey demonstrated the presence of AMF fungal spores in deep soil layers down to 8 m in pure and mixed

Eucalyptus grandis and *Acacia mangium* plantations in Brazil (Pereira et al. 2018) and in the rhizosphere of *Faidherbia albida* in Senegal, where the AMF spores were detected as deep as at 34 m below soil level (Dalpé et al. 2000). Pereira et al. (2018) showed AMF root colonization of Eucalyptus and Acacia for the 0–100 cm soil with colonization rates between 6 and 25%; however they found only about 10% below 1 m and 5% or 6% below 3 m. This weak AMF colonization may be partly due to the plants being already over 1 year of age, a period in which commonly the AMF are substituted gradually by ECMF in Eucalyptus. Nevertheless, these results indicate that one should take into account mycorrhizal symbiosis in the deeper layers of soil, and not only in the first 30 cm, as is most commonly done. The presence of ECMF propagules in deep soil has also been described (Santana et al. 2016), as well as the presence of typical ECM structures in Eucalyptus roots down to 6 m deep (Lambais et al. 2017).

The minirhizotron method is well suited for root observations and has already been used to study the dynamics of ECMF (McCormack et al. 2017), but it has aroused some doubts about the method itself perhaps being the reason for this finding on deep soil mycorrhization, due to contamination of deeper regions by surface soil. Recently, a new sampling method for deep soil layers was developed by sampling fine roots and ECM root tips during the digging of pits, layer by layer, to avoid contaminations between two continuous layers. This research demonstrated for the first time the presence of already well-developed ECM symbioses on deep roots (Robin A. personal data). In this study, the authors showed the mycorrhizal presence of *Pisolithus* on eucalyptus at a depth of 4 m. Visual observations were confirmed by molecular sequencing. The diversity study (by ITS Illumina sequencing) showed a strong impact of the depth on the intraspecific diversity of the fungus *Pisolithus*, with the presence of a reservoir of biodiversity associated with the deeper roots. The observation of ectomycorrhizal Acacia roots (by *Pisolithus* or *Scleroderma*) down to 1 m deep (Santana 2017) highlights the importance of evaluating deeper layers equally for acacia mycorrhizal colonization. The stratification between eucalyptus and acacia roots as a function of depth found in mixed plantations (Laclau et al. 2013; Germon et al. 2018) possibly could detect deep root mycorrhization, perhaps with different mycorrhizal communities when evaluating pure or mixed plantations between these two tree species.

7.6 Understanding the Concept of Common Mycorrhizal Networks (CMN) in Mixed Plantations of Eucalyptus and Acacia: Prospects for Future Research

The concept of mycorrhizal networks, defined as a common mycorrhizal mycelium linking the roots of at least two plants (Simard et al. 2012), is perfectly applicable in mixed plantations of eucalyptus and acacia. Common mycorrhizal networks have been shown to facilitate the transfer of carbon (Simard et al. 1997), nitrogen (He et al. 2009), and water (Egerton-Warburton et al. 2007; Prieto et al. 2016).

Concerning the nitrogen cycling by CMN, normally we find an overwhelming occurrence of the transfer of N compounds from the leguminous N-fixing tree to the non-fixing eucalyptus. This has also been studied for annual plants (Moyer-Henry et al. 2006; Jalonen et al. 2009), although nitrogen transfers were demonstrated between *Casuarina cunninghamiana* and *Eucalyptus* (He et al. 2005), highlighting its great value for eucalyptus, one of the most outstanding species in forestry worldwide. Transfers through CMN between Acacia and Eucalyptus have been studied involving AMF (Meng et al. 2015) or ECMF (He et al. 2005). Many studies were conducted under controlled conditions, for example in the greenhouse, while in situ in the field direct demonstrations of transfer are more difficult. An in situ study using fungicides showed ^{15}N transfer via mycorrhizal networks (Montesinos-Navarro et al. 2016). In eucalyptus and acacia plantations, N transfer was demonstrated in situ (Paula et al. 2015). An interesting experiment with a leguminous tree from the Mimosa group in the Atlantic Forest, nodulated by the beta-Rhizobium Burkholderia and inoculated by AMF, also demonstrated the synergic action of these two agents together (Lammel et al. 2015).

Mendes Filho et al. (2010) set up a pot experiment in a greenhouse, with exotic and native leguminous trees using cassiterite mine spoil as substrate. The substrate was very poor in microorganisms and even poorer in energetic organic matter, since it showed an almost absence of respiration when tested using a direct soil respiration test, but still somewhat more when applying the glucose-induced respiratory test. To this substrate, distributed in 2 L pots, we applied two main treatments, with or without organic compost (OM), and all pots were inoculated with a Rhizobium or Burkholderia strain with specificity for the corresponding leguminous tree. Then, each one of those main treatments was subdivided into four sub-treatments, i.e., inoculation of AM fungi, fertilization with P, or application of both, besides a control without any further treatment, resulting in eight treatments altogether (AMF, P, AMF + P, and control (OM+) and AMF, P, AMF + P, and control (OM−)), with four replicates. Right afterwards the seeds of the corresponding tree were planted in each pot.

After 1 month of growth under a constant watering regime, we took photos of the experiment, which demonstrate that some growth only occurred in the presence of compost. *Acacia mangium* (exotic) grew vigorously in AMF (OM+) and AMF + P (OM+) and showed a much poorer growth in control and P (OM+). In the treatments without compost (OM−) *A. mangium* showed only a little growth in the sub-treatment AMF + P and still less in AMF. *Mimosa caesalpiniaefolia* (native tree) responded in a similar way, but a real satisfactorily growth pattern only appeared in AMF (OM+) and in AMF + P (OM+), with a little growth in P (OM+). The minimal growth in some other sub-treatments should be neglected because the seedlings will never develop satisfactorily (Fig. 7.6).

These results show that in a soil of low fertility, especially when highly eroded, decapitated, or containing residues of heavy metals, organic matter would be the most fundamental ingredient for the development of such trees, but we only detected a real adequate development when they also were provided with AMF. In the case of *A. mangium*, even when P was not given, the mycorrhiza alone scavenged all necessary P from the depleted substrate. Nevertheless, an incipient growth was observed when growing with compost and receiving P fertilization, as well as in the

Fig. 7.6 The legumes *A. mangium* (exotic) and *M. caesalpiniaefolia* (native), inoculated with the respective rhizobia, were planted on a very poor and exhausted mine spoil with or without organic compost (OM), AMF, P or both, and a control (Mendes-Filho et al. 2010; Cardoso and Andreote 2016; Cardoso et al. 2017)

substrate without compost, but when receiving P and AMF inoculation. The Mimosa, however, seems to be more sensitive to negative soil conditions, with real adequate development only in the presence of AMF + P or AMF (OM+). In the absence of mycorrhiza, but in the presence of P (OM+) growth was just incipient.

Obviously, trees of the family Fabaceae will only develop satisfactorily when inoculated with the right rhizobia. Since this fact is well known already for a long time by microbiologists, we did not use rhizobial inoculation as a treatment. Instead, the adequate rhizobial strain was inoculated together with the seeds in each pot, since this is by far the most fundamental condition for legume growth, especially in poor substrates. In this study, however, we proved that mycorrhizae are also an indispensable factor for the rhizobia, because they cannot nodulate plant seedlings in the absence of P. Yet OM acts simultaneously as a chemical, physical, and biological factor of soil fertility. It acts as a chemical factor, contributing with some macro- and micronutrients for the plants and regulating the pH; as a physical factor, providing a more adequate structure of the substrate; and finally as a biological factor, furnishing the energetic substances necessary for the nutrition and multiplication of microorganisms, since these are also indispensable for soil and plant health, breaking down the complex molecules, and providing mineral nutrients for the plants.

7.7 Final Remarks

The study of mycorrhizal fungi in pure and mixed eucalyptus and acacia plantations is still incipient; however the results have been promising, especially for mixed plantations with both species. We have shown that the mycorrhizal root colonization of these plants differs when in intercropping. These effects are also dependent on several edaphic-climatic factors and can be affected by plant age, since their type and intensity are variable from site to site. Likewise, we have yet an important knowledge gap regarding the functional aspects involved in the mycorrhizal symbiosis, mainly for nutrient transfer between trees. For example, although Paula et al. (2015) showed the N transfer from acacia to eucalyptus roots, it would be difficult to guarantee its mediation through the mycorrhizal hyphal system, and the mechanisms remain to be studied to know if a part of this transfer could be due to mycorrhizal networks. For the future, we suggest a major focus on the fungus morphotype analysis for ECM symbiosis, as well as on the eventual successional processes from AM to ECM in the later stages of forest plantations. We also need to go beyond diversity approaches and to study more the functional aspects of these symbioses, and not only taking into account the first few centimeters of soil, with researches in the total depth of the soil profile. Implementing a holistic view of the mycorrhizal community in pure and mixed Eucalyptus and Acacia plantation will help the producers to get a more effective and sustainable production, reducing agrochemicals and other external inputs.

References

Agerer R (1995) Anatomical characteristics of identified ectomycorrhizas: an attempt towards a natural classification. In: Varma A, Hock B (eds) Mycorrhiza. Springer Verlag, Berlin, pp 685–734

Agerer R (2001) Exploration types of ectomycorrhizae—a proposal to classify ectomycorrhizal mycelial systems according to their patterns of differentiation and putative ecological importance. Mycorrhiza 11:107–114

Aguilar-Trigueros CA, Powell JR, Anderson IC et al (2014) Ecological understanding of root-infecting fungi using trait-based approaches. Trends Plant Sci 19:432–438

Anderson IC, Cairney JWG (2004) Diversity and ecology of soil fungal communities: increased understanding through the application of molecular techniques. Environ Microbiol 6:769–779

Barros FM, Bradi RM, Reis MS (1978) Micorriza em eucalipto. Rev Árv 2:130–140

Bellei MD, Garbaye J, Gil M (1992) Mycorrhizal succession in young Eucalyptus-Viminalis plantations in Santa-Catarina (southern Brazil). Forest Ecol Manag 54:205–213

Berbara RLL, Souza FA, Fonseca HMAC (2006) III—Fungos Micorrízicos Arbusculares: Muito Além da Nutrição. In: Nutrição Mineral de Plantas. http://www.ufrrj.br/amfoods/arquivos/arq_publicacao/20_ARQ.pdf. Accessed 18 Jun 2019

Bethlenfalvay GJ, Brown MS, Stafford AE (1985) Glycine-Glomus-rhizobium symbiosis: II. Antagonistic effects between mycorrhizal colonization and nodulation. Plant Physiol 79:1054–1058

Bini D (2012) Atributos microbianos e químicos do solo e da serapilheira em plantios puros e mistos de Eucalyptus e Acacia mangium. Thesis, Universidade de São Paulo

Bini D et al (2018) Intercropping Acacia mangium stimulates AMF colonization and soil phosphatase activity in Eucalyptus grandis. Sci Agric 75:102–110. https://doi.org/10.1590/1678-992x-2016-0337

Brundrett M (2004) Diversity and classification of mycorrhizal associations. Biol Rev 79:473–495

Brundrett MC (1991) Mycorrhizas in natural ecosystems. In: Macfayden A, Begon M, Fitter AH (eds) Advances in ecological research. Academic, London, pp 171–313

Brundrett MC (2002) Coevolution of roots and mycorrhizas of land plants. New Phytol. 154:275–304

Campos DTDS, Silva MDCSD, Luz JMRD et al (2011) Colonização micorrízica em plantios de eucalipto. Rev Árvore 35:965–974

Caproni AL et al (2005) Fungos micorrízicos arbusculares em estéril revegetado com Acacia mangium, após mineração de bauxita. Rev Árv 29:373–381

Cardoso EJBN (1985) Effect of Mycorrhiza and rock phosphate on growth and production of the Symbiosis soybean-rhizobium. Rev Bras Ci Solo 9:125–130

Cardoso EJBN et al (2010) Micorrizas arbusculares na aquisição de nutrientes pelas plantas. In: Siqueira JO et al (eds) Micorrizas: 30 anos de pesquisas no Brasil. Editora UFLA, Lavras, pp 153–215

Cardoso EJBN, Andreote FD (2016) Microbiologia do Solo (recurso eletrônico), 2.ed. ESALQ, Piracicaba-SP, 221 p

Cardoso EJBN, Nogueira MA, Zangaro W (2017) Importance of Mycorrhizae in tropical soils. Diversity and benefits of microorganisms from the tropics 245–267. https://doi.org/10.1007/978-3-319-55804-2_11

Carvalho TS, Moreira FMS (2010) Simbioses leguminosas, fungos micorrízicos e bactérias fixadoras de nitrogênio nodulíferas. In: Siqueira JO et al (eds) Micorrizas: 30 anos de pesquisa no Brasil. Editora UFLA, Lavras, pp 383–414

Cavagnaro TR, Sokolow SK, Jackson LE (2007) Mycorrhizal effects on growth and nutrition of tomato under elevated atmospheric carbon dioxide. Funct Plant Biol 34:730–736

Chen YL et al (2014) Use of mycorrhizal fungi for forest plantations and mine site rehabilitation. In: Solaiman ZM et al (eds) Mycorrhizal fungi: use in sustainable agriculture and land restoration. Springer, Berlin, Heidelberg, pp 325–355

Chen YL, Liu S, Dell B (2007) Mycorrhizal status of Eucalyptus plantations in South China and implications for management. Mycorrhiza 17:527–535

Chilvers GA, Pryor LD (1965) The structure of Eucalyptus mycorrhizas. Aust J Bot 13:245–259

Chiquete AAS (2011) Diversidade de Fungos em Solos de Florestas Plantadas de Eucalipto. Universidade Federal de Viçosa, Tese

Churchland C, Grayston SJ (2014) Specificity of plant-microbe interactions in the tree mycorrhizosphere biome and consequences for soil C cycling. Front Microbiol 5:261. https://doi.org/10.3389/fmicb.2014.00261

Clasen BE, Silveira ADO, Baldoni DB et al (2018) Characterization of Ectomycorrhizal species through molecular biology tools and morphotyping. Sci Agr 75(3):246–254

Craig GF, Atkins CA, Bell DT (1991) Effect of salinity on growth of four strains of rhizobium and their infectivity and effectiveness on two species of Acacia. Plant Soil 133(2):253–262

Dalpé Y, Diop TA, Plenchette C, Gueye M (2000) Glomales species associated with surface and deep rhizosphere of Faidherbia albida in Senegal. Mycorrhiza 10(3):125–129

Dickson S (2004) The Arum–Paris continuum of mycorrhizal symbioses. New Phytol 163(1):187–200

Dommergues YR (1987) The role of biological nitrogen fixation in agroforestry. Agroforestry, p 245

Egerton-Warburton LM, Querejeta JI, Allen MF (2007) Common mycorrhizal networks provide a potential pathway for the transfer of hydraulically lifted water between plants. J Exp Bot 58(6):1473–1483

Frank AB, Trappe JM (2005) On the nutritional dependence of certain trees on root symbiosis with belowground fungi (an English translation of A.B. Frank's classic paper of 1885). Mycorrhiza 15:267–275

Futai K, Taniguchi T, Kataoka R (2008) Ectomycorrhizae and their importance in forest ecosystems. In: Siddiqui ZA et al (eds) Mycorrhizae: sustainable agriculture and forestry. Springer, Dordrecht, pp 241–285
Gallaud J (1905) Étude sur les mycorrhizes endotrophes. Rev Gn Bot 17:5–500
Garbaye J (1991) Biological interactions in the mycorrhizosphere. Experientia 47:370–375
Gasparotto FA, Navarrete AA, Souza FA et al (2010) Técnicas moleculares aplicadas ao estudo das micorrizas. In: Siqueira JO (ed) Micorrizas: 30 anos de pesquisa no Brasil. Editora UFLA, Lavras, pp 551–582
Germon A, Guerrini IA, Bordron B et al (2018) Consequences of mixing *Acacia mangium* and *Eucalyptus grandis* trees on soil exploration by fine-roots down to a depth of 17 m. Plant Soil 424:203–220
Gocke MI, Huguet A, Derenne S et al (2017) Disentangling interactions between microbial communities and roots in deep subsoil. Sci Total Environ 575:135–145
Gomes SP, Trufem SFB (1998) Fungos micorrízicos arbusculares (Glomales, Zygomycota) na Ilha dos Eucaliptos, Represa do Guarapiranga, São Paulo, SP. Acta Bot Bras 12:393–401
Harley JLH, Smith SE (1983) Mycorrhizal symbiosis. Academic, New York, p 483
He X, Critchley C, Ng H, Bledsoe C et al (2005) Nodulated N_2-fixing *Casuarina cunninghamiana* is the sink for net N transfer from non-N_2-fixing *Eucalyptus maculata* via an ectomycorrhizal fungus Pisolithus sp using (NH^{4+})-N-15 or (NO^{3-})-N-15 supplied as ammonium nitrate. New Phytol 167:897–912
He X, Xu M, Qiu G et al (2009) Use of N-15 stable isotope to quantify nitrogen transfer between mycorrhizal plants. J Plant Ecol 2:107–118
Helgason T, Daniell TJ, Husband R et al (1998) Ploughing up the wood-wide web? Nature
Jalonen R, Nygren P, Sierra J (2009) Transfer of nitrogen from a tropical legume tree to an associated fodder grass via root exudation and common mycelial networks. Plant Cell Environ 32:1366–1376
Kasuya MCM, Costa MD, Araújo EF et al (2010) Ectomicorrizas no Brasil: biologia e nutrição de plantas. In: Siqueira JO et al (eds) Micorrizas: 30 anos de pesquisa no Brasil. Editora UFLA, Lavras, pp 615–643
Laclau JP, Nouvellon Y, Reine C et al (2013) Mixing Eucalyptus and Acacia trees leads to fine root over-yielding and vertical segregation between species. Oecologia 172:903–913
Lambais GR, Jourdan C, Piccolo MD et al (2017) Contrasting phenology of *Eucalyptus grandis* fine roots in upper and very deep soil layers in Brazil. Plant Soil 421:301–318
Lambais MR (2010) Sinalização e transdução de sinais em micorrizas arbusculares. In: Siqueira JO et al (eds) Micorrizas: 30 anos de pesquisas no Brasil. Editora UFLA, Lavras, p 119
Lambais MR et al (2005) Diversidade microbiana nos solos: definindo novos paradigmas. In: Vidal-Torrado P et al (eds) Tópicos em ciência do solo. SBCS, Viçosa, pp 43–84
Lammel DR, Cruz LM, Mescolotti DLC et al (2015) Woody Mimosa species are nodulated by Burkholderia in ombrophylous forest soils and their symbioses are enhanced by arbuscular mycorrhizal fungi (AMF). Plant Soil 393(1–2):123–135
Levisohn J (1958) Effects of mycorrhiza on tree growth. Soils Fertil 21:73–82
Li CH, Yan K, Tang LS et al (2014) Change in deep soil microbial communities due to long-term fertilization. Soil Biol Biochem 75:264–272
Liese R, Lubbe T, Albers NW et al (2018) The mycorrhizal type governs root exudation and nitrogen uptake of temperate tree species. Tree Physiol 38:83–95
Malajczuk N, Linderman RG, Kough J et al (1981) Presence of vesicular-arbuscular mycorrhizae in *Eucalyptus* spp. and *Acacia* sp., and their absence in *Banksia* sp. after inoculation with *Glomus fasciculatus*. New Phytol 87:567–572
McCormack ML, Fernandez CW, Brooks H et al (2017) Production dynamics of Cenococcum geophilum ectomycorrhizas in response to long-term elevated CO_2 and N fertilization. Fungal Ecol 26:11–19
Mendes Filho PF, Vasconcellos RLF, Paula AM et al (2010) Evaluating the potential of forest species under "microbial management" for the restoration of degraded mining areas. Water Air Soil Pollut 208:79–89

Meng LB, Zhang AY, Wang F et al (2015) Arbuscular mycorrhizal fungi and rhizobium facilitate nitrogen uptake and transfer in soybean/maize intercropping system. Front Plant Sci 6:10

Midgley MG, Brzostek E, Phillips RP (2015) Decay rates of leaf litters from arbuscular mycorrhizal trees are more sensitive to soil effects than litters from ectomycorrhizal trees. J Ecol 103:1454–1463

Molina R, Massicotte H, Trappe JM (1992) Specificity phenomena in mycorrhizal symbioses: community-ecological consequences and practical implications. In: Allen MF (ed) Mycorrhizal functioning: an integrative plant–fungal process. Chapman & Hall, New York, pp 357–423

Montesinos-Navarro A, Verdu M, Querejetac JI et al (2016) Soil fungi promote nitrogen transfer among plants involved in long-lasting facilitative interactions. Perspect Plant Ecol 18:45–51

Moreira FS, Siqueira JO (2006) Microbiologia e bioquimica do solo. Editora UFLA, Lavras

Mortimer PE, Pérez-Fernández MA, Valentine AJ (2008) The role of arbuscular mycorrhizal colonization in the carbon and nutrient economy of the tripartite symbiosis with nodulated Phaseolus vulgaris. Soil Biol Biochem 40:1019–1027

Moyer-Henry KA, Burton JW, Israel D et al (2006) Nitrogen transfer between plants: a N-15 natural abundance study with crop and weed species. Plant Soil 282:7–20

Nogueira MA, Cardoso EJBN (2006) Plant growth and phosphorus uptake in mycorrhizal Rangpur lime seedlings under different levels of phosphorus. Pesq Agrop Brasileira 41:93–99

Oliveira VL, Schmidt VDB, Bellei MM (1997) Patterns of arbuscular- and ecto-mycorrhizal colonization of *Eucalyptus dunnii* in southern Brazil. Ann Sci Forest 54:473–481

Paula RR, Bouillet J-P, Ocheuze Trivelin PC et al (2015) Evidence of short-term belowground transfer of nitrogen from *Acacia mangium* to *Eucalyptus grandis* trees in a tropical planted forest. Soil Biol Biochem 91:99–108

Pereira APA (2015) Influência da profundidade do solo e do manejo de *Eucalyptus grandis* e *Acacia mangium* na estrutura das comunidades microbianas do solo. Dissertação, Universidade de São Paulo

Pereira APD, de Andrade PAM, Bini D et al (2017) Shifts in the bacterial community composition along deep soil profiles in monospecific and mixed stands of *Eucalyptus grandis* and *Acacia mangium*. PLoS One 12

Pereira APD, Santana MC, Bonfim JA et al (2018) Digging deeper to study the distribution of mycorrhizal arbuscular fungi along the soil profile in pure and mixed *Eucalyptus grandis* and *Acacia mangium* plantations. Appl Soil Ecol 128:1–11

Peterson RL, Bonfante P (1994) Comparative structure of vesicular-arbuscular mycorrhizas and ectomycorrhizas. Plant Soil 159(1):79–88

Prieto I, Roldán A, Huygens D et al (2016) Species-specific roles of ectomycorrhizal fungi in facilitating interplant transfer of hydraulically redistributed water between Pinus halepensis saplings and seedlings. Plant Soil 406(1–2):15–27

Read DJ (1991) Mycorrhizas in ecosystems. Experientia 47(4):376–391

Requeña N, Perez-Solis E, Azcon-Aguilar C, Jeffries P, Barea JM (2001) Management of indigenous plant– microbe symbioses aids restoration of desertified ecosystems. Appl Environ Microbiol 67:495–498

Requeña N, Serrano E, Ocón Z, Breuninger M (2007) Plant signals and fungal perception during arbuscular mycorrhiza establishment. Phytochemistry 68(1):33–40

Santana MC (2017) Análise da comunidade de fungos em áreas de monoculturas e consórcio de *Eucalyptus grandis* e *Acacia mangium*. Dissertation, Universidade de São Paulo

Santana MC, Pereira APA, Forti VA, Cardoso EJBN (2016) Eucalypt as trap plant to capture associative fungi in soil samples from great depth. Int J Environ Agric Res 2:191–194

Santos JC, Finlay RD, Tehler A (2006) Molecular analysis of arbuscular mycorrhizal fungi colonising a semi-natural grassland along a fertilisation gradient. New Phytol 172(1):159–168

Schenck NC, Perez Y (1990) Manual for the identification of VA mycorrhizal fungi, vol 286. Synergistic, Gainesville

Shah F, Nicolás C, Bentzer J, Ellström M, Smits M, Rineau F et al (2016) Ectomycorrhizal fungi decompose soil organic matter using oxidative mechanisms adapted from saprotrophic ancestors. New Phytol 209(4):1705–1719

Simard SW, Beiler KJ, Bingham MA, Deslippe JR, Philip LJ, Teste FP (2012) Mycorrhizal networks: mechanisms, ecology and modelling. Fungal Biol Rev 26(1):39–60
Simard SW, Perry DA, Jones MD, Myrold DD, Durall DM, Molina R (1997) Net transfer of carbon between ectomycorrhizal tree species in the field. Nature 388(6642):579
Siqueira JO, Lambais MR, Stürmer SL (2002) Fungos micorrízicos arbusculares: origem e características dos fungos Glomaleanos. Biotecnol Ciên Desen 25:12–21
Smith SE, Read DJ (2008) The symbionts forming arbuscular mycorrhizas. Mycor Symb 2:13–41
Smith SE, Read DJ (2010) Mycorrhizal symbiosis. Academic, Cambridge
Smith SE, Smith FA (2011) Roles of arbuscular mycorrhizas in plant nutrition and growth: new paradigms from cellular to ecosystem scales. Annu Rev Plant Biol 62:227–250
Souza LABD, Bonnassis PAP, Silva Filho GN, Oliveira VLD (2008) New isolates of ectomycorrhizal fungi and the growth of eucalypt. Pesqui Agropecu Bras 43(2):235–241
Souza VC, Silva RA, Cardoso GD, Barreto AF (2006) Estudos sobre fungos micorrízicos. R Bras Eng Agríc Ambiental 10(3):612–618
Steffen RB, Antoniolli ZI, Steffen GPK, Eckhardt DP (2010) Micorrização das mudas de *Eucalyptus grandis* Hill ex Maiden comercializadas no município de Santa Maria, RS. Ciên Natura 32(1):25–35
Stürmer SL (2012) A history of the taxonomy and systematics of arbuscular mycorrhizal fungi belonging to the phylum Glomeromycota. Mycorrhiza 22(4):247–258
Suz LM, Azul AM, Morris MH, Bledsoe CS, Martín MP (2008) Morphotyping and molecular methods to characterize ectomycorrhizal roots and hyphae in soil. In: Molecular mechanisms of plant and microbe coexistence. Springer, Berlin, pp 437–474
Taylor MK, Lankau RA, Würzburger N (2016) Mycorrhizal associations of trees have different indirect effects on organic matter decomposition. J Ecol 104(6):1576–1584
Tedersoo L, Brundrett MC (2017) Evolution of ectomycorrhizal symbiosis in plants. In: Biogeography of mycorrhizal symbiosis. Springer, Cham, pp 407–467
Torrecillas E, Alguacil MM, Roldán A (2012) Host preferences of arbuscular mycorrhizal fungi colonizing annual herbaceous plant species in semiarid Mediterranean prairies. Appl Environ Microbiol 78(17):6180–6186
van Der Heijden MG, Martin FM, Selosse MA, Sanders IR (2015) Mycorrhizal ecology and evolution: the past, the present, and the future. New Phytol 205(4):1406–1423
Vandenkoornhuyse P, Ridgway KP, Watson IJ, Fitter AH, Young JPW (2003) Co-existing grass species have distinctive arbuscular mycorrhizal communities. Mol Ecol 12(11):3085–3095
Vesk PA, Ashford AE, Markovina AL, Allaway WG (2000) Apoplasmic barriers and their significance in the exodermis and sheath of *Eucalyptus pilularis*—*Pisolithus tinctorius* ectomycorrhizas. New Phytol 145(2):333–346
Wang B, Qiu YL (2006) Phylogenetic distribution and evolution of mycorrhizas in land plants. Mycorrhiza 16(5):299–363
Willis A, Rodrigues BF, Harris PJ (2013) The ecology of arbuscular mycorrhizal fungi. Crit Rev Plant Sci 32(1):1–20
Yin H, Wheeler E, Phillips RP (2014) Root-induced changes in nutrient cycling in forests depend on exudation rates. Soil Biol Biochem 78:213–221
Zheng L, Zhao X, Zhu G, Yang W, Xia C, Xu T (2017) Occurrence and abundance of ammonia-oxidizing archaea and bacteria from the surface to below the water table, in deep soil, and their contributions to nitrification. Microb Ope 6(4):e00488

Chapter 8
Mesofauna and Macrofauna in Soil and Litter of Mixed Plantations

Maurício Rumenos Guidetti Zagatto, Luís Carlos Iuñes Oliveira Filho, Pâmela Niederauer Pompeo, Cintia Carla Niva, Dilmar Baretta, and Elke Jurandy Bran Nogueira Cardoso

8.1 General Introduction

The soil fauna comprises the invertebrate community that lives permanently or at least in one of their development stages in soil or litter (Zagatto et al. 2017). These invertebrates may be classified according to the body diameter as microfauna, mesofauna, and macrofauna (Swift et al. 1979; Baretta et al. 2011). The soil microfauna comprises those invertebrates with body diameter smaller than 0.2 mm, while mesofauna lies between 0.2 and 2.0 mm, and macrofauna comprises the larger invertebrates with body diameter between 2 and 20 mm (Lavelle 1997; Oliveira Filho et al. 2018).

The soil macrofauna comprises the larger invertebrates, called "ecosystem engineers," since they affect the soil structure, building galleries and enlarging pores, as the earthworms and termites, while many of them mix organic matter with the mineral superficial layers of the soil profile. Still others move tons of soil while building their nests sometimes connected by kilometers of underground galleries, as ants and termites (Bardgett and Van der Putten 2014; Brown et al. 2015; Pereira et al. 2017b).

The soil mesofauna comprises mainly mites and springtails, besides several insects, myriapods, oligochaeta, crustacea, and others, which live mainly in the litter or on the soil surface. These small invertebrates actively participate in the initial

M. R. G. Zagatto (✉) · E. J. Bran Nogueira Cardoso
Department of Soil Science, University of São Paulo, "Luiz de Queiroz" College of Agriculture, Piracicaba, SP, Brazil

L. C. I. Oliveira Filho · P. N. Pompeo · D. Baretta
Santa Catarina State University, Chapecó, SC, Brazil

C. C. Niva
Embrapa Cerrados, Brazilian Agricultural Research Corporation, Planaltina, DF, Brazil

E. Cardoso et al. (eds.), *Mixed Plantations of Eucalyptus and Leguminous Trees*,
https://doi.org/10.1007/978-3-030-32365-3_8

fragmentation of organic matter and, consequently, favor microbial decomposition, nutrient cycling, and soil aggregation (Barbercheck et al. 2009; Lavelle et al. 1997). They also may be fungivorous, vegetarian, coprofagous, predators, parasitic, etc.

Soil fauna abundance and diversity are modulated by vegetation, weather, moisture, temperature, and agrochemical and fertilizer inputs, among other factors. These characteristics make them excellent bioindicators, since the presence or absence of several invertebrate groups and their quantity may reflect soil health or environmental degradation (Anderson 2009; Cardoso et al. 2013). The native ecosystems present great diversity and low dominance and, consequently, higher resistance and resilience than agricultural ecosystem (Zagatto et al. 2019b). In natural ecosystems, certain species present a greater tolerance to environmental stress and the higher species diversity in such ecosystems increases the functional redundancy (McCann 2000).

Due to the several functions performed by the invertebrates in soil, they are related to ecosystem processes and services as C sequestration, greenhouse gas mitigation, pest and disease control, and soil water storage. These aspects modify plant production, the habitat function of soil for other organisms, and the air and water flows in soil (Van Der Putten et al. 2004; Wagg et al. 2014; Creamer et al. 2016).

Eucalyptus cultivation is favored by tropical climate and this genus is planted in many regions of the world because of its rapid growth cycle and production of multipurpose wood (Ibá 2015). However, intensive cultivation leads to the depletion of N and other soil nutrients, requiring a high input of mineral fertilizers to avoid a decrease in the production rates (Gonçalves et al. 2008). Thus, the consortium of Eucalyptus with N-fixing trees allows for a decrease in fertilizer use, enriching the soil with N due to high biological nitrogen fixation (Forrester et al. 2006; Laclau et al. 2008, Chap. 6). Recently, in Brazil, intercropped plantations of *Eucalyptus grandis* (Eucalyptus) and *Acacia mangium* (Acacia) demonstrated high sustainability in comparison with Eucalyptus monocultures, favoring the microbial community, especially the phylum Proteobacteria, which contains most species involved in biological nitrogen fixation, as Rhizobium with *A. mangium* (Pereira et al. 2017a).

Moreover, mixed plantations of *E. grandis* and *A. mangium* deposit great amounts of litter on the soil surface (Chap. 3), which may serve as habitat and food for microorganisms and soil fauna. The formation of a litter layer helps to avoid soil heating and soil erosion and consists of a great organic reservoir that releases several nutrients to the soil solution, modulated mainly by weather, soil fauna, and microorganisms (Bachega et al. 2016).

Therefore, this chapter intends to clarify the effect of N-fixing tree introduction in Eucalyptus monocultures (especially of the mix *E. grandis* and *A. mangium*), on the soil faunal community. To our knowledge, there are no studies on soil and mesofauna in Brazil, especially not in mixed cultures of Eucalyptus and Acacia.

8.2 Effect of Forest Plantations on Soil Macrofauna

In forest management, some specific practices can cause changes in soil invertebrate biodiversity and consequently in the ecosystem performance, because they participate in ecosystem processes at different scales, both temporal and spatial, including organic matter decomposition, regulation of soil and nutrient losses, or bioturbation. Different forest management systems influence the soil macrofauna, which is important to maintain soil properties, because they are responsible for the creation of biogenic structures that promote changes in soil physical characteristics, as well as the availability of resources to other organisms (Brown et al. 2015; Pereira et al. 2015). Similarly, tree plantations may vary in structure and composition, and these variations may alter soil macrofauna communities (Warren and Zou 2002).

Forest and agricultural practices can have profound effects on population levels and species composition of various groups of organisms. Among the most affected by plantations of exotic species are insects such as beetles and ants, given their high sensitivity to changes in ecosystems (Aliaga et al. 2017). In this sense, many studies have evaluated these and other groups of soil macrofauna in different forest and agricultural land-use systems and demonstrated how land use and soil management impact these communities.

Warren and Zou (2002) evaluated the abundance and biomass of soil macroinvertebrates in *Leucaena leucocephala*, *Casuarina equisetifolia*, and *E. robusta* plants, being 9 years old, at a degraded site in Puerto Rico. The nutrient concentrations and the permanent litter stocks on the forest floor were also determined to examine the relationship between litter chemistry and soil macroinvertebrates. *Leucaena* plantations had greater abundances and biomass of millipede species than *Casuarina* and *Eucalyptus*. The biomass of the earthworms did not differ between the plantings. Millipede biomass was highly correlated with the N concentration and C/N ratio of the litter. The results found by the authors suggest that plantations of tree species differ in their influence on soil macrofauna, and the biomass and abundance of soil fauna can be regulated through the careful selection of tree species for planting in degraded tropical lands.

Rosa et al. (2015), studying the Plateau region of Santa Catarina (Brazil), aimed to evaluate the effect of land-use systems on the distribution of soil macrofauna and its relationship with soil chemical and physical attributes. In that study, native forest, *Eucalyptus* plantation, and perennial pasture favored edaphic biodiversity and were considered more stable than the crop-livestock integration and no-tillage systems with greater anthropogenic intervention, which reduced the macrofauna groups of the soil. The authors collected the fauna using soil monoliths (area 0.25 m by 0.25 m), which were excavated following the standard Tropical Soil Biology and Fertility (TSBF) sampling protocol (Anderson and Ingram 1993), using a sampling grid with nine points in each area and three true replicates of each land-use system. The fauna group that contributed the most to separate the eucalyptus plantation from the crop-livestock integration and no-tillage was Isoptera (termites) and from the native forest was Formicidae. The diversity measured by the Shannon-Wiener index was higher in the native forest, followed by the eucalyptus plantation, both in winter and summer.

Souza et al. (2016) carried out a similar study with the soil macrofauna in the Eastern region of Santa Catarina and found similar results, where they evidenced the eucalyptus plantation as a system with greater stability for the biodiversity of the soil fauna, together with the native forest and perennial pasture, when compared to systems with annual crops. As a result, greater densities of individuals in eucalyptus plantation and perennial pasture (in summer) were found than in other systems, resulting in higher values of diversity. The most important groups of soil macrofauna to differentiate land-use systems were Formicidae, Coleoptera, and Oligochaeta.

Kamau et al. (2017) evaluated the soil macrofauna at sites after different periods of conversion of primary forests to cropland, considering dominant tree species in the region (*Croton megalocarpus*, *E. grandis,* and *Zanthoxylum gilletii*), in Kapchorwa, Uganda, and demonstrated the importance of the diversity of tree cover in agricultural landscapes for soil macrofauna conservation. The authors sampled the fauna by the soil monolith method to verify if the spatial variation in the soil macrofauna abundance is affected by the age of cultivation, tree species, and distance from the trunk of the tree. The results showed a greater abundance of macrofauna in the soil after 16–62 years of cultivation than in the first 10 years, although this varied with the tree species and macrofaunal group. The abundance of earthworms was higher below the canopy of *Z. gilletii*; beetles were found in higher numbers under *E. grandis* and *C. megalocarpus* than under *Z. gilletii*; higher numbers of termites and centipedes were found under *E. grandis* after 16 years of cultivation. The quality of organic residues from trees has shown an important effect on macrofauna abundance and spatial distribution, indicating that the increasing diversity of tree species in agroecosystems can play an important role in maintaining biodiversity and ecosystem services (Kamau et al. 2017).

As observed in these studies, some groups of the soil macrofauna present greater contribution to demonstrate changes in the environments, in spite of seasonal effects. When evaluating families of Coleoptera they found low abundances in eucalyptus plantations; however, higher values of the diversity index of Shannon-Wiener were found in this system (Pompeo et al. 2016). The authors sampled Coleoptera using the same sampling technique in five agricultural and forest systems in the Plateau region of Santa Catarina. As seen in Fig. 8.1, some families were more associated with the eucalyptus plantation, and in winter the eucalyptus and native forest systems were related to a larger number of families, with Curculionidae and Chrysomelidae being the most associated ones. The families Staphylinidae and Phalacridae were related to the native forest. In summer, the native forest system was different from the other systems, and showed relations with a greater number of families, especially Staphylinidae, Chrysomelidae, and Leiodidae; in the eucalyptus plantation system, there was an association with Tenebrionidae, Phalacridae, and Curculionidae.

Bartz et al. (2014) studying the same land-use systems in municipalities of the West and Plateau regions of Santa Catarina evaluated the richness of earthworm species collected by soil monoliths and qualitative random samplings using alleatory excavations. In all systems, 24 species were identified, with 19 native species,

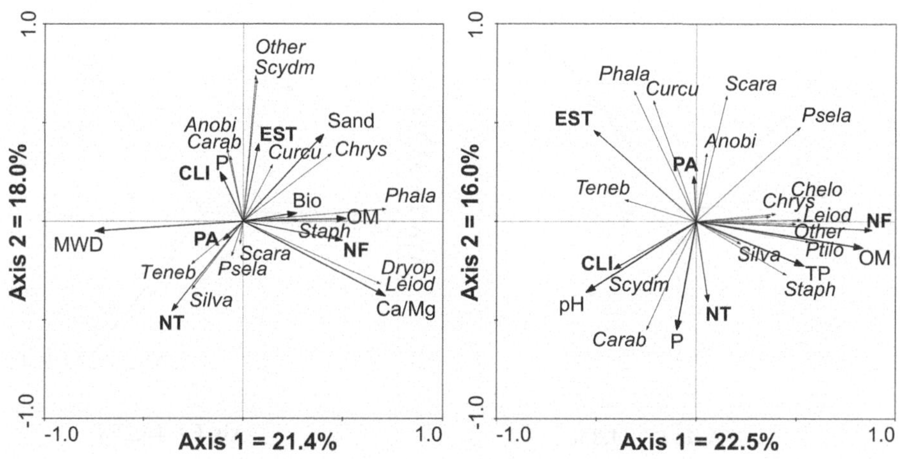

Fig. 8.1 Principal component analysis of Coleoptera families distinguishing land-use systems in winter (left) and summer (right), and environmental variables, used as explanatory variables, in the Southern Santa Catarina Plateau (Pompeo et al. 2016). *NF* native forest, *EST* Eucalyptus stands, *PA* perennial pasture, *CLI* crop-livestock integration, *NT* no-tillage, *OM* organic matter, *TP* total porosity, *MWD* mean weight diameter, *Bio* biopores, *Anobi* Anobiidae, *Carab* Carabidae, *Scydm* Scydmaenidae, *Curcu* Curculionidae, *Chrys* Chrysomelidae, *Phala* Phalacridae, *Staph* Staphylinidae, *Dryop* Dryopidae, *Leiod* Leiodidae, *Scara* Scarabaeidae, *Psela* Pselaphidae, *Silva* Silvanidae, *Teneb* Tenebrionidae, *Chelo* Chelonariidae, *Ptilo* Ptilodactylidae

including several ones that were new to science. Considering both regions together, native species accounted for 90%, 89%, and 80% of the species richness in native forest, eucalyptus plantation, and perennial grassland systems, respectively, while in annual cropping systems native species accounted for 65% and 60% of the total wealth in crop-livestock integration and no-tillage. In the western region, native forests and no-tillage assemblages were composed exclusively of native species, while in grazing and crop-livestock integration they represented 75% and 58%. The Plateau assemblies were composed of 100% native species in eucalyptus plantation and pasture, 88% in native forest, and only 67% and 54% in crop-livestock integration and no-tillage. These results highlight the importance of systematic surveys because prior to this study little was known about the impacts of forest and agricultural land use on earthworm populations in this region of Brazil.

8.3 Mesofauna Community in Mixed Plantations of Eucalyptus and Acacia: Effects of Soil and Litter Quality

The soil and litter samples for mesofauna assessment in pure and mixed plantations of *E. grandis* and *A. mangium* were collected at the Experimental Station of Forest Sciences located in Itatinga, São Paulo state, Brazil (23°02′01”S and 48°97′30”W).

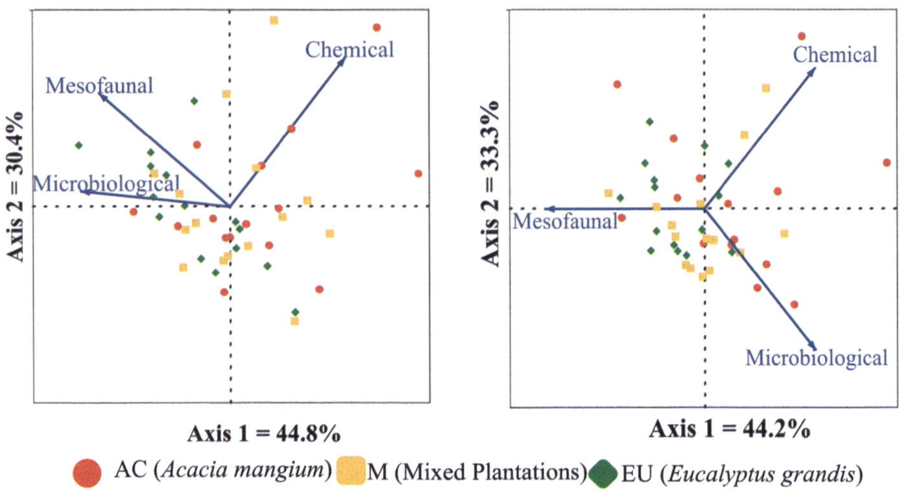

Fig. 8.2 Principal component analysis for soil quality variables based on microbiological, physicochemical, and mesofaunal attributes in pure and mixed plantations of *Acacia mangium* and *Eucalyptus grandis* under lower soil moisture (left—water soil content = 4.8%) and higher soil moisture (right—water soil content = 8.6%)

The soil was classified as a Geric Rhodic Ferralsol (FAO) with a texture of 84% sand, 4% silt, and 12% clay (Bini et al. 2013). The climate in this region is Cfa (Köppen classification), with an annual rainfall of 1350 mm, mostly (75%) concentrated between March and October (Laclau et al. 2008).

In soil under mixed plantations of *Acacia mangium* and *Eucalyptus grandis*, we can find the coexistence of some organisms originally belonging exclusively to *Acacia mangium* or to *Eucalyptus grandis*, as is the case of Enchytraeidae, Isopoda, and Thysanoptera (Zagatto et al. 2019a). This finding suggests that pure plantations of *Acacia mangium* and mixed plantations of Acacia and Eucalyptus present better soil conditions for soil fauna. The order Thysanoptera for example comprises some predators which require a well-established food chain for their development (Blasi et al. 2013; Melloni and Varanda 2015; Mound 2005). Enchytraeidae and Isopoda, on their turn, actively comminute the organic matter and stimulate the microbial activity and, consequently, the nutrient release to the soil solution (Paoletti and Hassall 1999; Van Vliet et al. 2004; Filser et al. 2016).

Recent research shows that soil moisture is one of the main abiotic factors responsible for the development of those invertebrates. In periods of severe water scarcity, a positive correlation was found between the abundance of soil mesofauna groups and soil microbiological attributes under pure and mixed *Eucalyptus grandis* and *Acacia mangium* plantations. In the period of higher rainfall, a great independence of the microbiological, mesofauna, and soil chemical variables was the rule (Fig. 8.2). Therefore, we postulate that periods of water scarcity provide a profound interaction between soil mesofauna and microorganisms, which favors their survival. During water stress there is also a great feeding difficulty because the

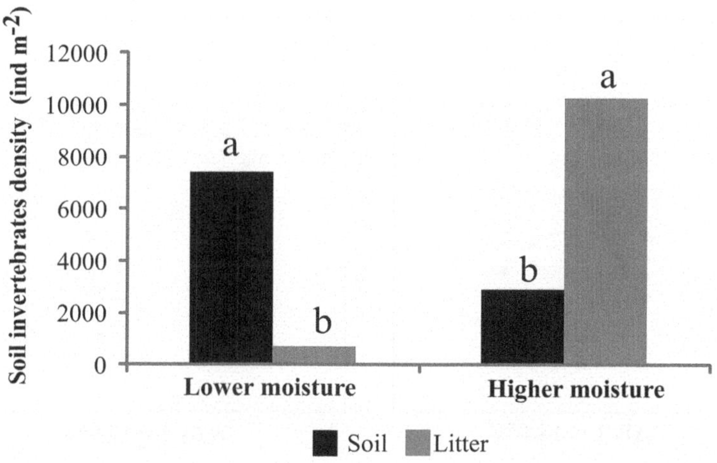

Fig. 8.3 Mesofaunal density in soil and litter. Values followed by different letters indicate differences between soil and litter at the same sampling date (Tukey's test at 5% significance)

decomposition of soil organic matter and the metabolic activities become much slower than in periods of water abundance.

Microorganisms and fauna in the soil probably build up important interactions for the ecosystem functioning; however little is known about this relation (Creamer et al. 2016), when studying the network of biotic interactions in the soil of several land-use systems in Europe. The authors found strong correlations between nematodes and arbuscular mycorrhizal fungi (AMF) in arable soils, between nematodes and Archaea and Archaea and bacteria in grasslands and between enchytraeids and AMF in forests. They also reported changes in these synergic relations due to changes in soil organic carbon contents and pH. Thus, in acid soils with less than 2% of organic carbon enchytraeids, collembolans and archaea showed strong connections, while in soils close to neutrality (pH 5–7) there was a greater correlation among mycorrhizae, archaea, and bacteria.

In periods of severe water scarcity, a higher density of mesofauna was present on the soil surface, below the litter layer, while in the moist period a much higher density of invertebrates inhabited the litter itself in forest plantations of *Eucalyptus grandis* and *Acacia mangium* (Fig. 8.3). These data suggest the preference of the mesofauna for soil below the litter where there is some moisture preservation during drought while the litter is already straw-dry. The litter however is chosen whenever it contains enough moisture. In periods of drought and high solar irradiation, the litter covers and protects the soil against excessive desiccation and maintains soil temperature in suitable ranges for the invertebrates' development in soil (Choi et al. 2006; Derpsch et al. 2010; Peña-Peña and Irmler 2016). In addition, our data suggest that there is a great interaction between soil mesofauna and microorganisms in soil during a drought season (under low soil and low litter moisture) (Fig. 8.2). This condition allowed higher mesofauna abundance in soil than in litter at this same sampling period (Fig. 8.3). We were surprised that such a small difference in moisture, with very low values, made a huge

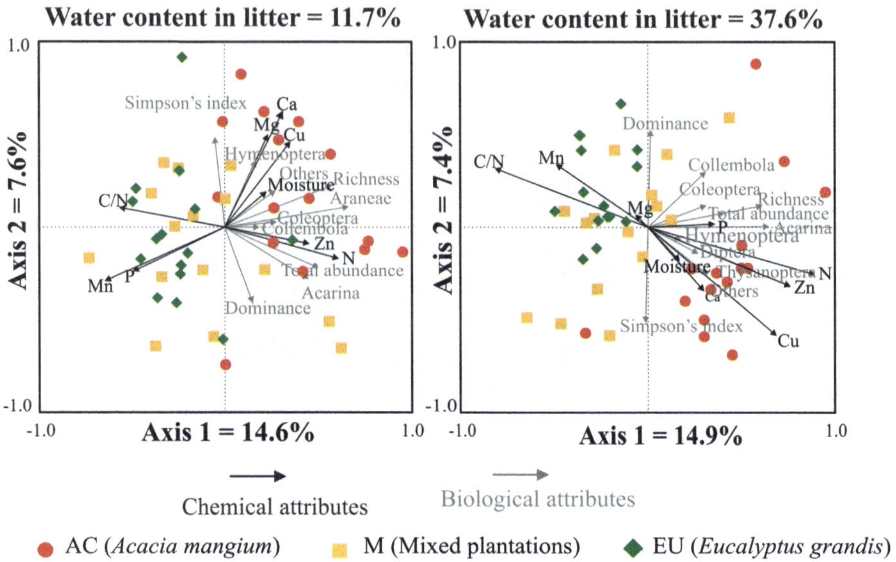

Fig. 8.4 Redundancy analysis (RDA) between response variables (biological litter attributes—grey vectors) and explanatory variables (chemical litter attributes—black vectors) in pure *E. grandis* (E) and *A. mangium* (AC) plantations, and a mixed system between them (M)

difference in the soil mesofauna habitat preference. Apparently, these invertebrates are extremely sensitive to weather conditions, since the values for soil moisture typify a dry soil in both cases.

The litter of *Eucalyptus grandis* and of *Acacia mangium* show deep differences in their chemical constitutions (Pereira et al. 2018), with Acacia litter presenting a lower C/N ratio and higher contents of several other nutrients, making it more adequate and palatable for the mesofauna (Kaneda and Kaneko 2011) (Fig. 8.4). Conversely, the higher C/N ratio and Mn contents of Eucalyptus litter correlate negatively with all groups of fauna sampled. Therefore, the more easily decomposable organic material of Acacia litter favors the soil faunal community, considering also that coprophagy between the different faunal groups is very intense (Kautz et al. 2002; Teuben and Verhoef 1992).

The rhizosphere of *Eucalyptus grandis* can exude some volatile compounds, which result in toxic effects for soil fauna (Zhiqun et al. 2017). In the Eucalyptus rhizosphere, many such chemical groups were identified, as 2,4-dimethyl heptane, 2,2,4,6-pentamethyl heptane, and 2,4-di-tert-butylphenol, which cause changes in the acetylcholinesterase (enzyme related to neuronal synapses); superoxide dismutase (related to superoxide radical changes in H_2O_2 and O_2); and glutathione-S transferase (related to tissue protection against oxidative stress) (Zhiqun et al. 2017). Hence, it is possible that these compounds are also responsible for the lower mesofauna abundance and diversity found in pure *Eucalyptus grandis* plantations (Fig. 8.3). Nevertheless, more studies are necessary in order to verify if such compounds are also emitted in the rhizospheres of mixed plantations and in pure plantations of *Acacia mangium*.

8.4 Springtails and Beetles in Forest and Agricultural Systems: An Approach to Morphotypes

Springtails or collembolans are small arthropods belonging to the class Hexapoda. These organisms usually are less than 2 mm wide and live in moist habitats, eating fresh organic matter, especially in the topsoil layers. On the other hand, beetles are larger organisms, much related to chemical and physical soil attributes, since they mix organic matter with soil clay particles (Filho et al. 2016; Brown et al. 2015).

Springtails and beetles are good bioindicators, since these faunal groups are conditioned by management, weather, and soil chemical, physical, and biological attributes. In addition, these organisms comminute the plant wastes and the soil organic matter, accelerating the decomposition processes. Therefore, they are considered soil quality indicators due to their capacity to modify the soil properties and due to their great sensitivity to changes in the soil (Nichols et al. 2008; Korasaki et al. 2012; Filho and Baretta 2016).

The predominant use and management of the soil in Brazil involve a drastic modification of the native forest by its fragmentation and conversion into agricultural production systems and forest residues. Consequently, after intervention, the functions of soil organisms and ecosystem services (as nutrient cycling, water quality, biodiversity maintenance, food production, among others) are affected.

An assessment of the functional and structural biodiversity of some of the soil fauna groups, considering the impact of different management systems, is necessary, in the interest of biodiversity preservation and ecosystem services provided by organisms (Van Capelle et al. 2012). Considering the growing interest in understanding the ecosystem functions where springtails and beetles act, a major limitation is the lack of taxonomists available. The differentiation of organisms using specific traits can be useful, especially due to the lack of taxonomists able to identify the biodiversity, constituting an opportunity to minimize the taxonomic insufficiency of these taxa in Brazil.

Thus, an alternative approach is morphotyping, which analyzes morphological traits and has been adapted for soil invertebrates (Oliveira Filho et al. 2016; Pey et al. 2014; Pompeo et al. 2017; Santos et al. 2018). The objective is to group organisms of each species according to their degree of adaptation to the soil, classifying the eco-morphological groups according to their habitat (edaphic, hemiedaphic, and epigeic) (Fig. 8.5) and morphotypes (life form) within each group. Due to the difficulties imposed by the lack of knowledge, an approach using functional characteristics has great possibilities of helping to understand both the functional role in the ecosystems and the effects of habitat modifications on the community structure (Fountain-Jones et al. 2015).

Thus, several studies have been carried out with this type of approach in Europe involving different land-use systems and soil fauna groups (Gardi et al. 2008; Menta et al. 2018a, b; Parisi 2001; Parisi et al. 2005). In Brazil, this approach has been used to verify differences between forests and agricultural systems (Machado et al. 2019; Pompeo et al. 2017; Santos et al. 2018).

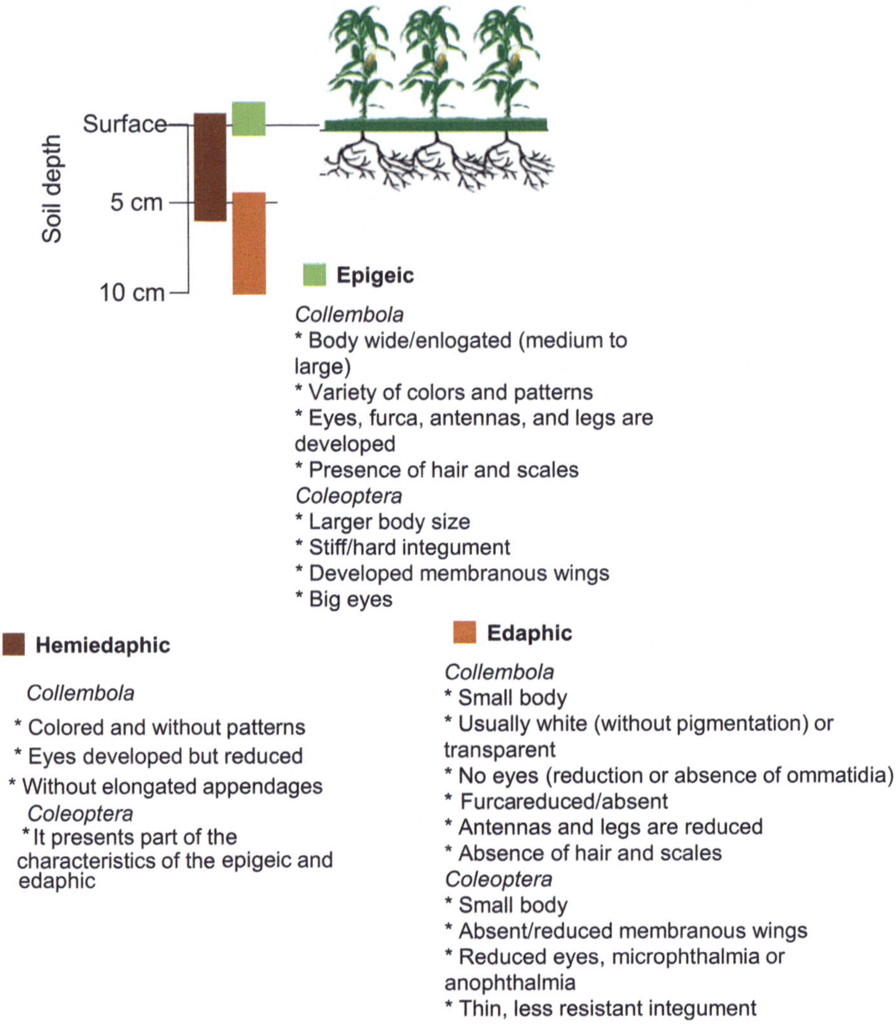

Fig. 8.5 Change of eco-morphological groups of springtails and beetles living inside the soil or living on the soil surface (modified from Filho and Baretta (2016)). Colored bars indicate the exact habitat (or depth of the soil) to which epigeic (green bar), hemiedaphic (dark brown), and edaphic (light brown) specimens are best adapted. Although the specimens have a preferred depth that best fits their needs, they may sometimes migrate vertically in the soil profile for some abiotic factors (e.g., humidity and temperature) and biotic factors (e.g., lack of prey or other food)

According to Machado et al. (2019) and Santos et al. (2018), evaluating the springtail communities by monitoring their morphological traits was efficient for the comparison of the different land-use systems in two regions of Santa Catarina state (Brazil), showing correlations between their communities and physical and chemical soil properties. These two studies presented 16 edaphic morphotypes, 25 hemiedaphic, and 5 epigeic, totaling 46 morphotypes collected in the two regions.

In general, both studies showed that the eucalyptus plantation presented the lowest values of abundance and richness of morphotypes when compared to areas of native forest, pasture, integrated crop-livestock, and no-tillage systems. Higher springtail abundance and richness were determined in the land-use systems without anthropological actions and soil management due to better ecological equilibrium, with increases of the food sources due to a higher diversity of plants. Thus, the low quality of the litter and the exudates of the Eucalyptus rhizosphere can be an important factor that limits the abundance and richness of morphotypes (Machado et al. 2019; Santos et al. 2018).

These studies show that soil attributes may explain the presence of several morphotypes in the areas (as determined by pH and potential acidity (H + Al), biopores, bulk density, C/N, Ca/Mg and Mg/K ratio, metabolic quotient, micro- and macroporosity, microbial biomass carbon, microbial basal respiration, organic matter, soil moisture, total organic carbon, and total porosity). However, only on the plateau, microporosity did explain the abundance of the morphotypes in eucalyptus plantations. In this case, this correlation may account for the identified morphotypes being mainly edaphic and hemiedaphic. This variable is important to explain the activity and establishment of the various hemiedaphic and epigeic morphotypes at this site.

Pompeo et al. (2017), when carrying out a study with the soil coleopteran morphotypes in forest and agricultural land uses in the Santa Catarina Plateau, in areas of Mixed Ombrophylous Forest (Atlantic Forest), evaluated the diversity of Coleoptera and their relationships with the land-use systems and edaphic properties in winter and summer. In this case, the native forest presented greater richness and abundance of morphotypes, as well as greater diversity in summer, proving to be the most stable among the studied land uses. This system was associated with a greater amount of morphotypes, and the soil properties related to carbon dynamics contributed to explain this distribution.

The *Eucalyptus* stands (EST) presented lower diversity and abundance of coleopteran morphotypes than native forest (Fig. 8.6). Though presenting smaller values than the other land-use systems, they did not differ significantly from pasture, crop-livestock integration, and no-tillage (Pompeo et al. 2017). Thus, the authors correlate the low density of beetles in EST to the lower plant diversity of these sites, since *Eucalyptus* is an exotic species, usually grown in monoculture. *Eucalyptus* stands also provide forest litter of lower quality than that of native forests, as already discussed above, therefore providing a less attractive environment for some beetles.

8.5 Enchytraeids in Forest and Agricultural Sites: The Dramatic Difference Between the Dataset Among Tropical and Temperate Regions

Enchytraeids are close relatives of the earthworms but not as well known worldwide. They live in all types of soils as long as they have a minimum of moisture, organic matter, and oxygen (Schmelz et al. 2013). Their small body size, generally

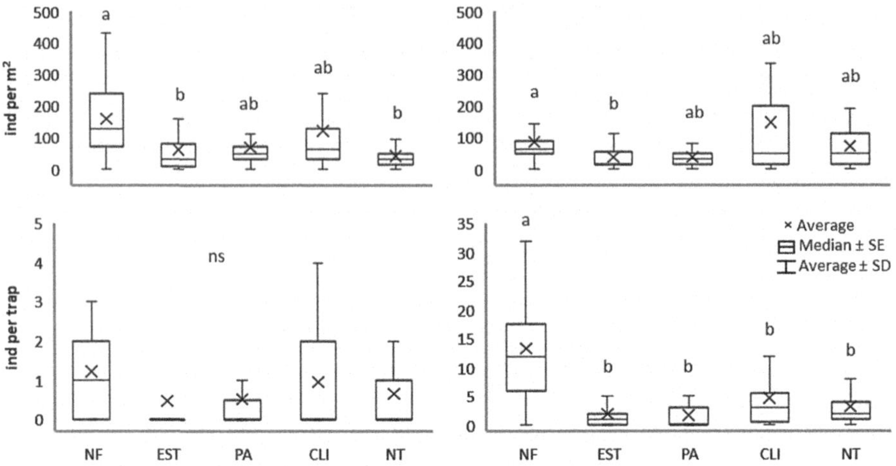

Fig. 8.6 Coleoptera density (individuals per m^2) and abundance (individuals per trap) in native forest (NF), Eucalyptus stands (EST), perennial pasture (PA), crop-livestock integration (CLI), and no-tillage (NT) in winter (left) and summer (right) on the Southern Santa Catarina Plateau (Pompeo et al. 2017). Mean values followed by the same letter are similar by the Kruskal-Wallis test ($p < 0.05$; $n = 135$); ns: nonsignificant difference

not more than 4 cm, and their whitish semitransparent body make them not so evident to be seen in soil with the naked eyes. When in high abundance in crop soils it is not rare that unsuspected farmers take enchytraeids as a pest, confounding them with nematodes, leaving them highly concerned about the health of their crops. However, their roles in soil are correlated with specific attributes of their larger relatives (earthworms), contributing to the decomposition of organic matter, building up of soil porosity and soil formation, mixing of organic matter, and regulation of microbial activity (Didden 1993), though in a lower scale because of the small body size, at least when in low abundance (Pelosi and Römbke 2016).

These small organisms play key roles in some particular natural ecosystems, such as coniferous forests, peatlands, moorlands, and inselbergs whose soils are acidic, rich in organic matter, and generally humid, either in the Northern or in the Southern hemisphere (Vaçulik et al. 2004; Schmelz et al. 2013; Carrera and Briones 2013). In these ecosystems, they may show extremely abundant populations reaching hundreds of thousands of individuals per square meter. The few data available on enchytraeids in the tropics suggest that their abundance can be much lower (>10,000) than in temperate regions (Römbke 2007; Schmelz et al. 2013). Conversely, some studies showed enchytraeid abundance reaching 44,000 individuals per square meter in a mixed Araucaria forest in Southern Brazil and a maximum average of 12,000, when determined by wet extraction (Schmelz et al. 2013; Niva et al. 2015).

The high abundance of enchytraeids reported in temperate regions is often associated to a community dominated by *Cognettia sphagnetorum*, a species, which proliferates rapidly with asexual reproduction by fragmentation of the body and regeneration of the lost parts (Nurminen 1967; Huhta et al. 1986; Schlaghamerský 2013). These forests with acidic soil very often present low occurrence of earthworms (Huhta 1984; Graefe and Beylich 2003) and may harbor enchytraeid communities composed in a habitat with more than 90% of *Cognettia* (Huhta 1984). The data available from South America so far show that the community of enchytraeids is quite different from the ones in the Northern hemisphere. For example, the genus *Cognettia*, which is common in European countries, has never been sampled in Brazil, and many species of *Guaranidrilus* and *Hemienchytraeus* are abundant in South America (Schmelz et al. 2013; Niva et al. 2015; Pelosi and Römbke 2016) but almost not found in Europe (Schmelz and Collado 2010) or North America (Schlaghamerský 2013). On the other hand, *Enchytraeus*, *Fridericia*, *Marionina*, and *Achaeta* are present in both continents (Schlaghamerský 2013; Schmelz et al. 2013; Niva et al. 2015; Pelosi and Römbke 2016). According to Pelosi and Römbke (2016), some species of the first two genera may be cosmopolitan or peregrine, and *Fridericia* seems to occur more in anthropogenic landscapes, such as pastures or croplands, and less in forests. It was also reported that the density and the distribution of enchytraeids do change with the different successional stages of the Atlantic Forest and soil types (cambisoil vs. gleysoil) (Römbke et al. 2007). The genus *Enchytraeus*, for example, was less frequent in primary forests than in less advanced stages of succession or pasture, while *Guaranidrilus* was slightly more frequent in primary forests. Another study, in the Araucaria mixed forest on cambisoil, shows that 34% of their population consisted of *Guaranidrilus*, the dominant genus.

It seems that, in Brazil, enchytraeid abundance can reach numbers similar to those found in temperate regions, however with a distinct species and genus composition (Schmelz et al. 2013). Silva et al. (2006) found extremely contrasting densities of enchytraeids when comparing an area with natural Cerrado to areas with conventional, no-tilling, integrated crop-livestock, and pasture systems. In one of the sampling dates, while in Cerrado there were more than 3000 enchytraeids per square meter; in the other systems there were less than 10, using the TSBF method for sampling. However, in general, the no-till, pasture, and crop-livestock systems favored enchytraeid populations reaching not more than 270 individuals per square meter. Assis (2015) found that enchytraeid density and richness were higher in a native mixed Araucaria forest than in conventional maize crop, while organic and conventional horticulture areas were somewhat in an intermediary position between the two. This author found *Guaranidrilus* and *Tupidrilus* only in the forest, while *Enchytraeus* and *Fridericia* were the most abundant ones in cultivated areas, either organic or conventional, confirming the hypothesis that these genera are more common in anthropogenic areas, and he also found a possible negative relation between phosphorus soil content and genus diversity.

8.6 Conclusions and Outlook

We believe that the soil fauna correlates with physical-chemical and microbiological soil processes, since soil invertebrates continuously fragment the soil organic matter and plant wastes, which facilitates many other processes in soil, such as microbial decomposition, nutrient cycling, and water-holding capacity; however the interaction between mesofauna and microorganisms is strongly conditioned by soil moisture. Eucalyptus plantations may exert some allelopathic effect in soil, once many researchers show an inhibition in the development of a diversified and abundant invertebrate community. In this sense, more research is needed in order to find out effectively which allelopathic substances are produced and what is the magnitude of this effect in relation to the provision of soil ecosystem services. Mixed plantations of *Eucalyptus grandis* and *Acacia mangium* share faunal groups which originate either exclusively from pure plantations of *Acacia mangium* or from *Eucalyptus grandis*, improving the soil quality, although the richness and diversity of such plantations differ from native forests. Therefore, more diversified mixed plantations may be the key for a higher soil fauna diversity and, consequently, for a better soil health, although perhaps these diversified systems are generally adopted in restoration sites. For production sites and silvicultural management however it may require greater adaptations.

References

Aliaga R, Fuentes AH, Clericus JEA (2017) Effect of post-harvest forestry residue management practices on the diversity of epigeal coleopterans. Rev Fac Nac Agron 70:8069–8075

Anderson JM (2009) Why should we care about soil fauna? Pesqui Agropecu Bras 44:835–842

Anderson JM, Ingram JSI (1993) Tropical soil biology and fertility: a handbook on methods, 2nd edn. CAB International, Wallingford

Assis O (2015) Enquitreídeos (Enchytraeidae, Oligochaeta) como indicadores do manejo do solo e em ensaios ecotoxicológicos. Master's Dissertation, Universidade Tecnológica do Paraná

Bachega LR, Bouillet JP, Picollo MC, Saint-André L, Bouvet JM, Nouvellon Y, Gonçalves JLM, Robin A, Laclau JP (2016) Decomposition of *Eucalyptus grandis* and *Acacia mangium* leaves and fine roots in tropical conditions did not meet the Home Field Advantage hypothesis. For Ecol Manag 359:33–43

Barbercheck ME, Neher DA, Anas O, El-Allaf SM, Weicht TR (2009) Response of soil invertebrates to disturbance across three resource regions in North Carolina. Environ Monit Assess 152:283–298

Bardgett RD, Van der Putten WH (2014) Belowground biodiversity and ecosystem functioning. Nature 515:505–511

Baretta D, Santos JCP, Segat JC, Geremia EV, Oliveira Filho LCI, Alves MV (2011) Fauna edáfica e qualidade do solo. In: Tópicos em Ciência do Solo. SBCS, Viçosa, pp 119–170

Bartz MLC, Brown GG, Rosa MG, Klauberg Filho O, James SW, Decaëns T, Baretta D (2014) Earthworm richness in land-use systems in Santa Catarina, Brazil. Appl Soil Ecol 83:59–70

Bini D, Santos CA, Bouillet JP, Gonçalves JLM, Cardoso EJBN (2013) *Eucalyptus grandis* and *Acacia mangium* in monoculture and intercropped plantations: evolution of soil and litter microbial and chemical attributes during early stages of plant development. Appl Soil Ecol 63:57–66

Blasi S, Menta C, Balducci L, Conti FD, Petrini E, Piovesan G (2013) Soil micro-arthropod communities from Mediterranean forest ecosystems in Central Italy under different disturbances. Environ Monit Assess 185:1637–1655

Brown GG, Niva CC, Zagatto MRG, Ferreira S, Nadolny H, Cardoso GX, Santos A, Martinez G, Pasini A, Bartz MLC, Sautter KD, Thomazini MJ, Baretta D, Silva E, Antoniolli ZI, Decaëns T, Lavelle P, Sousa JP, Carvalho F (2015) Biodiversidade da Fauna do solo e sua contribuição para os serviços ambientais. In: Parron LM, Garcia JR, Oliveira EB, Brown GG, Prado RB

(eds) Serviços Ambientais em Sistemas Agrícolas e Florestais do Bioma Mata Atlântica. EMBRAPA, Brasília, pp 113–154
Cardoso EJBN, Vasconcellos RLF, Bini D, Miyauchi MYH, Santos CA, Alves PRL, Paula AM, Nakatani AS, Pereira JM, Nogueira MA (2013) Soil health: looking for suitable indicators. What should be considered to assess the effects of use and management on soil health? Sci Agric 70:274–289
Carrera N, Briones MJI (2013) Arthropod community structure and diversity from galician upland peatlands. In: Riosmena-Rodriguez R (ed) Invertebrates: classification evolution and biodiversity. Nova Science, New York, p 251
Choi WI, Moorhead DL, Neher DA, MII R (2006) A modeling study of soil temperature and moisture effects on population dynamics of Paronychiuruskimi (Collembola: Onychiuridae). Biol Fertil Soils 43:69–75
Creamer RE, Hannula SE, JPV L, Stone D, Rutgers M, Schmelz RM, Ruitter PC, Hendriksen NB, Bolger T, Bouffaud ML, Buee M, Carvalho F, Costa D, Dirilgen T, Francisco R, Griffiths BS, Griffiths R, Martin F, Martins da Silva P, Mendes S, Morais PV, Pereira C, Philippot L, Plassart P, Redecker D, Römbke J, Sousa JP, Wouterse M, Lemanceau P (2016) Ecological network analysis reveals the inter-connection between soil biodiversity and ecosystem function as affected by land use across Europe. Appl Soil Ecol 97:112–124
Derpsch R, Friedrich T, Kassam A, Hongwen L (2010) Current status of adoption of no-till farming in the world and some of its main benefits. Int J Agric Biol Eng 3:1–25
Didden WAM (1993) Ecology of terrestrial Enchytraeidae. Pedobiologia 37:2–29
Filho LCIO, Baretta D (2016) Por que devemos nos importar com os colêmbolos edáficos? Sci Agrár 17:21–40
Filser J, Faber JH, Tiunov AV, Brussaard L, Frouz J, Deyn G, Uvarov AV, Berg MP, Lavelle P, Loreau M, Wall DH, Querner P, Eijsackers H, Jiménez JJ (2016) Soil fauna: key to new carbon models. Soil 2:565–582
Forrester DI, Bauhus J, Cowie AL, Vanclay JK (2006) Mixed-species plantations of Eucalyptus with nitrogen-fixing trees: A review. For Ecol Manag 233:211–230
Fountain-Jones NM, Baker SC, Jordan GJ (2015) Moving beyond the guild concept: developing a practical functional trait framework for terrestrial beetles. Ecol Entomol 40:1–13
Gardi C, Menta C, Leoni A (2008) Evaluation of the environmental impact of agricultural management practices using soil microarthropods. Fresenius Environ Bull 17:1165–1169
Gonçalves JLM, Stape JL, Laclau JP, Bouillet JP, Ranger J (2008) Assessing the effects of early silvicultural management on long-term site productivity of fast-growing eucalypt plantations: the Brazilian experience. South Forest 70:105–118
Graefe U, Beylich A (2003) Critical values of soil acidification for annelid species and decomposer community. Newsletter Enchytraeidae 8:51–55
Huhta V, Hyvönen R, Koskenniemi A, Vilkamaa P, Kaasalainen P, Sulander M (1986) Response of soil fauna to fertilization and manipulation of pH in coniferous forests. Acta Forest Fenn 195:1–30
Huhta V (1984) Response of Cognettia sphagnetorum (Enchytraeidae) to manipulation of pH and nutrient status to coniferous forest soil. Pedobiologia 27:245–260
IBÁ (2015) Anuário estatístico—Ano base 2014
Kamau S, Barrios E, Karanja NK, Ayuke FO, Lehmann J (2017) Soil macrofauna abundance under dominant tree species increases along a soil degradation gradient. Soil Biol Biochem 112:35–46
Kaneda S, Kaneko N (2011) Influence of collembola on nitrogen mineralization varies with soil moisture content. Soil Sci Plant Nutr 57:40–49
Kautz G, Zimmer M, Topp W (2002) Does Porcellioscaber (Isopoda: Oniscidea) gain from coprophagy? Soil Biol Biochem 34:1253–1259
Korasaki V, Lopes J, Brown GG, Louzada J (2012) Using dung beetles to evaluate the effects of urbanization on Atlantic Forest biodiversity. Insect Sci 00:1–14
Laclau JP, Bouillet JP, Gonçalves JLM, Silva EV, Jourdan C, Cunha MCS, Moreira MR, Saint-André L, Maquère V, Nouvellon Y, Ranger J (2008) Mixed-species plantations of *Acacia*

mangium and *Eucalyptus grandis* in Brazil: 1. Growth dynamics and aboveground net primary production. For Ecol Manag 255:3905–3917

Lavelle P (1997) Faunal activities and soil processes: adaptive strategies that determine ecosystem function. Adv Ecol Res 27:93–132

Lavelle P, Bignell D, Lepage M, Wolters W, Roger P, Ineson P, Heal OW, Dhillion S (1997) Soil function in a changing world: the role of invertebrate ecosystem engineers. Eur J Soil Biol 33:159–193

Machado JS, Oliveira Filho LCI, Santos JCP, Paulino AT, Baretta D (2019) Morphological diversity of springtails (Hexapoda: Collembola) as soil quality bioindicators in land use systems. Biota Neotrop 19:e20180618

McCann KS (2000) The diversity–stability debate. Nature 405:228–233

Meloni F, Varanda EM (2015) Litter and soil arthropod colonization in reforested semi-deciduous seasonal Atlantic forests. Restor Ecol 23:690–697

Menta C, Conti FD, Pinto S (2018a) Microarthropods biodiversity in natural, seminatural and cultivated soils—QBS-ar approach. Appl Soil Ecol 123:740–743

Menta C, Conti FD, Pinto S, Bodini A (2018b) Soil Biological Quality index (QBS-ar): 15 years of application at global scale. Ecol Indic 85:773–780

Mound LA (2005) Thysanoptera: diversity and interactions. Annu Rev Entomol 50:247–269

Nichols E, Spector S, Louzada J, Larsen T, Amezquita S, Favila ME (2008) Ecological functions and ecosystem services provided by Scarabaeinae dung beetles. Biol Conserv 141:1461–1474

Niva CC, Cezar RM, Fonseca PM, Zagatto MRG, Oliveira EM, Bush EF, Clasen LA, Brown GG (2015) Enchytraeid abundance in Araucaria mixed forest determined by cold and hot wet extraction. Brazilian J Biol 75:169–175

Nurminen M (1967) Ecology of enchytraeids (Oligochaeta) in Finnish coniferous forest soil. Ann Zool Fenn 4:147–157

Oliveira Filho LCI, Baretta D, Pereira JM, Maluche-Baretta CRD, Pompeo PN, Cardoso EJBN (2018) Fauna edáfica em ecossistemas florestais. In: Ciências Ambientais, pp 10–48

Oliveira Filho LCI, Klauberg Filho O, Baretta D, Tanaka CAS, Sousa JP (2016) Collembola community structure as a tool to assess land use effects on soil quality. Rev Bras Cienc Solo 40:1–18

Paoletti MG, Hassall M (1999) Woodlice (Isopoda: Oniscidea): their potential for assessing sustainability and use as bioindicators. Agric Ecosyst Environ 74:157–165

Parisi A (2001) The biological soil quality, a method based on microarthropods (in Italy). Acta Nat L'Ateneo Parm 37:97–106

Parisi V, Menta C, Gardi C, Jacomini C, Mozzanica E (2005) Microarthropod communities as a tool to assess soil quality and biodiversity: a new approach in Italy. Agric Ecosyst Environ 105:323–333

Pelosi C, Römbke J (2016) Are Enchytraeidae (Oligochaeta: Anellida) good indicators of agricultural management practices? Soil Biol Biochem 100:255–253

Peña-Peña K, Irmler U (2016) Moisture, seasonality, soil fauna, litter quality and land use as a driver of decomposition in Cerrado soils in SE–Mato Grosso, Brazil. Appl Soil Ecol 107:124–133

Pereira APA, Andrade PAM, Bini D, Durrer A, Robin A, Bouillet JP, Andreote FD, Cardoso EJBN (2017a) Shifts in the bacterial community composition along deep soil profiles in monospecific and mixed stands of *Eucalyptus grandis* and *Acacia mangium*. PLoS One 12:1–15

Pereira APA, Zagatto MRG, Brandani CB, Mescolotti DL, Cotta SR, Gonçalves JLM, Cardoso EJBN (2018) Acacia changes microbial indicators and increases C and N in soil organic fractions in intercropped Eucalyptus plantations. Front Microbiol 9:1–13

Pereira JDM, Segat JC, Baretta D, Leandro R (2017b) Soil Macrofauna as a soil quality indicator in native and replanted Araucaria angustifolia forests. Rev Bras Ciênc Solo 41:1–15

Pereira JM, Baretta D, Cardoso EJBN (2015) Fauna edáfica em floresta de Araucária. In: Cardoso EJBN, Vasconcellos RLF (eds) Floresta Com Araucária: Composição Florística e Biota Do Solo. Editora FEALQ, Piracicaba, pp 153–180

Pey B, Nahmani J, Auclerc A, Capowiez Y, Cluzeau D, Cortet J, Decaëns T, Deharveng L, Dubs F, Joimel S, Briard C, Grumiaux F, Laporte MA, Pasquet A, Pelosi C, Pernin C, Ponge JF, Salmon

S, Santorufo L, Hedde M (2014) Current use of and future needs for soil invertebrate functional traits in community ecology. Basic Appl Ecol 15:194–206

Pompeo PN, Oliveira Filho LCI, Filho OK, Mafra AL, Baretta CRDM, Baretta D (2016) Coleoptera diversity (Arthropoda: Insecta) and soil properties under soil management systems in the highlands of Santa Catarina state, Brazil. Sci Agrár 17:16–28

Pompeo PN, Oliveira Filho LCI, Santos MAB, Mafra AL, Klauberg Filho O, Baretta D (2017) Morphological diversity of Coleoptera (Arthropoda: Insecta) in agriculture and forest systems. Rev Bras Cienc Solo 41:e0160433

Römbke J (2007) Enchytraeidae of tropical soils: state of the art, with special emphasis on Latin America. Folia Facultatis Scientiarium Naturalium Universitatis Masarykianae Brunensis. Biologia 110:157–181

Römbke J, Collado R, Schmelz RM (2007) Abundance, distribution and indicator potential of enchytraeid genera (Enchytraeidae, Clitellata) in secondary forests and pastures of the Mata Atlântica. Acta Hydrobiol Sin 31:139–150

Rosa MG, Klauberg Filho O, Bartz MLC, Mafra AL, Sousa JPFA, Baretta D (2015) Macrofauna edáfica e atributos físicos e químicos em sistemas de uso do solo no planalto catarinense. Rev Bras Cienc Solo 39:1544–1553

Santos MAB, Oliveira Filho LCI, Pompeo PN, Ortiz DC, Mafra AL, Klauberg Filho O, Baretta D (2018) Morphological diversity of springtails in land use systems. Rev Bras Cienc Solo 41:e0170277

Schlaghamerský J (2013) Enchytraeid assemblages (Annelida: Clitellata: Enchytraeidae) of two old growth forests in the Porcupine Mountains (Michigan, USA). Soil Organisms 85:85–96

Schmelz RM, Collado R (2010) A guide to European terrestrial and freshwater species of Enchytraeidae (Oligochaeta). Soil Organisms 82:1–176

Schmelz RM, Niva CC, Römbke J, Collado R (2013) Diversity of terrestrial Enchytraeidae (Oligochaeta) in Latin America: current knowledge and future research potential. Appl Soil Ecol 69:13–20

Silva RF, Aquino AM, Mercante FM, Guimarães MF (2006) Soil invertebrate macrofauna under different production systems in a Hapludox in the Cerrado Regional. Pesqui Agropecu Bras 41(4):697–704

Souza ST, Cassol PC, Baretta D, Bartz MLC, Klauberg Filho O, Mafra AL, Rosa MG (2016) Abundance and diversity of soil macrofauna in native forest, eucalyptus plantations, perennial pasture, integrated crop-livestock, and no-tillage cropping. Rev Bras Cienc Solo 40:e0150248

Swift MJ, Heal OW, Anderson JM (1979) Decomposition in terrestrial ecosystems. Blackwell Scientific, Oxford

Teuben A, Verhoef HA (1992) Direct contribution by soil arthropods to nutrient availability through body and faecal nutrient content. Biol Fertil Soils 14:71–75

Vaçulik A, Kounda-Kiki C, Sarthou C, Ponge JF (2004) Soil invertebrate activity in biological crusts on tropical inselbergs. Eur J Soil Sci 55:539–549

Van Capelle C, Schrader S, Brunotte J (2012) Tillage-induced changes in the functional diversity of soil biota—a review with a focus on German data. Eur J Soil Biol 50:165–181

Van Der Putten WH, De Ruiter PC, Bezemer TM, Harvey JA, Wassen M, Wolters V (2004) Trophic interactions in a changing world. Basic Appl Ecol 5:487–494

Van Vliet PCJ, Beare MH, Coleman DC, Hendrix PF (2004) Effects of enchytraeids (Annelida: Oligochaeta) on soil carbon and nitrogen dynamics in laboratory incubations. Appl Soil Ecol 25:147–160

Wagg C, Bender F, Widmer F, van der Heyden MGA (2014) Soil biodiversity and soil community composition determine ecosystem multifunctionality. Proc Natl Acad Sci U S A 111:5266–5270

Warren MW, Zou X (2002) Soil macrofauna and litter nutrients in three tropical tree plantations on a disturbed site in Puerto Rico. For Ecol Manag 170:161–171

Zagatto MRG, Niva CC, Thomazini MJ, Baretta D, Santos A, Nadolny H, Cardoso GBX, Brown GG (2017) Soil invertebrates in different land use systems: how integrated production systems and seasonality affect soil Mesofauna communities. J Agric Sci Technol B 7:150–161

Zagatto MRG, Pereira APA, Souza AJ, Fabri RF, Baldesin LF, Pereira CM, Lopes RV, Cardoso EJBN (2019a) Interactions between mesofauna, microbiological and chemical soil attributes in pure and intercropped *Eucalyptus grandis* and *Acacia mangium* plantations. For Ecol Manag 433:240–247

Zagatto MRG, Zanão Júnior LA, Pereira APA, Estrada-Bonilla G, Cardoso EJBN (2019b) Soil mesofauna in consolidated land use systems: how management affects soil and litter invertebrates. Sci Agric 76(2):165–171

Zhiqun T, Jian Z, Junli Y, Chunzi W, Danju Z (2017) Chemosphere Allelopathic effects of volatile organic compounds from *Eucalyptus grandis* rhizosphere soil on Eisenia fetida assessed using avoidance bioassays, enzyme activity, and comet assays. Chemosphere 173:307–317

Chapter 9
Bioindicators of Soil Quality in Mixed Plantations of *Eucalyptus* and Leguminous Trees

Arthur Prudêncio de Araujo Pereira, Daniel Bini, Emanuela Gama Rodrigues, Maiele Cintra Santana, and Elke Jurandy Bran Nogueira Cardoso

9.1 Soil Quality Indicators: Definitions, Applications and Challenges

9.1.1 Definitions

Soil quality has an important role in both society and environment considering several ecosystem services provisioned through soils (e.g., food, feed, fiber, climate moderation through C cycling, waste disposal, water filtration and purification, elemental cycling) (Lal 2015). Soil quality is commonly defined as the capacity of a soil to function within ecosystem and land-use boundaries to sustain biological productivity, maintain environmental quality, and promote plant and animal health (Doran and Parkin 1994, 1996). This concept has been revised and it is proposed to discuss soil use rather than soil functions. Therefore, soil quality assessment would

A. P. A. Pereira (✉)
Department of Soil Science, University of São Paulo, "Luiz de Queiroz" College of Agriculture, Piracicaba, SP, Brazil

Soil Science Department (Pici Campus), Federal University of Ceará, Fortaleza, Ceará, Brazil
e-mail: arthur.prudencio@usp.br; arthur.prudencio@ufc.br

D. Bini
State University of the Central West, Guarapuava, PR, Brazil

E. G. Rodrigues
North Fluminense State University, Campos dos Goytacazes, RJ, Brazil

M. C. Santana · E. J. Bran Nogueira Cardoso
Department of Soil Science, University of São Paulo, "Luiz de Queiroz" College of Agriculture, Piracicaba, SP, Brazil

E. J. Bran Nogueira Cardoso et al. (eds.), *Mixed Plantations of Eucalyptus and Leguminous Trees*, https://doi.org/10.1007/978-3-030-32365-3_9

provide the scientific tools for evaluating the management of soil resources, also considering the societal demands of the various benefits that soils can provide to humankind if well managed. Therefore, the responsibility to maintain the quality of the soil could be clearly assigned to the user of the soil (Bünemann et al. 2018).

Additionally, several authors have proposed a link between soil functions and ecosystem services (ES) as the capacity of soils to deliver ES is determined largely by the soil functions (Greiner et al. 2017; Adhikari and Hartemink 2016). This link is a concept that has been considered a challenging yet promising approach for fostering the communication of nature's capital (Drobnik et al. 2018). According to Lal (2015), the soil quality must be preserved or restored to guarantee these services and also to enhance long-term productivity and improve the environment. Thus, this author suggests that the strategy is to produce "more from less" by reducing losses and increasing soil-, water-, and nutrient-use efficiency.

Measuring soil quality is an exercise in identifying inherent and dynamic soil properties which are responsive to management, are capable of being precisely measured within certain technical and economic constraints, and also are defined with respect to the delivery of ecosystem services (Bünemann et al. 2018).

In order to achieve an improvement in productivity and sustainability of pure or mixed *Eucalyptus* plantations, a different management is needed, based on a novel approach to measure soil quality which demonstrates the interrelationship between soil biota diversity and activity, improvements in nutrient cycling and consequently soil fertility, increase in the organic matter quality, and increase in wood productivity. All these aspects are necessary as a guide in the search for sustainable practices (Pereira et al. 2018a; Bini et al. 2013a; Gama-Rodrigues et al. 2008).

The role of soil biota for the functioning, integrity, and long-term sustainability of natural and managed terrestrial ecosystems is slowly increasing towards an adequate recognition, since these organisms are essential components of litter decomposition, nutrient cycling, soil aggregation, and growth of plant communities (Bender et al. 2016; Lal 2015). Soil microbial biomass is a labile fraction of the soil organic matter (SOM) and plays a crucial role in the maintenance of soil fertility and availability of plant nutrients (Jenkinson 1981). The microbial biomass is a sensitive indicator of organic matter dynamics because the microbial fraction changes comparatively fast and differences are detectable before they occur in total organic matter (Cardoso et al. 2013; Gama-Rodrigues and Gama-Rodrigues 1999). Although microbial biomass only constitutes an average of 2–5% of the soil organic C and 1–5% of the total soil N, it is the most important component of SOM, controlling the nature and rate of organic matter transformations (Jenkinson 1981; Smith et al. 1990). Moreover, it plays a critical role in soil C cycling and accounts for roughly half of the soil surface CO_2 efflux through heterotrophic soil respiration (Hanson et al. 2000; Högberg et al. 2001).

The litter-soil system comprises the habitat for the majority of species living on the planet, while the horizontal and vertical heterogeneity of both soil and litter can boost spatial variability in the distribution and activity of this biota (Coleman and Whitman 2005). The relationship between environmental attributes and biota of the litter-soil system becomes extremely important because any change in the soil or litter attributes affects the biota, and consequently the development of the plant community (Bini et al. 2013b; Zaia et al. 2012; Zagatto et al. 2019).

9.1.2 Applications

In this study, we are going to show some soil quality assessments with a focus on the provision of ecosystem services in the litter-soil system of pure and/or mixed Eucalyptus plantations.

Studying the interrelationships between microbial and soil chemical attributes Pereira et al. (2018a) were able to explore the microbial influence on C and N cycling and how it can discriminate between intercropped and pure Eucalyptus plantations. The study showed high efficiency of the microbial biomass in incorporating C and N, and improved organic matter (OM) cycling in mixed systems with *Acacia mangium*. It also showed the potential of longer maintenance of Acacia residues in the labile fraction of soil OM when compared to Eucalyptus treatments. Thus, the study concluded that mixed plantations promote a more efficient use of C and N by microbial communities, thereby increasing the plant nutrient availability in soils with low levels of OM. The same authors undertook an investigation to evaluate interactions between the bacterial community and biological functions involved in C and N cycles in the soil and litter layers resulting from pure or mixed *Eucalyptus grandis* and *Acacia mangium* plantations. This study showed a significant increase of bacterial community diversity and functional gene abundance, which improved C and N cycling in the soil and in the litter interface of a pure Acacia plantation, as well as in the intercropped plantation. Thus, the authors concluded that mixed plantations are a better alternative than using mineral N fertilizers for long-term soil health, as mineral N can reduce the abundance of functional genes, bacterial diversity, and microbial activities (Pereira et al. 2019).

The interrelationship between soil C, N, and P; litter C, N, P, lignin content, and polyphenol content; and microbial biomass and activity was examined by Bini et al. (2013a) in pure and mixed plantations of *Eucalyptus grandis* and *Acacia mangium* before and after senescent leaf drop. This study showed a stronger relationship between litter contents and microbiological soil attributes, as well as the important role played by the maintenance and quality of litter in regulating microbial biomass and activity in soils. The authors concluded that the synergism between the two tree species in the intercropped plantation established a new equilibrium in the soil microbiota after 20 months, maintaining and stimulating biogeochemical cycling as requirement for the sustainability of the intercropped plantations.

Changes in forest litter and soil where the native forest was replaced by eucalyptus plantations in four southeastern areas of Brazil were studied by Gama-Rodrigues et al. (2008). The authors observed that the interrelationship between litter and soil microbial attributes was sensitive to show the dissimilarity between eucalyptus sites and native forest. The study of these interactions also enabled observing that the impact of native forest conversion into eucalyptus stands varied in accordance with the site-specific characteristics that had been analyzed.

Another study related to the interaction between microbial and chemical soil attributes showed that the soil organic C and total N stock were more relevant to explain the dissimilarity between eucalyptus stands of different ages than the soil microbial attributes. On the other hand, both the microbial litter attributes and cellulose, lignin, and N content of litter were relevant to show the differences between those eucalyptus stands. Thus, the litter quality had a direct influence on litter microbial activity and microbial biomass C and N, which suggests a close relationship between C and N immobilization or mineralization and litter quality (Barreto 2008).

Zagatto et al. (2019) evaluated the density and diversity of soil mesofauna and its interaction with microbiological and chemical soil attributes in pure *Acacia mangium* (AC), *Eucalyptus grandis* (EU), and mixed *E. grandis* and *A. mangium* plantations (M). The authors found that the higher soil quality in pure Acacia plantations and in mixed plantations was due to the interaction between microbial activity and structure of the soil mesofauna community, which contributed to the increase in soil nutrients.

The interrelationship between arbuscular mycorrhizal fungi (AMF) root colonization, enzymatic activity, and soil and litter C, N, and P in both pure and mixed plantations of Acacia and Eucalyptus was evaluated by Bini et al. (2018). The results showed that the intercropped plantation increased the AMF colonization and the activity of acid and alkaline phosphatase. They also found negative correlations between root colonization and litter C/N and C/P ratios, and a positive correlation with soil acid phosphatase activity and soil N and P concentrations. Altogether, this means that intercropped systems with higher root colonization rates generate litter of better quality improving P cycling and P nutrition in soil and, therefore, the health and productivity of these forests. A study was carried out with the purpose to evaluate interactions between the structure and richness of soil bacteria, fungi, and archaea; the functional groups of nitrifying, denitrifying, and nitrogen-fixing bacteria; and their relationship with soil chemical changes in pure and mixed Eucalyptus and Acacia plantations. The results showed a distinct microbial community in mixed plantations with positive effects on soil phosphorus and nitrate content, which potentially reduces the demand for mineral fertilization (Rachid et al. 2013).

Additionally, Santos et al. (2018) studied the interaction between litter production, decomposition, and soil microbial activity in a pure and intercropped Eucalyptus and Acacia plantation. In this study, the more diverse litter composition in the mixed plantation provided a more balanced N and P supply, which in turn was able to sustain high microbial activity levels with positive consequences on litter decomposition and boosting nutrient cycling efficiency, being a sustainable option to offset the high nitrogen export from successive monoculture-based silvicultural systems.

9.1.3 Challenges

Seeking high productivity has made forest production a simplistic practice, with high fertilizer and pesticide applications producing detrimental environmental impacts. Nowadays, public concern in environmental issues is growing. This imposes the following question: How is it possible to achieve high ecological productivity by optimizing ecosystem services in low-input forest systems?

The interrelationship between soil and litter attributes (biological, microbial, chemical, and physical) has hitherto provided the fundamental context on how to improve ecosystem services. However, a mechanistic understanding of these relationships and to decide whether they are in relation to the soil as an ecosystem in itself or as part of a larger ecosystem in nature are undeniably complex and remain elusive.

A challenge in soil ecology is to develop multivariate hypotheses to describe not only interrelations between litter and soil attributes or between litter and soil fauna, but also the direct and indirect relations between attributes and fauna in the litter-soil system. There is a complex interconnection between the edaphic environment and productivity (Oliveira et al. 2018; Eisenhauer et al. 2015) (Fig. 9.1). This is a

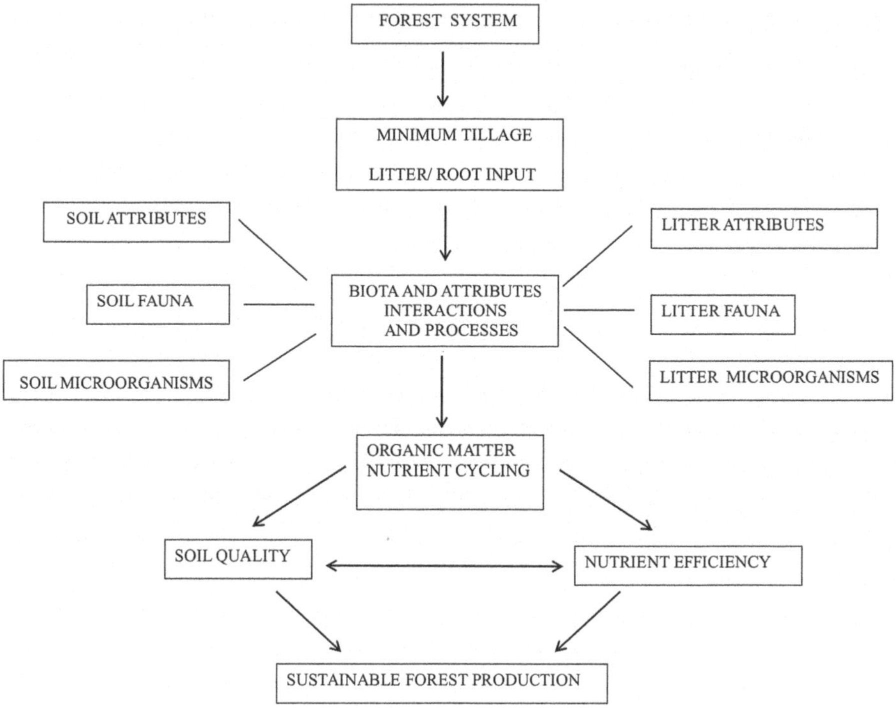

Fig. 9.1 Hypothetical model of interactions and processes mediated by soil biota for sustainable forest production

very important consideration. It even would justify a scientific study conducted by a multidisciplinary team to elucidate this intricate chain of interactions and the ecological interdependencies between land use, biodiversity, and ecosystem services, which will enable an increase in the productive capacity of soils. It could be the start of a basis of technological modifications for sustainable production, both economically and environmentally.

Thus, forest production would bring net benefits from ecosystem services, which are a source of revenue for producers who are attentive to sustainable business, and would guarantee the demands of future generations. It would increase competitive advantages, and finally it would remove trade barriers imposed because of environmental reasons.

9.2 Biological Properties in Forest Ecosystems: Why Are They Important for *Eucalyptus* Plantation?

9.2.1 Forest Habitats: A Brief Description

Natural forests are considered a specific ecosystem, representing high wood production and comprising huge habitats that support the microbiome life, which is dynamic and quickly responds to anthropogenic and environmental changes (Baldrian 2017). Moreover, reactions to the microbiome metabolism can occur in the most diverse plant organs and locations, such as leaves, flowers, seeds, fruit, wood, inside (endophytic) or on the tree surface (phyllosphere), as well as belowground (soil, roots, rhizosphere, and mycorrhizosphere) (Baldrian 2017). Even more important, habitats differ in properties such as nutrient availability, major environmental conditions, processes, and dynamics, which together can alter the microbiome dynamics. The forest microbiome research has been highly focused on soil habitats, emphasizing tree roots and their symbionts, while litter and other habitats have been greatly underexplored, mainly in pure and mixed plantations (Pereira et al. 2018a, b).

The forest environment has specific properties that differentiate it from other (e.g., agricultural systems and implanted forest) (Navarrete et al. 2015). One of the most important features is the huge effect of the dominant trees on the surrounding habitat, which can regulate aboveground and belowground interactions (Wardle et al. 2004). The trees interact with microbial activities and composition, and this is mediated by bulk soil and litter chemistry (Augusto et al. 2015; Šnajdr et al. 2013; Urbanová et al. 2015). Here, we include organic matter contents, soil pH, nitrogen transformations, and other macro- and micronutrients (Fierer and Jackson 2006; Lauber et al. 2008; Prescott and Grayston 2013; Rousk et al. 2010; Tedersoo et al. 2016; Urbanová et al. 2015). These effects seem to be extremely dependent on forest management (Tedersoo et al. 2016), and all drivers are combined by stochastic effects on microbiome assembly (Bahram et al. 2016), which can contribute to the dynamics of microbiomes in different forest habitats (Štursová et al. 2016).

9.3 Native and Planted Forest Environments

Planted forests are subject to multiple modes of disturbance, such as insect attacks, fires, and nutritional imbalances in soil, among others. In addition, this system is also significantly changed by many anthropogenic factors, as climate change or environmental pollution, water deficit, and management practices, which together may easily shift the balance of carbon and nitrogen cycling processes (Trumbore et al. 2015).

Soils under native forest have characteristics that differentiate them in numerous aspects from the planted forestry and agricultural soils (Fig. 9.2). For example, around 50% of the C fixed by trees is allocated in the soil through their root activity (Högberg et al. 2001), while the litter layers are important organic matter sources for the system, governing important stages of ecosystem services and nutrient cycling (Baldrian 2017). Besides the C contents, one of the most notorious effects of tree influence is the low pH of the soil solution, potentiated by the release of organic acids through root system exudation (Motavalli et al. 1995). In addition, there is a large root extrusion of enzymes that degrade organic matter which makes biogeochemical cycling very active in this environment, with increases in C and N contents above- and belowground (Fig. 9.2, Chap. 2).

Natural forests can provide several ecosystem services that are fundamental for the maintenance of the surrounding environment, mainly in soil protection (Lal 2014), biogeochemical cycling of nutrients (Laclau et al. 2010), maintenance of microbial biodiversity, meso- and macrofauna (Cardoso et al. 2013), and organic

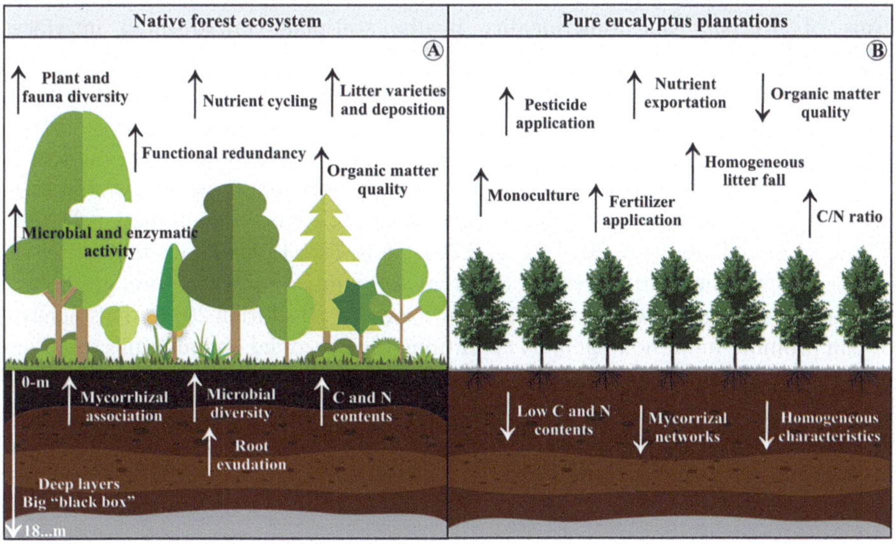

Fig. 9.2 Major differences between a natural forest and an implanted forest ecosystem. Arrows pointing upwards in (**a**) indicate better soil health than in (**b**), arrows pointing down, where the opposite is true

matter quality (Pereira et al. 2018a) among others (Fig. 9.2a). The preservation of natural forests and its functioning is becoming an evermore important subject, even for the common citizen. For example, in natural forests prevails a phenomenon that has been gently called the "Wood Wide Web," or either "The Forest Internet," responsible for permitting the existence of forests on our whole planet, a system that biologically interconnects all the trees in a forest, resulting from interactions of fungi, bacteria, and plants (including also some other macro- or microorganisms). Fungi and bacteria furnish nitrogen and phosphorus to the plants and receive in exchange carbon sources, moisture, and protection.

This is the result of millions of years of joint evolution, a real biological network which guarantees protection to all participating entities, which has been studied for over 30 years, starting even before there was a consolidation of the human Internet. Steidinger et al. (2019) showed the first map of these interconnections, demonstrating that, without this system, extensive forests would probably not exist. In tropical regions, where the predominating soils are generally very poor and acidic, with deficiencies in phosphorus and organic matter, such associations are even more fundamental for forest survival. This paper (about the "Wood Wide Web") includes about 200 scientists worldwide, among which seven Brazilian universities, with all earlier findings about the Wood Wide Web and its eventual risk of extinction due to deforestation and global warming (Steidinger et al. 2019). In planted pure forests, however, this web does not exist, while forest consortia composed of two, or even better multiple, tree species are prone to develop such a system, therefore being much more sustainable.

On the other hand, the conditions that occur in the planted forest ecosystems, such as in Eucalyptus plantations, differ strongly from their natural state, mainly in terms of diversity and functionality in the soil-plant-(micro)-biota interface (Fig. 9.2b).

In planted forests (monocultures) we find deposition of a unique litter type, which may present low nutrient availability (high C/N ratio) (Mercês et al. 2016; Snowdon et al. 2005). Moreover, when compared to natural systems, the depletion of mineralizable nutrients may occur over time (mainly due to wood exportation), and the type and quality of exudates secreted by the roots are extremely selective (Churchland and Grayston 2014) (Fig. 9.2b). Thus, forest plantations may differ in some soil properties and determine different soil temperatures, aeration gradients, porosity percentages, and soil water storage capacities (Baldrian 2017). This behavior can promote more homogenous conditions for microbial communities, making them less diverse and less efficient in the use of available resources, leading to negative plant-soil feedbacks (PSF) (Mariotte et al. 2017), or "soil fatigue" (Huang et al. 2013). In this sense, little emphasis was placed on studies to minimize PSF using intercropping systems to improve biological functions in different forest niches (Wang et al. 2017).

9.4 The Mixed or Consortiated Forest System with N_2-Fixing Trees: Brief Importance of Biological Functions for Soil, Plant Health, and Nutrient Cycling

There are around 3 trillion trees on planet Earth (Crowther et al. 2015), which are responsible for covering a large part of the soil surface covered by vegetation. This large volume of biomass is extremely important, especially in the regulation of the world's climate, and the health of soil and bodies of water (Kirilenko and Sedjo 2007). However, trees are closely dependent on the microbiome to survive, which provides nutrients essential for their development, such as N, P, and K (primary macronutrients), through organic matter cycling in the soil (Baldrian 2017). For example, it is estimated that N_2-fixing bacteria and mycorrhizal fungi are responsible for providing up to 75% of the nitrogen and 80% of the phosphorus that forests use during their life cycle (Van Der Heijden et al. 2008). In addition, all organic matter transformation steps depend on the activity of microorganisms (Singh 2018).

Acacia trees form symbiotic relationships with N_2-fixing bacteria (Chap. 6) and provide a key reservoir of N, C, and P for the surrounding ecosystem (Bini et al. 2013a; Paula et al. 2018; Pereira et al. 2018a; Taylor et al. 2017). In this sense, it is possible to integrate trees of high economic value (*Eucalyptus*) and trees of high ecological value (*Acacia*) in an intercropped system (Laclau et al. 2008; Pereira et al. 2017; Rachid et al. 2015) (Fig. 9.3) and improve the ecosystem services promoted by the plant-soil microbiome.

Recent studies have shown the N_2-fixing potential around 90–120 kg ha^{-1} $year^{-1}$, as well as the direct transfer of N of the roots of *A. mangium* to the *E. grandis* roots (Bouillet et al. 2008; Paula et al. 2015, 2018). In this sense, Eucalyptus plants would provide financial benefits and those of *A. mangium*, immeasurable ecological gains (Fig. 9.2). In this case, the availability of N can occur for *Eucalyptus* also after senescence of *Acacia* plant tissues (litter, fine roots, and nodules), root exudation, as well as cell death of organisms of the soil microbiota, providing N through mineralization processes (May and Attiwill 2003; He et al. 2003; Chalk et al. 2014). However, in spite of the diverse benefits of this association, studies evaluating the interactions at the soil-plant-microbiome interface in this type of forest management remain poorly understood.

The increase of N in the soil promoted by *A. mangium* sometimes is able to promote a significant increase in *E. grandis* productivity, even in the absence of the application of mineral fertilizers (Laclau et al. 2008), although this is very dependent on the climatic and edaphic conditions. In a review published by Forrester et al. (2006), a meta-analysis of 18 studies showed that several trials with mixed cultures were significantly more productive than monocultures, with fewer cases showing the opposite. For example, 11 years after the implantation of a mixed *E. globulus* and *A. mearnsii* plantation, mixed stands between species were more productive than monocultures in terms of aerial biomass, volume of wood produced, and C allocation in soil, with higher N and P cycling rates in litter (Forrester et al. 2004).

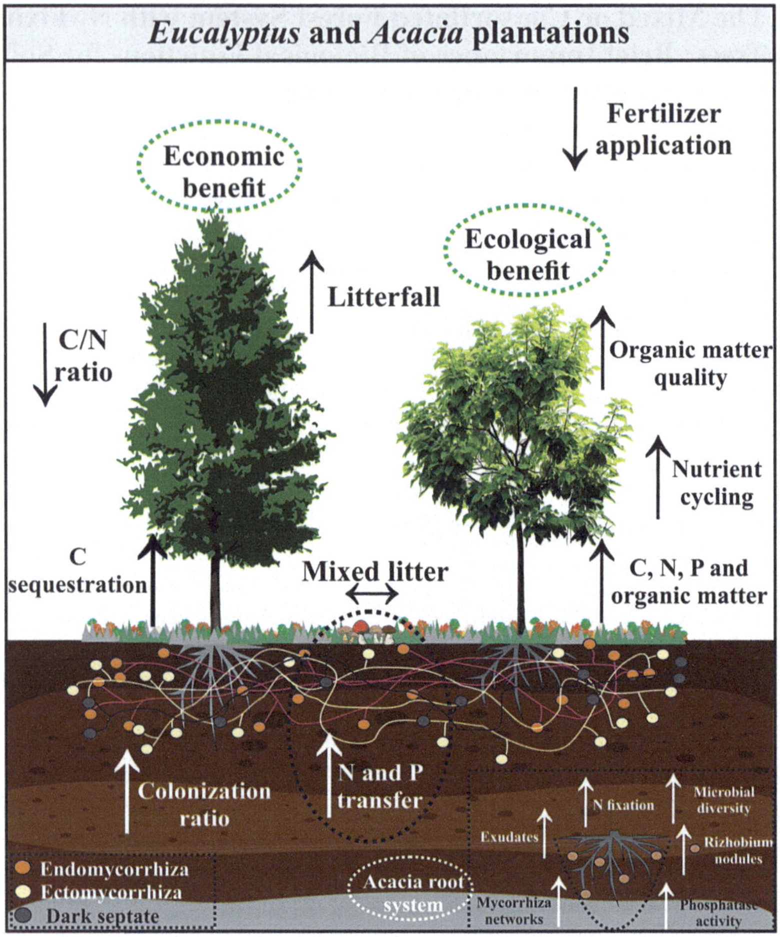

Fig. 9.3 Intercropped *Eucalyptus* and *Acacia* plantations. Upward arrows mean that soil health is better than in pure plantations, and the opposite is true for downward arrows. Belowground networks representing the mycorrhizal associations (endo- and ectomycorrhiza, and dark septate endophytes) and interactions between the two plants

9.5 Soil Microorganism Processes and Nutrient Cycling in Forest Plantations

Natural or planted forest sustainability shows a great dependence on geochemical, biochemical, and biogeochemical cycling. By definition, geochemical cycling is characterized by the inputs and outputs of mineral elements between the ecosystem and the environment. Biochemical cycling refers to the translocation of nutrients inside the plant, such as the process of nutrient translocation. Finally, biogeochemical cycling involves the processes of nutrient transfer between the soil and plant systems, with nutrient uptake by plants until their return to the soil via mineralization and decomposition processes or root exudation (Switzer and Nelson 1972). Thus, nutrient cycling is related with abiotic and biotic factors to maintain the ecosystem equilibrium. In general, nutrient inputs in ecosystems come from the air, rainfall, weathering process, biological N fixation (BNF), organic matter mineralization or decomposition, and throughfall. On the other hand, outputs are represented by erosion, runoff, volatilization, leaching, and nutrient removal during the forest harvest (Lavelle et al. 2005).

Commercial forests, as eucalyptus, are established commonly in poor soils and depend on nutrient cycling for their main sustainability. Appropriate management can promote the nutrient cycling, with ecological and economic benefits (Forrester et al. 2005b; Laclau et al. 2008). Thus, the ecological intensification promoted by mixed plantations favors processes of nutrient cycling and increases plant biomass and environmental sustainability (Forrester et al. 2005a). Leguminous trees intercropped with a nonleguminous tree (e.g., Eucalyptus) confer advantages to important microbial processes related to biogeochemical and geochemical cycling of C and nutrients, mainly N and P (Forrester et al. 2005b; Bini et al. 2013a, 2018; Baldrian 2017; Pereira et al. 2018b). In mixed plantations the microorganisms are the protagonists in nutrient cycling, based on three processes: BNF, mycorrhizal colonization, and decomposition or mineralization of organic matter (Ward and Jensen 2014; Bini et al. 2013a; Liang et al. 2017; Pereira et al. 2019).

Nitrogen-fixing bacteria present in leguminous trees promote BNF. These bacteria are of great value to promote N cycling in mixed plantations with nonlegume tree species. The fixed atmospheric N is first immobilized within the leguminous trees, and afterwards it is translocated and only then it becomes available for other trees (Parrotta et al. 1996; Khanna 1997; Forrester et al. 2006). The fixed nitrogen can be shared with nonlegume trees (a) via root exudation of the legume species; (b) by transfer of N through hyphal networks of mycorrhizal fungi that connect leguminous and nonleguminous plants; and (c) through the decomposition and mineralization of plant tissues of the legume species (Frey and Schüpp 1993; He et al. 2003; Forrester et al. 2006). This process is related to increases in N cycling in mixed stands. There is a high potential of BNF in mixed plantations, with approximately 20 g N m^{-2} $year^{-2}$, where more than 90% of N is derived from this process (Binkley 1992; Nygren et al. 2012). In mixed plantations of *A. mangium* and *E. urophylla* × *grandis*, Tchichelle et al. (2017b) found an amount of biologically fixed N four times higher than the total amount of commercial nitrogen fertilizer application at the beginning of the rotation. According to this author, 16% of the N present in the eucalyptus comes from the BNF promoted by Acacia. Moreover, N cycling tends to increase because there is a stimulation of the incorporation of N derived from the soil into the trees.

N and P cycling seems to be favored in mixed plantations. Of great biological importance, the available P needs special attention in tropical soils. In these soils there is inorganic P fixation onto iron or aluminum oxides (Hinsinger 2001), which makes planting of leguminous trees critical, since they demand great amounts of P to sustain BNF processes (Hinsinger 2001; Inagaki et al. 2011). This is the reason why legumes require more P than nonleguminous plants, such as eucalyptus (Binkley 1992; Koutika et al. 2016). A strategy for P acquisition in poor soils for legumes and eucalyptus is their great capacity of association with arbuscular mycorrhizal fungi (AMF) and ectomycorrhizae (Pagano and Scotti 2008; Mendes-Filho et al. 2009; Jimu et al. 2017; Bini et al. 2018). In general, these fungi can access P sources and other nutrients in soil through their hyphal network, even when the P ions are located further away from the plant roots, which cannot get access to them (He et al. 2003). In this sense, P cycling in mixed plantations seems to be dominated by the high mycorrhizal capacity of the leguminous species involved, which positively influences a higher colonization in the eucalyptus tree (Khanna 1997; Aggangan et al. 2010; Bini et al. 2018).

Pereira et al. (2018a) found colonization of eucalyptus roots by AMF at 0–50 cm depth, a fact stimulated by the presence of *A. mangium*. According to Bini et al. (2018), two strategies are important for P cycling in mixed plantations: mycorrhization by AMF and high activities of the phosphatase enzymes. *E. grandis* when intercropped with *A. mangium* has higher activity of acid and alkaline phosphatases than in pure plantations. However, P cycling is probably extremely fast, since it was not possible to detect significant differences between the P available in the soil in pure and mixed plantations, although there was a higher concentration of P in the trees (Bini et al. 2018). Thus, it is possible that other soil microorganisms also are active in improving fast P cycling. In addition, nutrient cycling is maximized by fungal networks that connect one plant to another, transferring nutrients between them without passing through the soil, a phenomenon that can occur between different plant species (Simard et al. 2003; Bini et al. 2018).

Among the three processes mentioned above, the process of decomposition and mineralization of organic matter is probably the main pathway for nutrient cycling in ecosystems and the most responsible for forest sustainability (Rahman et al. 2013). Soil organic matter results, largely, from the decomposition of animal and plant residues deposited on and under the soil. It is the main source of C, nutrients, and energy for soil microorganisms and plants (Brady and Weil 2009).

In tropical and subtropical soils, this is more evident, since it has a relationship with physical, chemical, and biological attributes of the soil, which makes the maintenance and management of organic matter fundamental for the productive capacity of forest soils in the long term (Switzer and Nelson 1972). Soil organic matter is produced during the fixation of C by photosynthesis, which generates organic compounds that can be comminuted by root exudates or deposits of fragments of senescent plants on the soil, called litter (Brady and Weil 2009; Rahman et al. 2013). For this reason, it is fundamental to create adequate edaphic conditions by stimulating biological processes as the degradation and mineralization of organic matter. Litter is first fragmented by soil mesofauna and, subsequently, by heterotrophic fungi and bacteria, which contribute to the formation of soil organic matter or humus, represented by stable or labile fractions, such as microbial biomass (Lavelle et al. 2006; Rahman et al. 2013).

Microbial biomass is the living part of the organic matter of the soil, being a source and sink of nutrients and, therefore, considered the organic matter that presents a fast cycling (Kaschuk et al. 2010). Bini et al. (2013a), in pure and mixed plantations of *A. mangium* and *E. grandis*, detected that the microbial biomass served as a sink of C and nutrients until 14 months of planting and after 20 months became the source of these elements. Litter microbial biomass can represent an important pool of C and nutrients, because it shows a greater capacity to be a nutrient sink than the microbial biomass of the soil (Gama-Rodrigues et al. 2011; Bini et al. 2013b).

Bini et al. (2013b) showed that the litter microbial biomass drained approximately 94% more N than the soil microbial biomass in pure and mixed plantations of *A. mangium* and *E. grandis*. Litter enters the decomposition and mineralization process by the action of several microorganisms by means of exuding many specific enzymes (phosphatases, cellulases, ligninases, ureases, etc.). Thus, part of the C is recycled into the atmosphere as CO_2, while N can be mineralized as NH^{4+} and then converted to NO^{3-}, while other elements such as P and S and several micronutrients can be transformed into mineral forms that are absorbed by plants. Only a small portion of this C is stabilized in the form of humus, which is still a source of C and nutrients, but whose mineralization rate is lower (Rahman et al. 2013).

It is important to highlight the fact that litter is the main source of soil organic matter, being a major component of the biogeochemical cycling process of nutrients in forest ecosystems (Rahman et al. 2013). Monocultures produce nondiversified litter, originated by only one type of plant residue. In contrast, mixed plantations produce more heterogeneous organic material (Binkley et al. 1992; Richards et al. 2010) with higher quality especially in mixes with legumes. Litter is the main vehicle to transfer C, N, P, and Ca from the trees to the soil; other elements as K are returned mainly through throughfall, and for Mg, it is variable for different forests (Cole and Rapp 1980; Bini et al. 2013b; Santos et al. 2017). However, there are variations in the rate of decomposition and mineralization of organic wastes. High levels of lignin, polyphenols, cellulose, and high C/N and C/P ratios characterize the recalcitrant residues, which make it difficult to recycle nutrients (Bini et al. 2013b; Rahman et al. 2013). Furthermore, the quantity and quality of the litter also depend mainly on the tree species and the soil attributes which govern nutrient availability.

In general, after closing of the canopy, eucalyptus produces a litter that is relatively poor in nutrients, due to high C/N and C/P ratios, with high percentages of cellulose (Gama-Rodrigues and Barros 2002; Paul et al. 2004; Bini et al. 2013b; Pereira et al. 2018b), contributing little to the replacement of soil nutrients. On the other hand, leguminous trees provide greater incorporation of organic matter and nutrients to the soil, due to their higher leaf quality and lower C/N and C/P ratios, favoring decomposition processes (Forrester et al. 2006; Richards et al. 2010; Bini et al. 2013a; Pereira et al. 2018a). However, although N-fixing legumes favor the increase of N, it is important to understand that the higher N and P contents do not always result in faster degradation of *A. mangium* residues, in comparison with *E. grandis* (Bini et al. 2013b). High lignin concentrations in the litter can reduce the decomposition rate of legumes (Wedderburn and Carter 1999; Prescott 2010; Bini et al. 2013b; Rahman et al. 2013).

According to Bini et al. (2013a) and Bachega et al. (2016), litter decomposition of *A. mangium* was slower than that of *E. grandis*, even though containing higher N and P concentrations. In mixed plantations, however, there are improvements in litter quality, since the lignin concentration of the mixture decreases (Bini et al. 2013b; Santos et al. 2017). Santos et al. (2017) reported that the deposition of N and K via litter was higher in stands of mixed species of Acacia and Eucalyptus than in Eucalyptus monocultures. In addition, P, Ca, and Mg depositions were even higher in mixed plantations than in Acacia monoculture. Tchichelle et al. (2017) found that soil N mineralization was higher in Acacia monocultures and mixed plantations, being 82% and 52% higher, respectively, than in *E. grandis* monoculture. These results suggest faster nutrient cycling in the mixture due to microbial decomposition processes (Pereira et al. 2019). Similar data were reported when higher N and P contents were observed in litter at mixed plantation (Li et al. 2001; Forrester et al. 2005b; Voigtlaender et al. 2012 Bini et al. 2013a, and other authors). With the increase of plant age the process of N translocation in leaves decreases and soil N contents increase (Richards et al. 2010; Santos et al. 2017).

In mixed stands mycorrhizae and BNF are important microbial processes for N and P inputs. The absorption of nutrients via roots or mycorrhizae increases plant biomass, and such nutrients are translocated and later reabsorbed from senescent tissues. Finally, after dropping back to the soil the plant residues are equivalent to litter deposition; this is when the initial decomposition and mineralization process is renewed, providing nutrients in the inorganic form to replenish the soil and to be reabsorbed by the plant. Thus, sustainability and nutrient cycling depend on microbial action. Microorganisms are involved in geochemical cycling and biogeochemistry, being the actors of decomposition and mineralization of organic matter, while BNF and mycorrhization complement nutrient cycling. Furthermore, in mixed plantations (and perhaps in most forestry environments) litter represents the main compartment to generate new mineral nutrients for plants. In any plan for forest implementation, this must be considered seriously since long-term nutrient losses can be economically and environmentally unfavorable or even calamitous for silviculture.

9.6 Final Remarks and Future Perspectives

Research-related bioindicators of soil quality with pure and mixed forests are only just a beginning. The achievement of sustainable Eucalyptus forest systems that meet the wood production demands and maintain an intimate relationship with biodiversity is an important challenge of the twenty-first century. Although our ability to describe major bioindicators in this type of forest remains incomplete, we have already created an integrated view of very important processes mediated by microorganisms and their diversity, mainly on nutrient cycling. We know very little about how to exploit the multifunctionality and multicomplexity of natural ecosystems and apply them to increase soil health, yield, and sustainability of forest ecosystems.

Understanding the dynamics of the bioindicator function is a complex but very important challenge because this relationship is one of intense mutual cooperation. There is a clear need for intensive studies in the setting up of experimental trials that consider multiple regions and to make relationships between them, even if these studies are initially descriptive. Focusing on the Eucalyptus and Acacia bioindicators of soil quality at the geographic scale will undoubtedly represent an important and valuable future field of work in the sustainability of these forest plantations. We must know the bioindicator ecology and applications, so we can make appropriate predictions of how forest ecosystems will respond to management changes, environmental changes, and climatic events in the coming decades.

References

Adhikari K, Hartemink AE (2016) Linking soils to ecosystem services—a global review. Geoderma 262:101–111

Aggangan NS, Moon HK, Han SH (2010) Growth response of Acacia mangium Willd. seedlings to arbuscular mycorrhizal fungi and four isolates of the ectomycorrhizal fungus Pisolithus tinctorius (Pers.) Coker and Couch. New For 39(2):215–230

Augusto L, De Schrijver A, Vesterdal L, Smolander A, Prescott C, Ranger J (2015) Influences of evergreen gymnosperm and deciduous angiosperm tree species on the functioning of temperate and boreal forests. Biol Rev 90:444–466

Bachega LR, Bouillet JP, Piccolo MC, Saint-André L, Bouvet JM, Nouvellon Y, Gonçalves JLM, Robin A, Laclau JP (2016) Decomposition of Eucalyptus grandis and Acacia mangium leaves and fine roots in tropical conditions did not meet the home field advantage hypothesis. For Ecol Manag 359:33–43

Bahram M, Kohout P, Anslan S, Harend H, Abarenkov K, Tedersoo L (2016) Stochastic distribution of small soil eukaryotes resulting from high dispersal and drift in a local environment. ISME J 10:885–896

Baldrian P (2017) Forest microbiome: diversity, complexity and dynamics. FEMS Microbiol Rev 41:109–130

Barreto PAB (2008) Activity, carbon and nitrogen of microbial biomass in eucalypt plantations in an age sequence. Rev Bras Cienc Solo 32:611–619

Bender SF, Wagg C, van der Heijden MGA (2016) An underground revolution: biodiversity and soil ecological engineering for agricultural sustainability. Trends Ecol Evol 31:440–452

Bini D, Santos CA, Bouillet JPP, Gonçalves JLM, Cardoso EJBN (2013a) Eucalyptus grandis and Acacia mangium in monoculture and intercropped plantations: evolution of soil and litter microbial and chemical attributes during early stages of plant development. Appl Soil Ecol 63:57–66. https://doi.org/10.1016/j.apsoil.2012.09.012

Bini D, Figueiredo AF, da Silva MCP, Vasconcellos RLF, Cardoso EJBN (2013b) Microbial biomass and activity in litter during the initial development of pure and mixed plantations of Eucalyptus grandis and Acacia mangium. Rev Bras Cienc Solo 37(1):76–85

Bini D, Santos CA, Silva MCP, Bonfim JA, Cardoso EJBN (2018) Intercropping Acacia mangium stimulates AMF colonization and soil phosphatase activity in Eucalyptus grandis. Sci Agric 75:102–110

Binkley D (1992) Mixtures of nitrogen-fixing and non-nitrogen-fixing tree species. In: Cannell MGR, Malcom DC, Robertson PA (eds) The ecology of mixed-species stands of trees. Blackwell Scientific Publications, Oxford, pp 99–123

Bouillet JP, Laclau JP, Gonçalves JLM, Moreira MZMR, Trivelin PCO, Jourdan C, Silva EV, Piccolo MC, Tsai SM, Galiana A, Bouillet JP, Gonçalves JLM, Silva EV, Jourdan C, Cunha

MCS, Moreira MZMR, Saint-André L, Maquère V, Nouvellon Y, Ranger J, Gonçalves JLM, Silva EV, Jourdan C, Cunha MCS, Moreira MZMR, Saint-André L, Maquere V, Nouvellon Y, Ranger J (2008). Mixed-species plantations of Acacia mangium and Eucalyptus grandis in Brazil. For Ecol Manag 255:3905–3917. https://doi.org/10.1016/j.foreco.2007.10.049

Bünemann EK, Bongiorno G, Bai Z, Creamer RE, De Deyn G, de Goede R, Pulleman M (2018) Soil quality–a critical review. Soil Biol Biochem 120:105–125

Brady NC, Weil RR (2009) Elements of the Nature and Properties of Soils. 3rd Ed. Pearson Education, Upper Saddle River, NJ, USA

Cardoso EJBN, Nogueira LR, Vasconcellos F, Bini D, Yumi M, Miyauchi H, Alcantara C, Roger P, Alves L, Paula AM, Nakatani AS, Vasconcellos RLF, Bini D, Miyauchi MYH, Santos CA, Alves PRL, Paula AM, Nakatani AS, Pereira JM, Nogueira MA (2013) Soil health: looking for suitable indicators. What should be considered to assess the effects of use and management on soil health? Sci Agric 70:274–289

Chalk PM, Peoples MB, McNeill AM, Boddey RM, Unkovich MJ, Gardener MJ, Silva CF, Chen D (2014) Methodologies for estimating nitrogen transfer between legumes and companion species in agro-ecosystems: a review of 15N-enriched techniques. Soil Biol Biochem 73:10–21

Churchland C, Grayston SJ (2014) Specificity of plant-microbe interactions in the tree mycorrhizosphere biome and consequences for soil C cycling. Front Microbiol 5:1–20

Cole DW, Rapp M (1980) Elemental cycling in forested ecosystems. In: Dynamic properties of forest ecosystems. Cambridge University, Cambridge, pp 341–409

Coleman DC, Whitman WB (2005) Linking species richness, biodiversity and ecosystem function in soil systems. Pedobiologia 49:479–497

Crowther TW, Glick HB, Covey KR, Bettigole C, Maynard DS, Thomas SM, Smith JR, Hintler G, Duguid MC, Amatulli G, Tuanmu MN, Jetz W, Salas C, Stam C, Piotto D, Tavani R, Green S, Bruce G, Williams SJ, Wiser SK, Huber MO, Hengeveld GM, Nabuurs GJ, Tikhonova E, Borchardt P, Li CF, Powrie LW, Fischer M, Hemp A, Homeier J, Cho P, Vibrans AC, Umunay PM, Piao SL, Rowe CW, Ashton MS, Crane PR, Bradford MA (2015) Mapping tree density at a global scale. Nature 525:201–205

Doran JW, Parkin TB (1994) Defining and assessing soil quality. In: Doran JW, Coleman DC, Bezdicek DF, Stewart BA (eds) Defining soil quality for a sustainable environment. SSSA, Madison, pp 3–21

Doran JW, Parkin TB (1996) Quantitative indicators of soil quality: a minimum data set. In: Doran JW, Jones AJ (eds) Methods for assessing soil quality. SSSA, Madison, pp 25–37

Drobnik T, Greiner L, Keller A, Grêt-Regamey A (2018) Soil quality indicators—from soil functions to ecosystem services. Ecol Indic 94:151–169

Eisenhauer N, Bowker MA, Grace JB, Powell JR (2015) From patterns to causal understanding: structural equation modeling (SEM) in soil ecology. Pedobiologia 58:65–72

Fierer N, Jackson RB (2006) The diversity and biogeography of soil bacterial communities. Proc Natl Acad Sci U S A 103:626–631

Forrester DI, Bauhus J, Khanna PK (2004) Growth dynamics in a mixed-species plantation of Eucalyptus globulus and Acacia mearnsii. For Ecol Manag 193:81–95

Forrester DI, Bauhus J, Cowie AL (2005a) On the success and failure of mixed species tree plantations: lessons learned from a model system of Eucalyptus globulus and Acacia mearnsii. For Ecol Manag 209:147–155

Forrester DI, Bauhus J, Cowie AL (2005b) Nutrient cycling in a mixed-species plantation of Eucalyptus globulus and Acacia mearnsii. Can J For Res 35:2942–2950

Forrester DI, Bauhus J, Cowie AL, Vanclay JK (2006) Mixed-species plantations of Eucalyptus with nitrogen-fixing trees: a review. For Ecol Manag 233:211–230

Frey B, Schüpp H (1993) A role of vesicular-arbuscular (VA) mycorrhizal fungi in facilitating interplant nitrogen transfer. Soil Biol Biochem 25:651–658

Gama-rodrigues AC, Barros NF (2002) Ciclagem de nutrientes em floresta natural e em plantios de eucalipto e de dandá no sudeste da Bahia, Brasil. R Árvore 26:193–207

Gama-Rodrigues EF, Gama-Rodrigues AC (1999) Biomassa microbiana e ciclagem de nutrientes. In: Santos GA, Silva LS, Canellas LP, Camargo FAO (eds) Fundamentos da matéria orgânica do solo: Ecossistemas tropicais e subtropicais. Gênesis, Porto Alegre, pp 159–1704

Gama-Rodrigues EF, Barros NF, Viana AP, Santos GA (2008) Microbial biomass and activity in soil and forest litter of eucalyptus plantations and native vegetation in Southeastern Brazil. Ver Bras Cienc Solo 32:1489–1499

Gama-Rodrigues EF, Gama-Rodrigues AC, Barros NF, Moço MKS (2011) The relationships between microbiological attributes and soil and litter quality in pure and mixed stands of native tree species in southeastern Bahia, Brazil. Can J Microbiol 895:887–895

Greiner L, Keller A, Grêt-Regamey A, Papritz A (2017) Soil function assessment: review of methods for quantifying the contributions of soils to ecosystem services. Land Use Policy 69:224–223

Hanson PJ, Edwards NT, Garten CT, Andrews JA (2000) Separating root and microbial contributions to soil respiration: a review of methods and observations. Biogeochemistry 48:115–146

He XH, Critchley C, Bledsoe C (2003) Nitrogen transfer within and between plants through common mycorrhizal networks (CMNs). Crit Rev Plant Sci 22:531–567

Hinsinger P (2001) Bioavailability of soil inorganic P in the rhizosphere as affected by root-induced chemical changes: a review. Plant Soil 237:173–195

Högberg P, Nordgren A, Buchmann N, Taylor AFS, Ekblad A, Högberg MN, Nyberg G, Ottosson-Löfvenius M, Read DJ (2001) Large-scale forest girdling shows that current photosynthesis drives soil respiration. Nature 411:789–792

Huang LF, Song LX, Xia XJ, Mao WH, Shi K, Zhou YH, Yu JQ (2013) Plant-soil feedbacks and soil sickness: from mechanisms to application in agriculture. J Chem Ecol 39:232–242

Inagaki M, Kamo K, Miyamoto K, Titin J, Jamalung L, Lapongan J, Miura S (2011) Nitrogen and phosphorus retranslocation and N:P ratios of litterfall in three tropical plantations: luxurious N and efficient P use by Acacia mangium. Plant Soil 341:295–307

Jenkinson DS (1981) Microbial biomass in soil: measurement and turnover. Soil Biochem 5:415–471

Jimu L, Kemler M, Mujuru L, Mwenje E (2017) Illumina DNA metabarcoding of Eucalyptus plantation soil reveals the presence of mycorrhizal and pathogenic fungi. Forestry Int J Forest Res 91(2):238–245

Kaschuk G, Alberton O, Hungria M (2010) Three decades of soil microbial biomass studies in Brazilian ecosystems: lessons learned about soil quality and indications for improving sustainability. Soil Biol Biochem 42(1):1–13

Khanna PK (1997) Comparison of growth and nutrition of young monocultures and mixed stands of Eucalyptus globulus and Acacia mearnsii. For Ecol Manag 94:105–113

Kirilenko AP, Sedjo RA (2007) Climate change impacts on forestry. Proc Natl Acad Sci U S A 104:19697–19702

Koutika LS, Mareschal L, Epron D (2016) Soil P availability under Eucalypt and acacia on Ferralic Arenosols, republic of the Congo. Geoderma 7(2):153–158

Laclau JP, Bouillet JP, Gonçalves JLM, Silva EV, Jourdan C, Cunha MCS, Moreira MR, Saint-André L, Maquere V, Nouvellon Y, Ranger J (2008) Mixed-species plantations of *Acacia mangium* and *Eucalyptus grandis* in Brazil. For Ecol Manag 255:3905–3917

Laclau JP, Ranger J, Gonçalves JLM, Maquère V, Krusche AV, M'Bou AT, Nouvellon Y, Saint-André L, Bouillet JP, Piccolo MC, Deleporte P (2010) Biogeochemical cycles of nutrients in tropical Eucalyptus plantations. Main features shown by intensive monitoring in Congo and Brazil. For Ecol Manag 259:1771–1785

Lal R (2014) Soil conservation and ecosystem services. Int Soil Water Conserv Res 2:36–47

Lal R (2015) Restoring soil quality to mitigate soil degradation. Sustainability 7:5875–5895

Lauber CL, Strickland MS, Bradford MA, Fierer N (2008) The influence of soil properties on the structure of bacterial and fungal communities across land-use types. Soil Biol Biochem 40:2407–2415

Lavelle P, Dugdale R, Scholes R, Berhe AA, Carpenter E, Codispoti L, Izac AM, Lemoalle J, Luizão F, Scholes M, Tréguer P, Ward B (2005) Nutrient cycling. In: Hassan RM, Scholes R, Neville A (eds) Millennium ecosystem assessment. Island Press, Washington, p 331

Lavelle P, Decaëns T, Aubert M, Barot S, Blouin M, Bureau F, Margerie P, Mora P, Rossi JP (2006) Soil invertebrates and ecosystem services. Eur J Soil Biol 42:S3–S15

Li ZA, Peng SL, Rae DJ, Zhou GY (2001) Litter decomposition and nitrogen mineralization of soils in subtropical plantation forests of southern China, with special attention to comparisons between legumes and non-legumes. Plant Soil 229:105–116

Liang C, Schimel JP, Jastrow JD (2017) The importance of anabolism in microbial control over soil carbon storage. Nat Microbiol 2:1–6

Mariotte P, Mehrabi Z, Bezemer TM, De Deyn GB, Kulmatiski A, Drigo B, Veen GF, van der Heijden MGA, Kardol P (2017) Plant-soil feedback: bridging natural and agricultural sciences. Trends Ecol Evol 33:129–142

May BM, Attiwill PM (2003) Nitrogen fixation by *Acacia dealbata* and changes in soil properties 5 years after mechanical disturbance or slash-burning following timber harvest. For Ecol Manag 181(3):339–355

Mendes-Filho PF, Vasconcellos RLF, Paula AM, Cardoso EJBN (2009) Evaluating the potential of forest species under "microbial management" for the restoration of degraded mining areas. Water Air Soil Pollut 208:79–89

Mercês E, Soares B, Silva IR, Barros NF, Teixeira RS (2016) Soil organic matter fractions under second-rotation Eucalyptus plantations in eastern Rio Grande do Sul. Rev Árvore 41(1). https://doi.org/10.1590/1806-90882017000100007

Motavalli PP, Palm CA, Parton WJ, Elliott ET, Frey SD (1995) Soil pH and organic C dynamics in tropical forest soils: Evidence from laboratory and simulation studies. Soil Biol Biochem 27:1589–1599

Navarrete AA, Tsai SM, Mendes LW, Faust K, Hollander M, Cassman NA, Raes J, Veen JA, Kuramae EE (2015) Soil microbiome responses to the short-term effects of Amazonian deforestation. Mol Ecol 24:2433–2448

Nygren P, Fernández M, Harmand JM, Leblanc H (2012) Symbiotic dinitrogen fixation by trees: an underestimated resource in agroforestry systems? Nutr Cycl Agroecosyst 94:123–160

Oliveira PHG, Gama-Rodrigues AC, Gama-Rodrigues EF, Sales MVS (2018) Litter and soil-related variation in functional group abundances in cacao agroforests using structural equation modeling. Ecol Indic 84:254–262

Pagano MC, Scotti MR (2008) Arbuscular and ectomycorrhizal colonization of two Eucalyptus species in semiarid Brazil. Mycoscience 49:379–384

Parrotta JA, Baker DD, Fried M (1996) Changes in dinitrogen fixation in maturing stands of Casuarina equisetifolia and Leucaena leucocephala. Can J For Res 26:1684–1691

Paul K, Polglase P, Bauhus J, Raison J, Khanna P (2004) Modeling change in litter and soil carbon following afforestation or reforestation: calibration of the FULLCAM 'BETA' model. National Carbon Accounting System Technical Report No. 40, Canberra, Australian Greenhouse Office

Paula RR, Bouillet JP, Trivelin PCO, Zeller B, Gonçalves JLM, Nouvellon Y, Bouvet JM, Plassard C, Laclau JP (2015) Evidence of short-term belowground transfer of nitrogen from Acacia mangium to Eucalyptus grandis trees in a tropical planted forest. Soil Biol Biochem 91:99–108

Paula RR, Bouillet JP, Gonçalves JLM, Trivelin PCO, Balieiro FC, Nouvellon Y, Oliveira JC, Júnior JCD, Bordron B, Laclau JP (2018) Nitrogen fixation rate of *Acacia mangium* Wild at mid rotation in Brazil is higher in mixed plantations with *Eucalyptus grandis* Hill ex Maiden than in monocultures. Ann For Sci 75:14

Pereira APA, Andrade PAM, Bini D, Durrer A, Robin A, Bouillet JP, Andreote FD, Cardoso EJBN (2017) Shifts in the bacterial community composition along deep soil profiles in monospecific and mixed stands of Eucalyptus grandis and Acacia mangium. PLoS One 12:e0180371

Pereira APA, Zagatto MRG, Brandani CB, Mescolotti DL, Cotta SR, Gonçalves JLM, Cardoso EJBN (2018a) Acacia changes microbial indicators and increases C and N in soil organic fractions in intercropped Eucalyptus plantations. Front Microbiol 9:1–13. https://doi.org/10.3389/fmicb.2018.00655

Pereira APA, Santana MC, Bonfim JA, de Lourdes Mescolotti D, Cardoso EJBN (2018b) Digging deeper to study the distribution of mycorrhizal arbuscular fungi along the soil profile in pure and mixed *Eucalyptus grandis* and *Acacia mangium* plantations. Appl Soil Ecol 128:1–11

Pereira APA, Durrer A, Gumiere T, Gonçalves JLM, Robin A, Bouillet JP, Wang J, Verma JP, Singh BK, Cardoso EJBN (2019) Mixed Eucalyptus plantations induce changes in microbial communities and increase biological functions in the soil and litter layers. For Ecol Manag 433:332–342

Prescott CE (2010) Litter decomposition: what controls it and how can we alter it to sequester more carbon in forest soils? Biogeochemistry 101:133–149

Prescott CE, Grayston SJ (2013) Tree species influence on microbial communities in litter and soil: current knowledge and research needs. For Ecol Manag 309:19–27

Rachid CTCC, Balieiro FC, Peixoto RS, Pinheiro YAS, Piccolo MC, Chaer GM, Rosado AS (2013) Mixed plantations can promote microbial integration and soil nitrate increases with changes in the N cycling genes. Soil Biol Biochem 66:146–153

Rachid CTCC, Balieiro FC, Fonseca ES, Peixoto RS, Chaer GM, Tiedje JM, Rosado AS (2015) Intercropped silviculture systems, a key to achieving soil fungal community management in eucalyptus plantations. PLoS One 10:1–13

Rahman MM, Tsukamoto J, Tokumoto Y, Shuvo MAR (2013) The role of quantitative traits of leaf litter on decomposition and nutrient cycling of the forest ecosystems. J For Environ Sci 29(1):38–48

Richards AE, Forrester DI, Bauhus J, Scherer-Lorenzen M (2010) The influence of mixed tree plantations on the nutrition of individual species: a review. Tree Physiol 30(9):1192–1208

Rousk J, Bååth E, Brookes PC, Lauber CL, Lozupone C, Caporaso JG, Knight R, Fierer N (2010) Soil bacterial and fungal communities across a pH gradient in an arable soil. ISME J 4:1340–1351

Santos FM, Chaer GM, Diniz AR, Balieiro FC (2017) Nutrient cycling over five years of mixed-species plantations of Eucalyptus and Acacia on a sandy tropical soil. For Ecol Manag 384:110–121

Santos FM, Balieiro FC, Fontes MA, Chaer GM (2018) Understanding the enhanced litter decomposition of mixed-species plantations of Eucalyptus and Acacia mangium. Plant Soil 423:141–155

Simard WS, Jones MD, Durall DM (2003) Carbon and nutrient fluxes within and between mycorrhizal plants. In: Van der Heijden MGA, Sanders IR (eds) Mycorrhizal ecology. Springer, Berlin, pp 34–74

Singh BK (2018) Soil carbon storage: modulators, mechanisms and modeling, 1st edn. Academic Press, London, p 340

Smith JL, Paul EA, Bollag JM, Stotzky G (1990) The significance of soil microbial biomass estimations. Soil Biochemistry 6:357–396

Šnajdr J, Dobiášová P, Urbanová M, Petránková M, Cajthaml T, Frouz J, Baldrian P (2013) Dominant trees affect microbial community composition and activity in post-mining afforested soils. Soil Biol Biochem 56:105–115

Snowdon P, Ryan P, Raison J (2005) National carbon accounting system technical report no. 45 Review of C:N ratios in vegetation, litter and soil under Australian native forests and plantations, p 72

Steidinger BS, Crowther TW, Liang J, Van Nuland ME, Werner GD, Reich PB, Nabuurs G, de-Miguel S, Zhou M, Picard N, Herault B, Zhao X, Zhang C, Routh D, Peay KG, GFBI consortium (2019) Climatic controls of decomposition drive the global biogeography of forest-tree symbioses. Nature 569:404–408

Štursová M, Bárta J, Šantrůčková H, Baldrian P (2016) Small-scale spatial heterogeneity of ecosystem properties, microbial community composition and microbial activities in a temperate mountain forest soil. FEMS Microbiol Ecol 92(12):fiw185

Switzer GL, Nelson LE (1972) Nutrient accumulation and cycling in Loblolly Pine (*Pinus taeda*) plantation ecosystems: the first 20 years. SSSA 36:143–147

Taylor BN, Chazdon RL, Bachelot B, Menge DNL (2017) Nitrogen-fixing trees inhibit growth of regenerating Costa Rican rainforests. Proc Natl Acad Sci U S A 114(33):8817–8822

Tchichelle SV, Epron D, Mialoundama F, Koutika LS, Harmand JM, Bouillet JP, Mareschal L (2017a) Differences in nitrogen cycling and soil mineralization between a eucalypt plantation and a mixed Eucalypt and Acacia mangium plantation on a sandy tropical soil. South For J For Sci 79(1):1–8

Tchichelle SV, Mareschal L, Koutika LS, Epron D (2017b) Biomass production, nitrogen accumulation and symbiotic nitrogen fixation in a mixed-species plantation of eucalypt and acacia on a nutrient-poor tropical soil. For Ecol Manag 403:103–111

Tedersoo L, Bahram M, Cajthaml T, Põlme S, Hiiesalu I, Anslan S, Harend H, Buegger F, Pritsch K, Koricheva J, Abarenkov K (2016) Tree diversity and species identity effects on soil fungi, protists and animals are context dependent. ISME J 10:346–362

Trumbore S, Brando P, Hartmann H, Gauthier S, Bernier P, Kuuluvainen T, Shvidenko AZ, Schepaschenko DG (2015) Boreal forest health and global change. Science 349:819–822

Urbanová M, Šnajdr J, Baldrian P (2015) Composition of fungal and bacterial communities in forest litter and soil is largely determined by dominant trees. Soil Biol Biochem 84:53–64

Van Der Heijden MGA, Bardgett RD, Van Straalen NM (2008) The unseen majority: soil microbes as drivers of plant diversity and productivity in terrestrial ecosystems. Ecol Lett 11:296–310

Voigtlaender M, Laclau JP, de Gonçalves JLM, Piccolo MC, Moreira MZ, Nouvellon Y, Ranger J, Bouillet JP (2012) Introducing *Acacia mangium* trees in *Eucalyptus grandis* plantations: consequences for soil organic matter stocks and nitrogen mineralization. Plant Soil 352(1-2):99–111

Wang GZ, Li HG, Christie P, Zhang FS, Zhang JL, Bever JD (2017) Plant-soil feedback contributes to intercropping overyielding by reducing the negative effect of take-all on wheat and compensating the growth of faba bean. Plant Soil 415:1–12

Ward BB, Jensen MM (2014) The microbial nitrogen cycle. Front Microbiol 5:2–3

Wardle DA, Bardgett RD, Klironomonas JN, Setala H, Van Der Putten WH, Wall DH (2004) Belowground biota ecological linkages between aboveground and belowground biota. Science 304:1629–1633

Wedderburn ME, Carter J (1999) Litter decomposition by four functional tree types for use in silvopastoral systems. Soil Biol Biochem 31:455–461

Zagatto MRG, Pereira AP, Souza AJ, Pereira RF, Baldesin LF, Pereira CM, Lopes RV, Cardoso EJBN (2019) Interactions between mesofauna, microbiological and chemical soil attributes in pure and intercropped *Eucalyptus grandis* and *Acacia mangium* plantations. For Ecol Manag 433:240–247

Zaia FC, Gama-Rodrigues AC, Gama-Rodrigues EF, Moço MKS, Machado RCR, Baligar VC (2012) Carbon, nitrogen, organic phosphorus, microbial biomass and N mineralization in soils under cacao agroforestry systems in Bahia, Brazil. Agrofor Syst 86:197–212

Chapter 10
Ecosystem Services in Eucalyptus Planted Forests and Mixed and Multifunctional Planted Forests

Fabiano de Carvalho Balieiro, Luiz Fernando Duarte de Moraes, Rachel Bardy Prado, Ciro José Ribeiro de Moura, Felipe Martini Santos, and Arthur Prudêncio de Araujo Pereira

10.1 A Brief State of the Art of Native and Planted Forests

Forests act as a source of food, fuel, and medicine for more than a billion people around the world (FAO 2018). In addition forests hold more than three-quarters of the world's terrestrial biodiversity, provide many products and services that contribute to the socioeconomic development, and are particularly important for hundreds of millions of people in rural areas (FAO 2018). According to this report (Global Forest Resource Assessment, FRA) the world's forest area decreased from 31.6 to 30.6% between 1990 and 2015, but at a slower pace in recent years (FAO 2018). On the other hand, planted forests have their area increased year after year, albeit at a slower pace in recent years. The average annual rate of increase between 1990 and 2000 was 3.6 million ha. The rate peaked at 5.9 million ha per year for the period 2000–2005 and slowed to 3.3 million ha per year between 2010 and 2015 (FAO 2018).

F. de Carvalho Balieiro (✉) · R. B. Prado
EMBRAPA Soils, Brazilian Agricultural Research Corporation, Rio de Janeiro, RJ, Brazil
e-mail: fabiano.balieiro@embrapa.br

L. F. D. de Moraes
EMBRAPA Agrobiology, Brazilian Agricultural Research Corporation, Seropédica, RJ, Brazil

C. J. R. de Moura
Federal University of Rio de Janeiro, Rio de Janeiro, RJ, Brazil

F. M. Santos
Federal Rural University of Rio de Janeiro, Seropédica, RJ, Brazil

A. P. de. Araujo Pereira
Department of Soil Science, University of São Paulo, "Luiz de Queiroz" College of Agriculture, Piracicaba, SP, Brazil

Soil Science Department (Pici Campus), Federal University of Ceará, Fortaleza, Ceará, Brazil

E. J. Bran Nogueira Cardoso et al. (eds.), *Mixed Plantations of Eucalyptus and Leguminous Trees*, https://doi.org/10.1007/978-3-030-32365-3_10

Land-use change leads to the destruction and fragmentation of forests, with negative impacts on the biogeochemical cycles of nutrients, increasing the risk of invasion of species and bringing significant losses of biodiversity (Myers et al. 2000; Rodrigues et al. 2009; Brockerhoff et al. 2013; Newbold et al. 2015). The interference of human activities that influence climate also pressures these natural ecosystems (Sala et al. 2000; Bonan 2008; Ballester et al. 2010; FAO 2018). A study by Mcneill and Mcneill (2003) observed that the loss of average abundance of the planet's original biodiversity was around 73% in 2002 and that this should reach a level of 84% by 2050. For Newbold et al. (2015), the loss of local species richness above 20% could substantially undermine the contribution of biodiversity to ecosystem function and services and, consequently, human well-being.

In Brazil, native forests are among the most biodiverse and threatened ecosystems on the planet (Myers et al. 2000). As part of the group of 24 biomes with an exceptional concentration of endemic species and with alarming habitat loss, the Atlantic Rainforest and Cerrado biomes, for example, were placed on the list of global *hotspots*. According to the authors, in only 1.4% of the terrestrial surface, these 24 biomes harbor more than 44% of all vascular plant species and 35% of all vertebrate species (mammals, birds, amphibians, and reptiles). Data from Ribeiro et al. (2009) indicate that the Atlantic Rainforest has only 11.4–16% of its original coverage. The Amazon, despite its huge area, has had its deforestation monitored by the National Institute for Space Research (INPE) since 1988. According to the records presented by the National Forestry Information System (2017), for the period 2016–2017, the increase in the deforested area was equal to 662,400 ha, contrary to the trend of decreasing deforestation in previous years (2002–2011). In 2016, land-use changes accounted for 51% of Brazilian total greenhouse gas emissions, equivalent to 1.17 billion Mg ($=10^6$ g) CO_2 equivalent (CO_2e) (Brandão Jr et al. 2018). Deforestation was the main source of emissions in relation to land-use changes, with the Amazon biome contributing 602 million Mg CO_2e (52%) of the sector's emissions in 2016, Cerrado 21% (~248 million Mg CO_2e), and the Atlantic Rainforest also 21%.

Despite the continuous degradation of native forest ecosystems, forest cover has increased in several countries as the result of regeneration in abandoned agricultural areas and forest plantations for commercial or restoration purposes (Chazdon et al. 2016). However, as pointed out by the author, based on the nature of the data and methodology used, he cannot infer about the return of biodiversity and ecosystem services lost with the conversion of forest to other land uses or degradation. Planted forests emerge as an alternative to biomass production since they occupy a reduced area (~2%) globally (FAO 2010). In Brazil alone, these forests occupy more than seven million ha (IBÁ 2017). Although heavily criticized on environmental aspects, these forests represent alternative sources of raw material, energy, and income for farmers. Thousands of direct and indirect jobs, investments in local infrastructure, and foreign exchange for the country are due to forestry business. In Brazil, about BRL 10 billion was generated for the communities around the business units of the sector (http://www.iba.org/statisticaldata).

About the relationship between ecosystem services (ES) and management of planted forests we can say that, on different scales, companies and research, extension, and education institutions have studied, monitored, and proposed management and land-use alternatives capable of reversing the processes of degradation of natural resources or even increasing the supply of goods and services provided by them (Brockerhoff et al. 2013; Ferraz et al. 2013; Gonçalves et al. 2013).

Water regulation, maintenance of soil fertility, regulation of climatic conditions, and erosion control are some of the ESs provided by planted forests when well managed. In most of the tropical regions, planted forests are highly productive monocultures due to the uniformity of plots and management (Gonçalves et al. 2013; Liu et al. 2018), but mixed (with poor diversity as reported in this book) and multi-diversified or multipurpose plantations can make the production of wood and fibers environmentally and socially fairer. Rural development, natural resource management, biodiversity conservation, and ecological restoration are concepts that should guide the activities of industrial or family enterprises involved in the forestry business (Lima et al. 2012a; Liu et al. 2018).

This chapter provides a conceptual basis and application of the ES approach to planted forests and forest plantations in three distinct Brazilian production environments: short-rotation intensive Eucalyptus plantations; low-diverse mixed plantations and high-diversity mixed plantations; or multipurpose plantations. These three systems differ in terms of area occupied, technological level adopted, and purpose.

The first system is represented by great paper, pulp, and coal companies in the forestry business, which occupy more than seven million ha and are present in at least 15 states. They play an important role in the economy and in the generation of jobs and income (IBÁ 2017). It is currently one of the most advanced agricultural activities in Brazil, thanks to investments in research in the areas of plant breeding, genetic improvement, and appropriated site management practice (Gonçalves et al. 2013).

The second management system is represented by less complex mixed plantations with low diversity of species. Within this category, the most studied plantations in Brazil are a mix of leguminous trees associated with diazotrophic bacteria (such as *Acacia mangium*) and non-N_2-fixing species (such as *Eucalyptus*), which are mentioned in this book, and the less diverse plantations for the purpose of restoring severely impacted degraded areas (Franco and Faria 1997; Parrotta and Knowles 1999; Chaer et al. 2011; Balieiro et al. 2018; Franco et al. 2018). In terms of area, these plantations are basically experimental plantations or small areas undergoing restoration in mining areas. The last commented category is mixed multipurpose plantations. This can be subdivided into two subclasses, for restoration purposes, with the possibility of using part of the plantation for commercial/extractive purposes (restricted-use areas) and agroforestry systems.

10.2 Ecosystem Services

10.2.1 Background and Conception

According to Hermann et al. (2011), the concept of ecosystem services dates back to the late 1960s and 1970s, highlighting the value of society over the roles of nature (King 1966; Helliwell 1969; Ehrlich and Ehrlich 1970; Dee et al. 1973; Ehrlich et al. 1977; Bormann and Likens 1979). In the same way, in the 1970s, 1980s, and 1990s, other scientists already drew the attention of society to the economic dependence on natural capital (Westman 1977; De Groot 1987; Daily 1997 and Costanza and Folke 1997), where natural capital is the natural stock of natural assets that generates a flow of goods or services that are useful or profitable to man over time (Costanza and Daly 1992). This concept has persisted up to now, with small variations in the scope of environmental economy. For Gómez-Baggethun and De Groot (2007), from an ecological perspective, *natural capital* cannot be conceived only as a stock or aggregation of natural elements, but as encompassing all ecosystem processes and interactions, which determine its integrity and ecological balance.

Daily (1997) was one of the first authors to approach the concept of ecosystem services as "the services provided by natural ecosystems and the species that compose them, in sustaining and fulfilling the conditions for the permanence of human life on Earth." The definition of Daily (1997) is similar to that of Millennium Ecosystem Assessment (MEA 2005), where ecosystem services are "the benefits that human beings derive from ecosystems," and it has been used in the literature in general, with small variations (Nicholson et al. 2009).

The ES approach has some advantages that can be highlighted: working on multiple scales, connection between science and politics, emphasizing social and economic aspects related to human well-being, aiming to promote the multifunctionality of ESs, and providing financial or nonfinancial compensation to those who work in favor of ecosystem services, among other aspects.

Furthermore, the literature on this theme has increased exponentially (Fisher et al. 2009), especially after the launch of the Millennium Ecosystem Assessment (MEA 2005), which proposed to evaluate ESs and the benefits derived directly and indirectly from ecosystems.

The MEA was requested by UN Secretary-General Kofi Annan in 2000 and was conducted between 2001 and 2005, involving more than 1300 scientists and 95 countries (TEEB 2010). It aimed to assess the consequences that changes in ecosystems bring to human well-being and the scientific basis for actions needed to improve the preservation and sustainable use of these ecosystems. This unique effort to systematize information on ecosystem services and its contribution to human well-being demonstrates that the international community recognizes the need for and urgency of adopting innovative measures to protect ecosystems, aligning preservation with economic development (Andrade and Romeiro 2019).

After 2005, several authors worked in this area; they produced data for the classification, evaluation, quantification, mapping, modeling, and valuation of ecosystem services, in order to subsidize decision-making in relation to ecosystems (Wilson and Carpenter 1999; Heal 2000; De Groot et al. 2002, 2010; MEA 2003, 2005; Turner et al. 2003; De Groot 2006; Fisher et al. 2009; Rounsevell et al. 2010; Dominati et al. 2010; Ferraz et al. 2013; among others).

There are several global initiatives to promote research, development, and public policies focused on the provision of ecosystem services, such as The Economics of Ecosystems and Biodiversity (TEEB), the Natural Capital Project, Intergovernmental Platform on Biodiversity and Ecosystem Services (IPBES), Ecosystem Services Partnership (ESP), Knowledge and Learning Mechanisms on Biodiversity and Ecosystem Services (EKLIPSE), Europe Ecosystem Research Network (Alter-Net), and Water Funds.

In Brazil, many studies have been and still are developed on biodiversity preservation and environmental conservation in the different biomes, due to the rich biodiversity and natural resources of the country and the processes of degradation due to different anthropogenic pressures. The number of publications and interest specifically on the ES theme are also increasing (see Ferreira et al. 2012; Brockerhoff et al. 2013; Ferraz et al. 2013, 2014; Prado et al. 2016; Periotto and Tundisi 2018).

10.2.2 Classification of Ecosystem Services

In terms of services provided by ecosystems, three categories are generally considered: regulation, provision (supply), and cultural (MEA 2003; Hein et al. 2006). In addition, another category of support services has been widely used in MEA (2003), but the latter has not been widely used in ecosystem assessment because of the double meaning or overlap with other service categories (Fisher and Turner 2008).

It should be mentioned that, based on the work on environmental accounting carried out by the European Environmental Agency (EEA), an international classification system for ESs has been developed since 2009, called the Common International Classification of Ecosystem Services (CICES), which is currently in its V5.1 version (Haines-Young and Potschin-Young 2018). This initiative, in direct contribution to the United Nations Statistics Division (UNSD), a review of the Environmental Economic Accounting System (SEEA), aimed to establish an internationally standardized ES classification system. The idea of establishing an international classification is from the need to standardize the description of the ESs so as to enable the establishment of methods of environmental accounting, mapping, and evaluation for ESs that may be replicable and comparable.

The CICES classification used as its starting point the typology suggested by the Millennium Ecosystem Assessment (MEA 2005), which, with the exception of the category of support services, considered the other three categories of services: provision, regulation, and cultural. The support services category was deliberately excluded because, as an intermediary service, relations between the ecosystem and environmental accounting in this case are not explicit. Thus, CICES, adopting a pragmatic view, chose to emphasize the final outputs of processes that effectively benefit and have direct and explicit value to people (Haines-Young and Potschin-Young 2018). However, as the authors themselves warn, the intermediary and support services should not be ignored or neglected. The classification presented below will be MEA (2005).

10.2.2.1 Support Services

For MEA (2005), support services are those required for the production of the other ESs. They differ from the basic categories insofar as their impacts on man are indirect or occur in the long run. Examples are primary production, atmospheric oxygen production, soil formation and retention, nutrient cycling, water cycling, and habitat provision. Forests are great natural assets that provide these services.

The cycles of several key nutrients for life support have been significantly altered by human activities over the past two centuries, with positive and negative consequences for other ecosystem services, as well as impacts on human well-being.

Forests are responsible for maintaining biodiversity. In Brazil, the ecosystem services were threatened by actions of deforestation and fires, associated with the dynamics of land use, agricultural and livestock expansion, and urban areas. According to the Ministry of the Environment, the loss of natural environments is estimated at between 15 and 18% in the Amazon biome; 50% in the Cerrado, Pampas, and Caatinga biomes; and 88% in the Atlantic Forest biome (Ferreira et al. 2012). Sparovek et al. (2010) estimated an environmental liability of 21–30 million hectares, which have to be restored in Brazil.

Natural ecosystems provide habitat and food requirements for a wide range of arthropod predators and parasitoids, insectivorous birds, and microbial pathogens that act as natural enemies of agricultural pests and thus provide biological control services (Tscharntke et al. 2005). An ecosystem service that has been greatly compromised by the suppression of forests is pollination by reducing habitats for birds and insects, compromising the ESs of regulation and provision.

10.2.2.2 Regulation Ecosystem Services

This relates to the regulatory characteristics of ecosystem processes, such as maintenance of air quality, climate regulation, erosion control, purification and regulation of water flow, self-purification of water (process of degradation of nutrients contained in water bodies due to sources of pollution, usually sewage), regulation of human diseases and pests in agriculture, pollination, and mitigation of natural damages. These services are derived almost exclusively from regulatory ecosystem purposes.

Unlike provisioning services, their assessment does not occur by their "level" of production or quantity available, but by the analysis of the ability of ecosystems to regulate certain services.

Forests play an essential role in regulating services, for it is through them that climate regulation, for example, is affected by deforestation and burning practices, drastically impacting climate change and its effects on the economy and the quality of life of society.

Forests also participate in water regulation and carbon sequestration. Water regulation is an ecosystem service that is highly related to the management of production systems, either through the direct use of water in irrigated systems or because of the changes they impose on the physical-biotic environment of the river basins that interfere with the water, hydrological, and climatic cycles. On the other hand, the deforestation caused for the implantation of agricultural systems can alter the regional precipitation regimes, through the changes in the evapotranspiration flows of clouds. The local climate can then become drier, not only impacting ecosystems but also compromising water security (Vergara and Scholz 2010).

Many studies estimated the potential of forests to sequester carbon, which is an ecosystem service regulator (Lal 2005; Jandl et al. 2007; Sedjo and Sohngen 2012). In times of climate change, they disagree on whether the carbon forest balance in future will be positive or negative (in photosynthesis or respiration) and heat up the backstage of research (Bonan 2008; Bellassen and Luyssaert 2014; Nottingham et al. 2015).

In this sense, planted forests, when well managed and with mixed plantations, can also contribute to the regulatory services mentioned here, as will be presented in the subsequent items.

10.2.2.3 Provision (Supply) Services

For MEA (2005), these services include products obtained directly from natural or seminatural ecosystems (agriculture), such as food and fiber; wood for fuel and other materials that serve as a source of energy; genetic resources; biochemical, medicinal, and pharmaceutical products; ornamental resources; and water.

Data from world food production illustrate the increase in the generation of provisioning services. According to MEA (2005), between 1961 and 2003 food production increased by more than 160%, with cereal production increasing 2.5 times, beef and sheep production increasing 40%, and production of pork and poultry meat increasing 60% and 100%, respectively.

Forests, in addition to the provision of water and food, are able to provide medicinal products, fiber, wood, and energy. Due to the richness of Brazilian biodiversity, native vegetation is a source of food resources in all Brazilian biomes. Many native plants are now domesticated and widely used in the country, such as palm heart, cassava, pepper, peanut, guaraná, pineapple, and cacao (Prado and Murrieta 2015), while others are more regional, such as pine nuts in the Araucaria Forests.

Products derived from plant extraction can be classified as timber and non-timber. In 2016, the value of non-timber products (BRL 1.9 billion) was 4.6% higher than in 2015 and 18% higher than in 2014 (SNIF 2017), with 86.5% (BRL 1.6 billion) corresponding to extractive activities in native forests. Non-timber products (as waxes, saponin, honey, or food products) generally are extracted by traditional populations and family farmers. Food products, such as açaí, native erva mate, and Brazil nut, generated in 2015, respectively, \$480, \$396, and \$107 million Brazilian Reals, while waxes (carnauba powder), oilseeds (babaçu nuts), and fibers (piassava) each generated more than \$195, \$107, and \$101 million Brazilian Reals in 2015 (SNIF 2016).

Brazil is the third largest exporter of forestry products (e.g., timber, pulp, paper, resins, tannins, gums), accounting for 3.64% of the total global market volume (FAO stat; data for 2016). Forest products generally rank fourth in the ranking of the value of national agribusiness exports, only behind soybeans, meat, and sugar-alcohol complex. Therefore, native and planted forests definitely contribute to this national scenario of ES provision.

10.2.2.4 Cultural Services

These include cultural diversity, as the very diversity of ecosystems influences the multiplicity of cultures, religious and spiritual values, generation of (formal and traditional) knowledge, and educational and aesthetic values, among others. These services are closely linked to human values and behaviors, as well as to social institutions and patterns, characteristics that make their perception different among groups of individuals, making it difficult to evaluate their provision (Andrade and Romeiro 2019).

Still, according to Andrade and Romeiro (2019), societies have developed an intimate interaction with the natural environment, which has shaped cultural diversity and human value systems. However, the transformation of natural ecosystems into cultivated landscapes with more homogeneous characteristics associated with economic and social changes, such as rapid urbanization, improvement and cost reduction of transportation conditions, and intensification of globalization, has weakened the links between ecosystems and cultural diversity/identity.

On the other hand, the use of ecosystems for recreation and tourism purposes has increased, mainly due to population growth and greater availability of time for leisure, mainly of the populations with greater purchasing power and greater access to infrastructure, which facilitate access to cultural services. Ecological tourism, for example, corresponds to one of the main sources of income for some countries that still have a large part of their ecosystems conserved.

Forests contribute effectively to cultural services, since they make up landscapes with a greater diversity of flora and fauna, which becomes an attraction for humans. They also contribute to water regulation and provision, as well as keeping water bodies such as rivers, lakes, waterfalls, and other pristin services, enabling recreation and tourism. In addition to these benefits, forests are associated with diverse believes religions, and spiritual aspects of traditional peoples such as indigenous people and others living in forests, factors that are related to cultural ecosystem services (ES). Kreye et al. (2017) present a discussion about forest-related, cultural ES.

It should also be highlighted that in order to show the ES provided by natural and planted forests, many studies have been carried out, based on different methodologies, aiming at their quantification and valuation. Masiero et al. (2019) present a manual with several methods for quantifying and valuating ESs from forests.

10.3 Ecosystem Services of Monoculture, Intensely Managed Planted Forests

Over the past 50 years, eucalypts went from being a risky investment in Brazilian silviculture to a global forestry success story, perhaps compared to the achievement of Henry Wickham, responsible for the expatriation of the rubber tree to Kew Gardens and then to its final scale in Malaysia.

Due to its geographical origin and anatomical characteristics that imply growth stresses that hinder drying, in addition to its wood being considered of moderate durability against rotting fungi and termites (Silva et al. 1995), *Eucalyptus* was viewed with distrust at the time of the introduction of the industrial plantations. But what is eucalypts for? This was a relevant question years ago, and it is with it that the book by Higa et al. (2000) entitled "Eucalyptus Plantation in Small Rural Properties" begins. The publication presents the genus *Eucalyptus* spp. for small properties, and argues:

> *Eucalyptus involves more than 600 species that are adapted to different climates and soils and can be used for different purposes. Eucalyptus can be planted as ornamental trees in parks and gardens; the leaves can be used in floral arrangements and for extracting oil and the flowers are used for honey production. The most common use is the use of wood such as firewood, poles, fences, rural buildings, production of saw wood, panel manufacturing and paper and pulp manufacturing.*

Eucalypt is the most planted forest crop in Brazil and is a raw material for industrial and domestic use. Currently, more than 91% of all wood produced for industrial purposes in Brazil comes from this essence, generating 510,000 jobs directly and 3.7 million indirectly, and contributing to the GDP with BRL 71 billion (IBÁ 2017).

Since its introduction in Brazil in 1909 by Edmundo Navarro de Andrade, in the state of São Paulo, *Eucalyptus* came to fulfill one of the four types of ecosystem services, the provisioning.

Its introduction and naturalization were a huge success and, in addition to supplying the Brazilian and world fiber market, eucalypt is also a source of energy and timber for Brazilian forestry companies, considerably reducing the pressure on the native forests, guaranteeing the conservation of the genetic patrimony associated with these forests.

It cannot be denied that in addition to provisioning, *Eucalyptus* provides other ecosystem services important to our society, regulating environmental conditions by carbon sequestration; providing cultural services such as spaces for recreation and a socioeconomic identity factor; and finally supporting services that influence the formation of soil and habitats, nutrient cycling, and oxygen production, among others.

Regarding support services, the forest sector currently protects and conserves almost six million hectares, in areas of permanent preservation (APPs, 30%), legal reserve areas (RL, 67%), and private reserve areas (RPPN, 3%), which contribute directly to the conservation of biodiversity, soil, and water (IBÁ 2017). According to a database of the main companies in the sector and organized in their 2017 report on the biomes of the Atlantic Forest and Cerrado, of the species threatened with extinction in the national territory, 38% of the mammals and 41% of the birds were found in areas belonging to Brazilian forest companies (IBÁ 2017). Around the world and in Brazil, planted forests represent an increasing proportion of the global forest area and partly compensate for the loss of natural forest in terms of forest area, habitat for biodiversity, and ecological function (Brockerhoff et al. 2013; Ferraz et al. 2013).

The performance of planted forests in terms of environmental benefits is highly dependent on their management plan, especially with respect to the hydrological regulation of micro-basins, since their growth and productivity are related to important social (hydrosolidarity and land-use conflicts) and environmental (climate change and biodiversity) issues (Lima et al. 2012b; Ferraz et al. 2013). Recently, Cassiano (2017) monitored the hydrological regime and water quality in three micro-basins (one micro-basin with age and species in mosaic management, two with conventional eucalypt management), demonstrating that the mosaic was more adequate in the regulation of the hydrological regime and water quality in relation to conventional management of eucalypt-planted forests. Ferraz et al. (2013) studied examples of forest management alternatives at macro- (theoretical thresholds for the management of evapotranspiration) and meso-scales (using data from a catchment experiment) that contribute to improve water conservation in forest landscape areas in Brazil. Their results suggest that for effective water conservation in *Eucalyptus* spp. plantations the scales of evaluation must be considered. At regional scale, the natural climatic constraints of water availability should drive the choice of more water-efficient species/varieties and forest management, while at the meso-scale the proportion of native forest in the landscape plays a crucial role in reduction and regulation of water use. In other words, mosaic management could stabilize flows from plantation areas. Figure 10.1 summarizes the theoretical potential of different forest landscapes to provide some hydric ecosystem services (adapted from Ferraz et al. 2013).

In micro-basins, generalized clear cutting can increase the concentration of nutrients and suspended sediments in the waters of the micro-basin, leading to losses of the natural nutritional capital of the soil, water quality, and storage capacity of the reservoirs and damages to productivity in medium and long terms (Lima et al. 2012b). The water quality indicators that were most affected by the clear cutting of eucalypt in monitored micro-basins are turbidity, color, and electrical conductivity (Câmara et al. 2000).

Water consumption in *Eucalyptus* plantations is not very different from the consumption of other species that have the same growth rate. That is, the greater the biomass production, the greater the water consumption to provide growth. However, there are differences within and between species in the efficiency of water use (biomass produced per unit of water consumed), due, for example, to physiological mechanisms that reduce transpiration as in cases of reduction in leaf area index in regions of greater water deficit (Fig. 10.2).

Management strategies considering selected genotypes and plant spacing can also increase the water security of commercial forests. Wood biomass at tighter spacing is generally higher (47–57 kg tree^{-1} at a planting density of 591 trees ha^{-1} vs. 13–24 kg tree^{-1} with 2949 trees ha^{-1}) but exhibited trees present lower leaf water potentials resulting in a trade-off between productivity and potential water stress (Hakamada et al. 2017). Of the genotypes tested, the *E. urograndis* clone presented the best performance from the lowest to highest planting density, while the hybrid *E. grandis* × *E. camaldulensis*, the worst. In other words, for both industry and family farmers they will be less vulnerable to climate change (water stress) understanding the interaction between planting density and genotypes. The authors expect that for lower water availability, regional potential water stress will be higher with

Fig. 10.1 Mosaic of eucalyptus forest plantations (on the upper plateau) and native forest in lower part of the landscape close to the stream (Internet photo)

total stand (stem wood) biomass productivity increase, and the opposite would be true in areas with higher water availability.

Another important factor to consider in the water balance is the fact that forests, despite being innate water consumers, also intercept rainfall and in the case of eucalypt, due to leaf characteristics such as leaf area, allow more water to reach the forest floor, when compared to the *Pinus* genus, for example (Paula Lima et al. 2013).

Companies are also striving to adopt more sustainable practices. Management techniques as leaving plant residues on the soil (ground or not), planting seedlings directly on the straw, and desiccating undergrowth rather than revolving the soil, all have led to minimized soil erosion. Likewise, rational use of pesticides, management of fertility and crop health, and planning and maintenance of roads and roadways within the areas also contributed to the mitigation of erosive processes, significant improvements in soil and water quality, and biodiversity conservation (Chaer and Tótola 2007; Brockerhoff et al. 2013; Gonçalves et al. 2013). All of these actions contribute to the environmental regularization of properties and forest certification by the Forest Stewardship Council (FSC) and the Program for the Endorsement of Forest Certification (PEFC), represented in Brazil by the National Forest Certification Program (Cerflor), with strict monitoring indicators and biodiversity management.

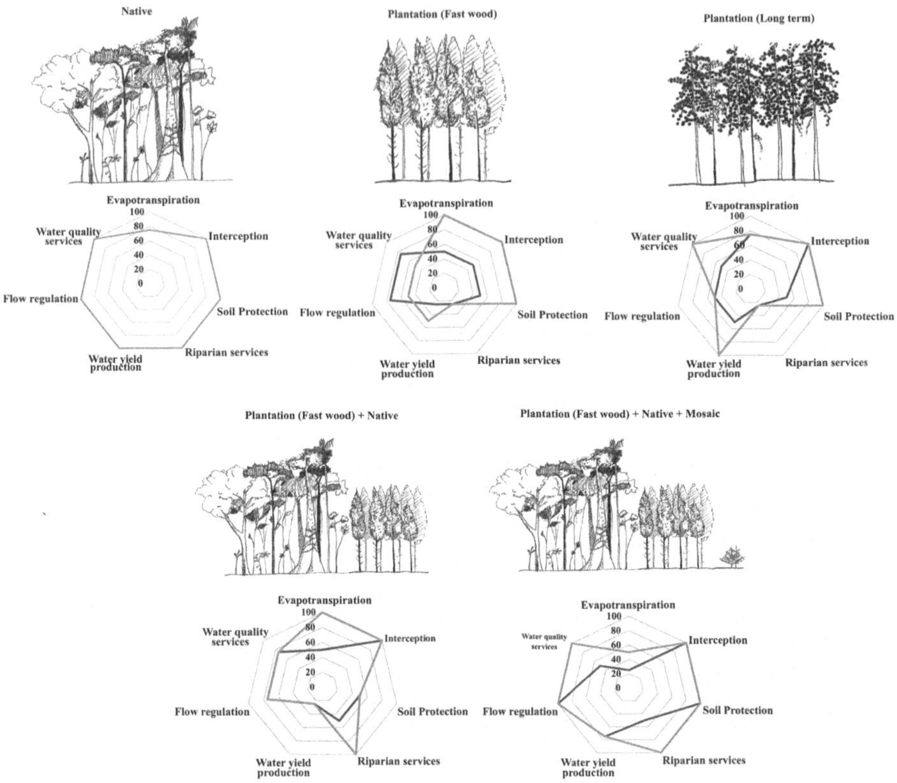

Fig. 10.2 Performance variation of expected ecosystem services linked to water conservation provided by different forest cover management systems (adapted from Ferraz et al. 2013; illustration by F.C. Balieiro)

Despite these advances, production of planted forests in detriment of other ecosystem services is advancing in some companies, but this vision must change, especially under the logic that increasing the area of planted forests will not compensate for loss of biodiversity and scenic or cultural services (for example) due to the loss of natural forest cover. To give an example, because it is extremely sensitive to weed competition, the control of weeds (Zen 1987; Tarouco et al. 2009) and invasive or pioneer species is difficult to give up given the sensitivity of *Eucalyptus* to resources in the initial stage of establishment. On the other hand, as detected by Stallings (1990) in the Atlantic Forest biome, more than 50% of the species of mammals and birds found in primary forest were found in eucalypt areas with understory vegetation and, on the contrary, none of them in plantations without this vegetation. In other words, the production logistics of the sector must be rethought, since the mixing of tree species (mixed plantations) can improve the performance of the Eucalyptus stand, with gains in several other ecosystem services.

An important aspect to be considered in relation to the environmental services of *Eucalyptus* as well as its impacts, is the scale at which species of this genus are planted. Another point concerns the species, since the choice of the appropriate species will influence the quality and quantity of the ecosystem service.

Regarding the ecosystem services of the soil associated with *Eucalyptus* plantations, it is worth mentioning that soil carbon stocks and organic matter increases have been comparable to natural ecosystems (Fialho and Zinn 2014). Although some changes in soil and climate behavior may be evident (Cook et al. 2016), forest ecosystems retain more of this soil asset than the Brazilian agricultural sector. According to the Brazilian Institute of Geography and Statistics, IBGE, agricultural activities cover an area of about 350 million hectares, of which approximately 172 million hectares are destined to pasture and with more than 50% of these areas in some state of degradation (Macedo et al. 2013). These degraded areas could be managed by the forest sector with many other environmental gains.

The management of *Eucalyptus* as well as any other silvicultural activity implies a permanent culture with cycles varying from 5 to 20 years of rotation, with fire being banned from these areas, unlike poorly managed pastures that sometimes burn in one cycle of fire-free cutting of *Eucalyptus*, thus avoiding emissions of greenhouse gases and soil degradation.

In summary, the planting of *Eucalyptus* may be the opportunity for the conversion of exclusively agricultural landscapes to agroforestry such as agrisilviculture, silvipastoral, and agrosilvopastoral systems. After all, eucalypt is recognized by farmers as a commodity, being adopted as a more familiar crop, as opposed to planting native species which, although more advantageous in providing ecosystem services, is not easily accepted in rural areas.

Thus, the adoption of *Eucalyptus* and appropriate management can promote the transition to natural forest restoration through the regeneration of pioneer and early native species in its understory, which may facilitate the succession of forests (Parrotta et al. 1997), and thus the provision of ecosystem services.

10.4 Ecosystem Services in Low-Diversity, Mixed-Planted Forests

Low-diversity (with up to five species), mixed-planted forests are common in programs for the restoration of severely degraded areas, such as mining areas (Franco and de Faria 1997; Parrotta and Knowles 1999; Singh et al. 2006; Franco et al. 2018), or in areas under environmental restoration where the availability of seedlings or seeds is low (Rodrigues et al. 2009). In Brazil and in other countries such as Australia, the USA, Congo, and Costa Rica, plantations that associate N_2-fixing trees and non-N_2-fixing tree species have gained importance in the last years. These mixed plantations have been studied under different experimental designs (Forrester et al. 2006; Kelty 2006), and most of them concentrate on the use of exotic species (Forrester et al. 2006; Liu et al. 2018; Marron and Epron 2019). Most of the results brought to this point are derived from the other chapters, especially from the experience with *Eucalyptus* and *Acacia mangium* in Brazil.

The provision of timber for energy or industrial purposes (paper and pulp) through mixed *Eucalyptus* and N_2-fixing species may reach higher levels than *Eucalyptus* monocultures (Binkley et al. 1992; Kaye et al. 2000; Forrester et al. 2013; Koutika et al. 2014; Santos et al. 2016). However, this fact cannot be generalized (Marron and Epron 2019). Based on data from experiments in five locations in Brazil and one in the Congo, the global production of timber from mixed plantations was higher in poorer and sandy soils and where the climate favored the development of the legume (*Acacia mangium*) (Santos et al. 2016). A very recent meta-analysis by Marron and Epron (2019), with 148 case studies around the world, showed that plant mixtures have a significantly positive global effect, with mixed-tree plantations being 18% more productive than the non-N_2-fixing monocultures, and this effect was significantly different from zero (null) under temperate conditions (24% more productive) but not under tropical conditions (12% more productive). They attributed these findings to nitrogen availability (generally less in temperate climate). However, as cited above, the marked success of the mixture is more evident in sites with low biomass production potential.

Comparing with the national average of eucalyptus productivity in monocultures (~20 Mg ha^{-1} $year^{-1}$) (Stape et al. 2010; Gonçalves et al. 2013), and the possibility of opening markets for other types of raw material, the rural producer and the industry would have more income and business options when handling mixed plantations. Considering that one half of the pasture area is in some degree of degradation (Oliveira et al. 2004), and that sandy soils cover around 10% of the Brazilian territory, these mixed plantations can provide additional gains in timber, and of several other ecosystem services for the property or countryside. The possibility of using timber or non-native species can also bring even greater benefits resulting from the conservation of native biodiversity.

One of the main concerns regarding successive rotations of tropical *Eucalyptus* plantations is the soil nitrogen (N) balance, which generally becomes negative with multiple rotations due to a combination of high N exports from timber harvests and low doses of N fertilizers that are typically applied (Corbeels et al. 2005; Laclau et al. 2010). Thus, N inputs are required to sustain satisfactory forest production, which is accomplished generally by nitrogen fertilizer applications (e.g., urea, ammonium sulfate, and ammonium nitrate). Plantation of *Eucalyptus* in consortium with N_2-fixing trees is a valuable and sustainable technique, because it can provide an increase in the N availability for *Eucalyptus* trees, obtain high timber productions even without the use of N fertilizers (Voigtlaender et al. 2012; Santos et al. 2016), with low greenhouse gas emissions, especially the nitrous oxide (N_2O), originating from N fertilizations (Silveira 2018). Tchichelle et al. (2017) showed that soil N mineralization was 82% higher under *Acacia mangium* monoculture and 53% higher under mixtures with *A. mangium* and *E. grandis* than in *E. grandis* monoculture, although differences in N stock and C:N ratio in the 0–25 cm soil layer between the two pure treatments were not significant. The higher N mineralization rate in pure legume tree plantations or in mixed plantations, when compared to *Eucalyptus* monocultures, has also been observed in other studies in the tropics (Bernhard-Reversat 1996; Parrotta 1999; Forrester et al. 2005).

In terms of N_2O emissions in mixed plantations established in the Southeast region, Santos et al. (in preparation) detected very low emissions of nitrous oxide from sandy soil (except in some periods of prolonged rainfall). This is due to the absence of mineral fertilization at the fifth year of the plantation and the fact that the soil in the experimental area has a low water retention capacity. However, when analyzing the accumulated emissions, these were higher in the *A. mangium* environment, in relation to *E. urograndis*, and with intermediate values for the mixed plantations (*E. urograndis* + *A. mangium*). In recently established tree plantations on medium-textured Oxisols (clay content ranging from 34 to 42%), in central-western Brazil, Silveira (2018) observed the same emission pattern, with low average flows in the dry period. This was explained by the low availability of inorganic nitrogen in the soil and due to the absence of microanaerobic sites within the microaggregates of the soil, and higher flows during the wet season. However, the application of N fertilizer (via urea) in the E + N treatment resulted in an increase in N_2O fluxes that began to be noticed a few days after application of N, coinciding with the highest rainfall event after application of N, when soil moisture content increased. It has been speculated that several years later, far from the initial phase of forest growth, when fertilization and soil-tillage operations cease and the N emissions are usually very low, the enrichment of N in the soil provided by the legume might raise the emissions of N_2O, in relation to *Eucalyptus* monocultures (Rachid et al. 2013). However, many more studies must be encouraged before this can be corroborated as a fact. In addition, we believe that the low nitrous oxide emission probably is compensated by higher C sequestration by the soil and no N fertilizer application in mixed plantations, but these balances have not been obtained yet.

The sequestration of atmospheric C by mixed-planted forests is another ecosystem service of these forests. This service is proportional to the growth rates of these plantations, which can vary with factors such as planting structure, composition, age, climatic condition, and soil type (Cook et al. 2016; Brandani et al. 2017; Marron and Epron 2019). The total amount of C sequestered by these plantations varies with the proportion of the total allocated in woody tissues and the end use of this biomass. Thus, uses for civil construction, wooden posts, furniture, and handicrafts increase the mitigating potential of the planted forests. On the other hand, if the use is for energy purposes, the sequestration becomes temporary. Carbon sequestration through soil is strongly related to the conversion to other uses for mixed plantings, or for more conservative management of soil and crop residues. Soil organic matter plays a role in a series of processes and properties from which several essential ecosystem services emerge, such as nutrient filtration, waste recycling, water filtration and storage, and flood control, among others (Dominati et al. 2010). As already presented in several chapters of this book, mixed plantations with non-N_2-fixing trees and N_2-fixing species can sequester more C than pure plantations of non-N_2-fixing trees, such as *Eucalyptus*. However, this effect is particularly dependent on the development and type of interactions of the species, as well as on the local edaphoclimatic conditions (Balieiro et al. 2008; Voigtlaender et al. 2012, 2019; Koutika et al. 2014; Rocha et al. 2019).

Erosion control and flood mitigation are ecosystem services provided by all kinds of forests. This includes mixed plantations, with planting on the remnants of the previous crop, because of the rapid covering of the soil by the plants (<2 years), and efficient exploitation of the soil mass resulting from complementarity and/or competition of aerial and underground niches (Chaer and Tótola 2007; Laclau et al. 2013; Silva et al. 2015; Santos et al. 2016). They protect the soil against the impact of raindrops, preserving and stabilizing the soil structure. Soil losses in *Eucalyptus* and *Acacia* plantations are usually low (<2.0 Mg ha^{-1}), reducing associated losses of nutrients from these plantations.

The expected improvements of the microbial activity, due to litter introduction of N_2-fixing legumes are conditioned by the structural quality of the litter produced by the species, the litter stoichiometry (N:P ratio), and the increase in the diversity of intercropping tree species, but divergent results can be found in the literature, in terms of decomposition, as discussed in Chap. 4.

Recent research has increased our understanding of the interactions that occur between microbiome-soil and microbiome-plant in mixed-plantation systems, particularly using *Acacia mangium* trees as a model plant (Rachid et al. 2015; Pereira et al. 2017; Fonseca et al. 2018; Bini et al. 2018). The search for specific roles in natural ecosystems and applying them in agriculture (Andreote and Silva 2017), as in the case of mixed-forest plantations, can increase the sustainability of planted forests. Studies with microorganisms associated with the soil-plant interface in *Eucalyptus* breeding programs have been neglected for years and we still know very little about this plant-microbiome interaction. In a forest system, trees are dependent on the microbiome to survive, which provides nutrients essential to their development, such as N, P, and K (primary macronutrients), by cycling organic matter in the soil. It was estimated that N_2-fixing bacteria and mycorrhizal fungi are responsible for furnishing up to 80% of all phosphorus and 75% of all nitrogen that forests use in their life cycle (van der Heijden et al. 2013).

A recent study evaluating a mixed-cultivation system between *E. grandis* and *A. mangium* on a 4-year-old and sandy-textured red-yellow Latosol (Ferralsol) showed strong influence of the composition of plants on the microbial community structure of the soil, both at the surface (0–20 cm) and at the subsurface (20–800 cm) (Pereira et al. 2017). In this study, the structure of the bacterial community was completely different when comparing with the pure treatment of *E. grandis* and the mixed treatment of the same species, suggesting that the acacia plants are the main modulators of the structure of the microbial community of the soil (Pereira et al. 2017). It is still unclear which effects are due to this behavior, but this may have occurred due to the production of root exudates by *A. mangium* and, mainly, by changes in soil chemical attributes that occur in the root region, especially high levels of carbon and nitrogen (Rachid et al. 2013; Pereira et al. 2018).

Similarly, a mixed plantation of *E. urograndis* and *A. mangium* at 3 years of age installed in a Haplic Planosol (0–10 cm) of sandy texture (90% sand) showed, through the PCR-DGGE technique, profiles of bands with significant differentiation in the soil bacterial community structure (Rachid 2013). In this study, the bacterial community showed the effect of each isolated plantation, which presented a community structure with its own characteristics, and community integration in the mixed treatment. In other words, the mixed treatment bacterial community had characteristics of both species, where the intermediate position leads to the belief that there is a balance in the influence of each species in relation to the community structure (Rachid et al. 2013). Thus, the influence of the plant composition of the mixed plantations between *Eucalyptus* and *Acacia* seems to be variable, becoming differentiated according to the edaphoclimatic factors of the place, the soil layer sampled, and the age of the stands. Because such studies were introduced relatively recently, learning about the changes in the soil microbial community in different regions, in contrasting soil and climate conditions, becomes necessary in future research.

Among the fungal types, two main classes are predominant in forest ecosystems, saprophytic and symbiotic fungi. In this sense, the roots of the trees can perform associations with the most varied types of fungi, such as arbuscular mycorrhizal (AMF) and ectomycorrhizal (ECM) fungi (Bonfanti and Anca 2009). Several types of mycorrhizal fungi present specificity for a single tree species (Lang et al. 2011). However, *Eucalyptus* and *Acacia* roots have the ability to associate with both AMF and ECM and, in some cases, with both types of symbionts at the same time (Pereira et al. 2018; Bini et al. 2018).

Bini et al. (2018) published possibly the first evaluation of AMF effects on a mixed cropping system. A study by Bini et al. (2018), on the dynamics of AMF during the first 20 months after planting pure and mixed stands of *E. grandis* and *A. mangium*, showed that the colonization by AMF in *E. grandis* roots was significantly higher, when the trees were grown in mixed systems with *A. mangium*. They also identified a strong correlation between AMF colonization rates and acid and alkaline phosphatase activities in soil, which are produced in greater quantity in forests containing *A. mangium*.

In this aspect, with the greater plant diversity, the materials deposited on and within the soil have the possibility of intensifying the nutrient cycling (Laclau et al. 2010).

10.5 Ecosystem Services of High-Diversity, Mixed Plantations: Multifunctional Planted Forest

Distinct types of forests—natural and planted forests or reforested areas—may provide ecosystem services at different levels; assessing the different aspects of forest state, related biodiversity, and landscape context becomes essential to monitor the forest and to estimate its contribution to the provision of ecosystem services (Chazdon et al. 2016). Since multiple-use forests (accounting for almost 20% of tropical forest area) are those that allow both production and conservation, in order to provide a great extent of ecosystem services, planted forests are expected to be multifunctional (Sloan and Sayer 2015).

As described in Chap. 11 of this book, to be considered multifunctional planted forests, high-diversity mixed-plantation forests must focus beyond forest structure or species composition, targeting also ecological functioning, which may enhance the provision of a wider variety of ecosystem services. Ecosystem multifunctionality requires greater numbers of species, since different tree species were found to influence different functions (Hector and Bagchi 2007). Besides providing ecosystem stability, the provision of ecosystem services by biodiversity also meets human society needs (Mori et al. 2013).

The evidence that the conservation of biodiversity is essential to provide ecosystem services has led to the discussion of changes on both production systems and rural landscapes. Deforestation and cropping remain the major causes of degradation of natural ecosystems (Rey Benayas and Bullock 2012; Sloan and Sayer 2015; Curtis et al. 2018), leading to the challenge of proposing land uses to provide a wide range of ecosystem services while conserving biodiversity in agricultural landscapes (Rey Benayas and Bullock 2012; Mori et al. 2013).

Agricultural intensification also causes biodiversity loss (Clough et al. 2011), and the consequent loss of ecosystem functions sustaining ecosystem services (Mori et al. 2013). In order to provide a full range of services, we need high-diversity-based production systems, planned to target the inclusion of ecosystem functions. In a perspective from the biodiversity and ecosystem functioning (BEF) theory, applying response-and-effect traits into the assembling of a planted forest will help for an optimal ecosystem multifunctionality (Mori et al. 2013; Laughlin 2014), leading to improved conservation priorities and a more resilient ecosystem to face negative effects of climate changes (Mori et al. 2013).

In order to have more multifunctionality in the rural landscape in the tropics, mixed-planted forests are so an alternative to traditional cropping and allow a good balance between production (as wood quality) and ecological benefits, like nutrient cycling; further, mixed plantations with higher diversity may contribute to a more efficient use of soil water (Amazonas et al. 2018). As commented above, adding plant functional traits to mixed-planted forests may be effective for increasing multifunctionality (Blesh 2018). The *Eucalyptus-Acacia* mixing model may supply nitrogen to the field, since Acacia species are N_2-fixing trees. Including biological nitrogen fixation (BNF) may supply nitrogen (N) to farm fields; in forest restoration, N_2-fixing cover crops have been used for weed control, contributing to enhance the multifunctionality of the ecosystem (Blesh 2018). When *Eucalyptus* is combined with a mixture of native trees in a high-diversity arrangement, the whole system was found to be more efficient in the use of soil water, which may suggest that such an agroecosystem may adapt to climate change conditions (Amazonas et al. 2018).

A meta-analysis showed that biodiverse agroforestry systems improve the provision of ecosystems in comparison to simple (low diversity) agroforestry systems and conventional production systems (Santos et al. 2018). Biodiverse agroforests were found to reduce biodiversity loss and negative impact on the provision of ecosystem services (Santos et al. 2018). The design and associated sustainable management of agroforests can avoid the removal of forest cover, optimizing both biodiversity and crop production benefits (Clough et al. 2011; Lescourret et al. 2015; Damour et al. 2018).

The diverse layer consisting of multiple-use tree species is the main factor responsible for the effective provision of ecosystem services (Lescourret et al. 2015).

Biodiverse agroforests are a type of high-diversity planted forests very highly recommended to improve biodiversity levels and provision of ecosystem services, supporting the planning of a biodiversity-friendly agricultural landscape (Santos et al. 2018; Clough et al. 2011). Agriculture intensification is a cause for landscape simplification, and land-use planning must seek multifunctional landscapes, providing goods and services, offering food security and actions for biodiversity conservation (O'Farrell and Anderson 2010).

Multifunctionality is the capacity of a landscape to provide multiple socioeconomic and ecological benefits (Hölting et al. 2019). Despite being unusual, it is important to understand how biodiversity in agricultural landscapes provides multiple ecosystem services, which may help enhancing landscape complexity (Birkhofer et al. 2018). Multifunctional, complex landscapes are expected to reduce biodiversity loss and maintain a balanced supply of ecosystem services (Frueh-Mueller et al. 2018), and multifunctional planted forests play an essential role in meeting those goals. Studies indicate that landscape management actions taken to conserve biodiversity will always promote an enhancement in the provision of ecosystem services (Birkhofer et al. 2018).

A strategy adopted to promote complex landscapes is the concept of land sparing, when biodiversity actions are allocated separately from production. On the other hand, land sharing may consist of biodiversity-based agricultural practices, with land uses or production systems that provide ecosystem services without decreasing agricultural production (Rey Benayas and Bullock 2012). High-diversity planted forests used for forest restoration is a good example of a land use based on the principle of land sharing used to reduce biodiversity loss and environmental health increase (Rey Benayas and Bullock 2012).

Beyond the arrangement of species in planted forests based on high biodiversity, the landscape approach may support the design of land management to achieve socioeconomic and ecological benefits in areas where land use traditionally competes with biodiversity conservation (Sayer et al. 2013). The achievement of multiple benefits depends on how stakeholders (farmers) are motivated to contribute to a wider range of ecosystem services (Frueh-Mueller et al. 2018). Incentive mechanism as payments for ecosystem services (PES) is a way to stimulate high and balanced provision of ecosystem services in productive landscapes, offering the opportunity for local stakeholders to participate in the decision-making process (Frueh-Mueller et al. 2018). PES may encourage farmers to keep the forest cover and enhance the permeability of the landscape (Tscharntke et al. 2011). Farmers have the chance to make choices for multispecies cropping system designs, selecting plant species and creating multifunctional agroecosystems, managing for crop yield and provision of ecosystem services (Damour et al. 2018). Multifunctional landscapes try to hold social and ecological dimensions in the management of a more complex landscape (Lescourret et al. 2015).

Multifunctional planted forests are able to provide economic besides ecological benefits, by either restoring or conserving biodiversity and related ecosystem services.

However, that approach presents some challenges as well, since it is necessary to ensure that, besides providing ecosystem services, multifunctional planted forests must support the needs of different stakeholders (Bullock et al. 2011). The assessment of studies focusing on the definition of multifunctionality indicated that about 33% of studies assessed ecological and socioeconomic variables in equal shares and integrated the perspectives of the stakeholders as well (Hölting et al. 2019).

High-diversity-based planted forests may ensure the provision of multiple services, since farmers have a clear understanding of how the ecosystem design and associated management practices can do that. In the landscape approach, the exchange of knowledge and experience among stakeholders is essential for a multifunctional landscape (Tscharntke et al. 2011). The success in using multifunctional planted forests to build a sustainable agricultural landscape is strongly based on achieving social and economic benefits. Therefore, the demand policies must consider a landscape scale to promote multifunctionality (Holt et al. 2016).

10.6 Final Comments

Forests may provide ecosystem services (ES) at different levels, but the implementation of a sustainable and multifunctional landscape requires focus on a range of issues and principles, as adaptive management, stakeholder involvement, and multiple objectives (O'Farrell and Anderson 2010; Sayer et al. 2013). Some constraints that must be overcome include institutional and governance concerns, transparent negotiation, and share of rights (O'Farrell and Anderson 2010; Sayer et al. 2013). Learning actions to bring together the multiple stakeholders in the construction of multifunctional landscapes and ways to foster synergy between ecosystem functioning and social dynamics are needed (O'Farrell and Anderson 2010; Lescourret et al. 2015).

Under local and regional scale the mosaic of exotic planted forests and native forests in the rural landscape matrix has brought immeasurable benefits to sustainability. The mixed-planted forests (with low or high diversity) emerge as an alternative to traditional cropping and allow a right balance between production (wood) and ecological benefits. Multifunctional landscapes are expected to reduce biodiversity loss and maintain a stable supply of ecosystem services, while under commercial level increased rotation length, multiple uses, alternative spacing and species or clone arrangements (as mixed with N_2-fixing trees), as well as conservative soil practices are necessary to mitigate the loss of natural capital.

References

Amazonas NT, Forrester DI, Silva CC, Almeida DRA, Rodrigues RR, Brancalion PH (2018) High diversity mixed plantations of Eucalyptus and native trees: an interface between production and restoration for the tropics. For Ecol Manag 417:247–256

Andrade D, Romeiro A (2019) Serviços ecossistêmicos e sua importância para o sistema econômico e o bem-estar humano. IE/UNICAMP, Campinas, p 155

Andreote FD, Silva MCP (2017) Microbial communities associated with plants: learning from nature to apply it in agriculture. Curr Opin Microbiol 37:29–34

Balieiro FC, Pereira MG, Alves BJR, Resende ASD, Franco AA (2008) Soil carbon and nitrogen in pasture soil reforested with eucalyptus and guachapele. Rev Bras Cienc Solo 32:1253–1260

Balieiro FC, Tonini H, Lima RA (2018) Produção científica brasileira (2007–2016) sobre *Acacia mangium* Willd.: estado da arte e reflexões. Cad Cienc Tecnol 35:37–52

Ballester MVR, Victoria RL, Krusche AV (2010) Agroenergia e sustentabilidade do solo e da água. In: Prado RB, Turetta APD, Andrade AG (org), Manejo e conservação do solo e da água no contexto das mudanças ambientais. Embrapa, Rio de Janeiro, vol. 1:215-236

Bellassen V, Luyssaert S (2014) Carbon sequestration: managing forests in uncertain times. Nat News 506:153

Bernhard-Reversat F (1996) Nitrogen cycling in tree plantations grown on a poor sandy savanna soil in Congo. Appl Soil Ecol 4:161–172

Bini D, Santos CA, Silva MCP, Bonfim JA, Cardoso EJBN (2018) Intercropping *Acacia mangium* stimulates AMF colonization and soil phosphatase activity in *Eucalyptus grandis*. Sci Agric 75:102–110

Binkley D, Dunkin KA, DeBell D, Ryan MG (1992) Production and nutrient cycling in mixed plantations of Eucalyptus and Albizia in Hawaii. For Sci 38:393–408

Birkhofer K, Andersson GKS, Bengtsson J, Bommarco R, Dänhardt J, Ekbom B, Ekroos J, Hahn T, Hedlund K, Jönsson AM, Lindborg R, Olsson O, Rader R, Rusch A, Stjernman M, Williams A, Smith HG (2018) Relationships between multiple biodiversity components and ecosystem services along a landscape complexity gradient. Biol Conserv 218:247–253

Blesh J (2018) Functional traits in cover crop mixtures: biological nitrogen fixation and multifunctionality. J Appl Ecol 55:38–48

Bonan GB (2008) Forests and climate change: forcings, feedbacks, and the climate benefits. Science 320:1444–1449. https://doi.org/10.1126/science.1155121

Bonfanti P, Anca LA (2009) Plants, mycorrhizal fungi, and bacteria: a network of interactions. Annu Rev Microbiol 63:363–383

Bormann FH, Likens GE (1979) Catastrophic disturbance and the steady-state in northern hardwood forests. Am Sci 67(6):660–669

Brandani CB, Rocha JHT, Godinho TO, Wenzel AVA, Gonçalves JLM (2017) Soil C and Al availability in tropical single and mixed-species of *Eucalyptus* sp. and *Acacia mangium* plantations. Geod Reg 10:85–92

Brandão Jr A, Barreto P, Lenti F, Shimbo J, Alencar A (2018) Emissões do setor de mudanças de uso da terra. SEEG, Sistema de Estimativa das Emissões de Gases de Efeito Estufa, Documento de Análise, p 56

Brockerhoff EG, Jactel H, Parrotta JA, Ferraz SFB (2013) Role of eucalypt and other planted forests in biodiversity conservation and the provision of biodiversity-related ecosystem services. For Ecol Manag 301:43–50

Bullock JM, Aronson J, Newton AC, Pywell RF, Rey-Benayas JM (2011) Restoration of ecosystem services and biodiversity: conflicts and opportunities. Trends Ecol Evol 26:541–549

Câmara CD, De Paula Lima W, Vieira SA (2000) Clear-cutting of a 50 year old Eucalyptus plantation: impacts on nutrient cycling in an experimental catchment. Sci For 57:99–109

Cassiano CC (2017) Efeitos hidrológicos da composição da paisagem em microbacias com florestas plantadas de Eucalyptus. PhD Thesis, Universidade de São Paulo

Chaer GM, Tótola MR (2007) Impact of organic residue management on soil quality indicators during replanting of eucalypt stands. Rev Bras Cienc Solo 31:1381–1396

Chaer GM, Resende AS, Campello EFC, Faria SM, Boddey RM (2011) Nitrogen-fixing legume tree species for the reclamation of severely degraded lands in Brazil. Tree Physiol 31(2):139–149

Chazdon RL, Brancalion PHS, Laestadius L, Bennett-Currt A, Buckingham K, Kumar C, Moll-Rocek J, Vieira ICG, Wilson SJ (2016) When is a forest a forest? Forest concepts and definitions in the era of forest and landscape restoration. Ambio 45:538–550

Clough Y, Barkmann J, Juhrbandt J, Kessler M, Wanger TC, Anshary A, Buchori D, Cicuzza D, Darras K, Putra DD, Erasmi S, Pitopang R, Schmidt C, Schulze CH, Seidel D, Steffan-Dewenter I, Stenchly K, Vidal S, weist M, Wielgoss AC, Tscharntke T (2011) Combining high biodiversity with high yields in tropical agroforests. Proc Natl Acad Sci U S A 108:8311–8316

Cook RL, Binkley D, Stape JL (2016) Eucalyptus plantation effects on soil carbon after 20 years and three rotations in Brazil. For Ecol Manag 359:92–98
Costanza R, Daly H (1992) Natural capital and sustainable development. Conserv Biol 6:37–46
Costanza R, Folke C (1997) Valuing ecosystem services with efficiency, fairness, and sustainability as goals. In: Daily GC (ed) Nature's services: societal dependence on natural ecosystems. Island Press, Washington DC, pp 49–70
Corbeels M, McMurtrie RE, Pepper DA, O'Connell AM (2005) A process-based model of nitrogen cycling in forest plantations: Part I. Structure, calibration and analysis of the decomposition model. Ecol Model 187(4):426–448
Curtis PG, Slay CM, Harris NL, Tyukavina A, Hansen MC (2018) Classifying drivers of global forest loss. Science 361:1108–1111
Daily GC (1997) Nature's services: societal dependence on natural ecosystems. Island Press, Washington DC, p 392
Damour G, Navas ML, Garnier E (2018) A revised trait-based framework for agroecosystems including decision rules. J Appl Ecol 55:12–24
De Groot RS (1987) Environmental functions as a unifying concept for ecology and economics. Environmentalist 7:105–109
De Groot RS (2006) Function-analysis and valuation as a tool to assess land use conflicts in planning for sustainable, multifunctional landscapes. Landscape Urban Plan 75(3-4):175–186
De Groot RS, Wilson MA, Boumans RMJ (2002) A typology for the classification, description and valuation of ecosystem functions, goods and services. Ecol Econ 41:393–408
De Groot RS, Alkemade R, Braat L, Hein L, Willemen L (2010) Challenges in integrating the concept of ecosystem services and values in landscape planning, management and decision making. Ecol Complex 7(3):260–272
Dee N, Baker J, Drobny N, Duke K, Whitman I, Fahringer D (1973) An environmental evaluation system for water resource planning. Water Resour Res 9:523–535
Dominati E, Patterson M, Mackay A (2010) A framework for classifying and quantifying the natural capital and ecosystem services of soils. Eco Econ 69:1858–1868
Ehrlich PR, Ehrlich AH (1970) Population, resources, environment: issues in human ecology, 2nd edn. W. H. Freeman, San Francisco, p 383
Ehrlich PR, Ehrlich AH, Holdren JP (1977) Ecoscience: population, resources, environment. W.H. Freeman, San Francisco, pp 546–547
FAO (2010) Global forest resources assessment. FAO, Rome, p 378
FAO (2018) The state of the world's forests 2018—forest pathways to sustainable development. FAO, Rome, p 139
Ferraz SFB, Lima WP, Rodrigues CB (2013) Managing forest plantation landscapes for water conservation. For Ecol. Manag. 301:58–66
Ferraz SFB, Ferraz KMPMB, Cassiano CC, Brancalion PHS, Luz DTA, Azevedo TN, Tambosi LR, Metzger JP (2014) How good are tropical forest patches for ecosystem services provisioning? Landsc Ecol 29:187–200
Ferreira J, Pardini R, Metzger JP, Fonseca CR, Pompeu PS, Sparovek G, Louzada J (2012) Towards environmentally sustainable agriculture in Brazil: challenges and opportunities for applied ecological research. J Appl Ecol 49:535–541
Fialho RC, Zinn YL (2014) Changes in soil organic carbon under Eucalyptus plantations in Brazil: a comparative analysis. Land Degrad Develop 25:428–437
Fisher B, Turner RK (2008) Ecosystem services: classification for valuation. Biol Conserv 141:1167–1169
Fisher B, Costanza R, Turner RK, Morling P (2009) Defining and classifying ecosystem services for decision making. Ecol Econ 68:643–653
Fonseca ES, Peixoto RS, Rosado AS, Balieiro FC, Tiedje JM, Rachid CTCC (2018) The Microbiome of Eucalyptus roots under different management conditions and its potential for biological nitrogen fixation. Microb Ecol 75:183–191
Forrester DI, Bauhus J, Cowie AL (2005) Nutrient cycling in a mixed-species plantation of Eucalyptus globulus and Acacia mearnsii. Can J Forest Res 35:2942–2950

Forrester DI, Bauhus J, Cowie AL, Vanclay JK (2006) Mixed-species plantations of Eucalyptus with nitrogen-fixing trees: a review. For Ecol Manag 233:211–230

Forrester DI, Pares A, O'Hara C, Khanna PK, Bauhus J (2013) Soil organic carbon is increased in mixed-species plantations of Eucalyptus and nitrogen-fixing Acacia. Ecosystems 16:123–132

Franco AA, Faria SM (1997) The contribution of N2-fixing tree legumes to land reclamation and sustainability in the tropics. Soil Bio Biochem 29:897–903

Franco AA, Resende AS, Campello EFC (2018) In: Alba JMF (ed) Recuperação de áreas mineradas, 3rd edn. Embrapa, Brasília, p 456

Frueh-Mueller A, Krippes C, Hotes S, Breuer L, Koellner T, Wolters V (2018) Spatial correlation of agri-environmental measures with high levels of ecosystem services. Ecol Indic 84:364–370

Gómez-Baggethun E, De Groot R (2007) Capital natural y funciones de los ecosistemas: explorando las bases ecológicas de la economía. Ecosistemas 16(3):4–14

Gonçalves JLM, Alvares CA, Higa AR, Silva LD, Alfenas AC, Stahl J, Ferraz SFB, Lima WP, Brancalion PHS, Hubner A, Bouillet JPD, Laclau JP, Nouvellon Y, Epron D (2013) Integrating genetic and silvicultural strategies to minimize abiotic and biotic constraints in Brazilian eucalypt plantations. For Ecol Manag 301:6–27

Haines-Young R, Potschin-Young M (2018) Revision of the Common International Classification for Ecosystem Services (CICES V5.1): A Policy Brief. One Ecosystem 3: e27108. https://doi.org/10.3897/oneeco.3.e27108

Hakamada R, Hubbard RM, Ferraz S, Stape JL, Lemos C (2017) Biomass production and potential water stress increase with planting density in four highly productive clonal Eucalyptus genotypes. South For J For Sci 79(3):251–257

Hein L, van Koppen K, de Groot RS, van Ierland EC (2006) Spatial scales, stakeholders and the valuation of ecosystem services. Ecological Economics. 57(2):209–228. https://doi.org/10.1016/j.ecolecon.2005.04.005

Heal G (2000) Valuing ecosystem services. Ecosystems 3(1):24–30

Hector A, Bagchi R (2007) Biodiversity and ecosystem multifunctionality. Nature 448:188–190

Helliwell DR (1969) Valuation of wildlife resources. Reg Stud 3:41–49

Hermann A, Schleifer S, Wrbk T (2011) The concept of ecosystem services regarding landscape research: a review. Living Rev Landscape Res 5:1–37

Higa RCV, Mora AL, Higa AR (2000) Plantio de eucalipto na pequena propriedade rural. Embrapa, Curitiba, p 24

Holt AR, Alix A, Thompson A, Maltby L (2016) Food production, ecosystem services and biodiversity: we can't have it all everywhere. Sci Tot Environ 573:1422–1429

Hölting L, Beckmann M, Volk M, Cord AF (2019) Multifunctionality assessments—more than assessing multiple ecosystem functions and services? A quantitative literature review. Ecol Indic 103:226–235

IBÁ (2017) Relatório 2017. IBÁ, p 80

Jandl R, Lindner M, Vesterdal L, Bauwens B, Baritz R, Hagedorn F, Johnson DW, Minkkinen K, Byrne KA (2007) How strongly can forest management influence soil carbon sequestration? Geoderma 137:253–268

Kaye JP, Resh SC, Kaye MW, Chimner RA (2000) Nutrient and carbon dynamics in a replacement series of Eucalyptus and Albizia trees. Ecology 81:3267–3273

Kelty MJ (2006) The role of species mixtures in plantation forestry. For Ecol Manag 233:195–204

King RT (1966) Wildlife and man. NY Conservationist 20(6):8–11

Koutika LS, Epron D, Bouillet JP, Mareschal L (2014) Changes in N and C concentrations, soil acidity and P availability in tropical mixed acacia and eucalypt plantations on a nutrient-poor sandy soil. Plant Soil 379:205–216

Kreye MM, Adams DC, Ghimire R, Morse W, Stein T, Bowker JM (2017) Forest ecosystem services: cultural values. General technical report SRS-226, vol 226. US Department of Agriculture Forest Service, Southern Research Station, Asheville, NC, pp 11–30

Laclau JP, Levillain J, Deleporte P, Nzila JD, Bouillet JP, André LS, Versini A, Mareschal L, Nouvellon Y, M'Bou AT, Ranger J (2010) Organic residue mass at planting is an excellent predictor of tree growth in Eucalyptus plantations established on a sandy tropical soil. For Ecol Manag 260:2148–2159

Laclau JP, Gonçalves JLM, Stape JL (2013) Perspectives for the management of eucalypt plantations under biotic and abiotic stresses. For Ecol Manag 301:1–5

Lal R (2005) Forest soils and carbon sequestration. For Ecol Manag 220:242–258

Lang C, Seven J, Polle A (2011) Host preferences and differential contributions of deciduous tree species shape mycorrhizal species richness in a mixed Central European forest. Mycorrhiza 21(4):297–308

Laughlin DC (2014) Applying trait-based models to achieve functional targets for theory-driven ecological restoration. Ecol Lett 17:771–784

Lescourret F, Magda D, Richard G, Adam-Blondon AF, Bardy M, Baudry J, Doussan I, Dumont B, Lefèvre F, Litrico I, Martin-Clouaire R, Montuelle B, Pellerin S, Plantegenest M, Tancoigne E, Thomas A, Guyomard H, Soussana F (2015) A social–ecological approach to managing multiple agro-ecosystem services. Curr Opin Environ Sust 14:68–75

Lima WP, Ferraz SFB, Rodrigues CB, Voigtlaender M (2012a) Assessing the hydrological effects of forest plantations in Brazil. In: River conservation and management. John Wiley & Sons, Ltd, Hoboken, NJ, pp 59–68

Lima WP, Laprovitera R, Ferraz SFB, Rodrigues CB, Silva MM (2012b) Forest plantations and water consumption: a strategy for hydrosolidarity. Int J Forest Res 2012:8

Liu CLC, Kuchma O, Krutovsky KV (2018) Mixed-species versus monocultures in plantation forestry: development, benefits, ecosystem services and perspectives for the future. Global Ecol Cons 15:e00419

Macedo MCM, Zimmer AH, Kichel AN, Almeida RG, Araújo AR (2013) Degradação de pastagens, alternativas de recuperação e renovação, e formas de mitigação. In: Encontro de Adubação de Pastagens da Scot Consultoria-Tec-Fértil. Scot Consultoria, Bebedouro, pp 158–181

Marron N, Epron D (2019) Are mixed-tree plantations including a nitrogen-fixing species more productive than monocultures? For Ecol Manag 441:242–252

Masiero M, Pettenella D, Boscolo M, Barua SK, Animon I, Matta R (2019) Valuing forest ecosystem services—a training manual for planners and project developers. FAO, Rome, p 216

Mcneill JR, Mcneill WH (2003) The human web: a bird's-eye view of world history. WW Norton & Company, New York, p 350

MEA (2003) Ecosystems and human well-being: a framework for assessment. Island Press, Washington, DC

MEA (2005) Ecosystems and human well-being: synthesis. Island Press, Washington, DC

Mori AS, Furukawa T, Sasaki T (2013) Response diversity determines the resilience of ecosystems to environmental change. Biol Rev 88:349–364

Myers N, Mittermeier RA, Mittermeier CG, Fonseca GA, Kent J (2000) Biodiversity hotspots for conservation priorities. Nature 403:853–858

Newbold T, Hudson LN, Hill SLL, Contu S, Lysenko I, Senior RB, Börger L, Bennett DJ, Choi es A, Collen B, Day J, Palma A, Díaz S, Echeverria-Londoño S, Edgar MJ, Feldman A, Garon M, Harrison MLK, Alhusseini T, Ingram DJ, Itescu Y, Kattge J, Kemp V, Kirkpatrick L, Kleyer M, Correia DLP, Martin CD, Meiri S, Novosolov M, Pan Y, Phillips HRP, Purves DW, Robinson A, Simpson J, Tuck SL, Weiher E, Whote HJ, Ewers RM, Mace GM, Scharlemann JPW, Purvis A (2015) Global effects of land use on local terrestrial biodiversity. Nature 520:45–50

Nicholson E, Mace GM, Armsworth PR, Atkinson G, Buckle S, Clements T, Ewers RM, Fa JE, Gardner TA, Gibbons J, Grenyer R, Metcalfe R, Mourato S, Muûls M, Osborn D, Reuman DC, Watson C, Milner-Gulland EJ (2009) Priority research areas for ecosystem services in a changing world. J Appl Ecol 46(6):1365–2664

Nottingham AT, Turner BL, Whitaker J, Ostle NJ, McNamara NP, Bardgett RD, Salinas N, Meir P (2015) Soil microbial nutrient constraints along a tropical forest elevation gradient: a belowground test of a biogeochemical paradigm. Biogeosciences 12:6071–6083

O'Farrell PJ, Anderson PM (2010) Sustainable multifunctional landscapes: a review to implementation. Curr Opin Environ Sustain 2:59–65

Oliveira OC, Oliveira IP, Alves BJR, Urquiaga S, Boddey RM (2004) Chemical and biological indicators of decline/degradation of Brachiaria pastures in the Brazilian Cerrado. Agr Eco Env 103:289–300

Parrotta JA (1999) Productivity, nutrient cycling, and succession in single- and mixed-species plantations of *Casuarina equisetifolia*, *Eucalyptus robusta*, and *Leucaena leucocephala* in Puerto Rico. For Ecol Manag 124:45–77

Parrotta JA, Knowles OH (1999) Restoration of tropical moist forest on bauxite mined lands in the Brazilian Amazon. Restor Ecol 7:103–116

Parrotta JA, Turnbull JW, Jones N (1997) Catalyzing native forest regeneration on degraded tropical lands. For Ecol Manag 99:1–7

Paula Lima WD, Barros Ferraz SFD, Barros Ferraz KMPMD (2013) Interações bióticas e abióticas na paisagem: uma perspectiva eco-hidrológica. In: Calijuri MC, Cunha DGF (eds) Engenharia Ambiental. Elsevier, Rio de Janeiro, pp 215–244

Pereira APA, Andrade PAM, Bini D, Durrer A, Robin A, Bouillet JP, Andreote FD, Cardoso EJBN (2017) Shifts in the bacterial community composition along deep soil profiles in monospecific and mixed stands of *Eucalyptus grandis* and *Acacia mangium*. PLoS One 12(7):e0180371

Pereira APA, Zagatto MRG, Brandani CB, Mescolotti DDL, Cotta SR, Gonçalves JLM, Cardoso EJBN (2018) Acacia changes microbial indicators and increases C and N in soil organic fractions in intercropped Eucalyptus plantations. Front Microbiol 9:655

Periotto NA, Tundisi JG (2018) A characterization of ecosystem services, drivers and values of two watersheds in São Paulo State, Brazil. Braz J Biol 78:397–407

Prado HM, Murrieta RSS (2015) Presentes do Passado. Ciência Hoje 326(55):32–37

Prado RB, Fidalgo ECC, Monteiro JMG, Schuler AE, Vezzani FM, Garcia JR, Oliveira AP, Viana JHM, Pedreira BCCG, Mendes IC, Reatto A, Parron LM, Clemente EP, Donagemma GK, Turetta APD, Simões M (2016) Current overview and potential applications of the soil ecosystem services approach in Brazil. Pesqui Agropecu Bras 51:1021–1038

Rachid CTCC (2013) Biodisponibilidade de nutrientes e estrutura microbiana do sistema solo-serapilheira em floresta plantada mista de *Eucalyptus urograndis e Acacia mangium*. PhD Thesis, Universidade Federal do Rio de Janeiro

Rachid CTCC, Balieiro FC, Peixoto RS, Pinheiro YAS, Piccolo MC, Chaer GM, Rosado AS (2013) Mixed plantations can promote microbial integration and soil nitrate increases with changes in the N cycling genes. Soil Bio Biochem 66:146–153

Rachid CTCC, Balieiro FC, Fonseca ES, Peixoto RS, Chaer GM, Tiedje JM, Rosado AS (2015) Intercropped Silviculture systems, a key to achieving soil fungal community management in Eucalyptus plantations. PLoS One 10:e0118515

Rey Benayas JM, Bullock JM (2012) Restoration of biodiversity and ecosystem services on agricultural land. Ecosystems 15:883–899

Ribeiro MC, Metzger JP, Martensen AC, Ponzoni FJ, Hirota MM (2009) The Brazilian Atlantic forest how much is left, and how is the remaining forest distributed implications for conservation. Bio Conserv 142:1141–1153

Rocha PV, Ataíde DHS, Lima JSS et al (2019) Preparo do solo e leguminosa arbórea fixadora de N2 consoricada com eucalipto afetam os estoques de carbono e nitrogênio de solo arenoso. In: Anais do Simpósio Brasileiro de Solos Arenosos. Campo Grande—MS

Rodrigues RR, Lima RAF, Gandolfi S, Nave AG (2009) On the restoration of high diversity forests: 30 years of experience in the Brazilian Atlantic Forest. Bio Conserv 142:1242–1251

Rounsevell MDA, Dawson TP, Harrison PA (2010) A conceptual framework to assess the effects of environmental change on ecosystem services. Biodivers Conserv 19(10):2823–2842

Sala OE, Chapin FS, Armesto JJ, Berlow E, Bloomfield J, Dirzo R, Huber-Sanwald E, Huenneke LF, Jackson R, Kinzig A, Leemans R, Lodge D, Mooney HA, Oesterheld M, Poff LT, Sykes M, Walker BH, Walker M, Wall D (2000) Global biodiversity scenarios for the year 2100. Science 287:1770–1774

Santos DG, Domingues AF, Gisler CVT (2010) Gestão de recursos hídricos na agricultura: O Programa Produtor de Água. In: Prado RB, Turetta APD, Andrade AG (eds) Manejo e Conservação do Solo e da Água no Contexto das Mudanças Ambientais. Embrapa Solos, Rio de Janeiro, pp 353–376

Santos FM, Balieiro FC, Ataíde DHS, Diniz AR, Chaer GM (2016) Dynamics of aboveground biomass accumulation in monospecific and mixed-species plantations of Eucalyptus and Acacia on a Brazilian sandy soil. For Ecol Manag 363:86–97028

Santos FM, Balieiro F de C, Fontes MA, Chaer GM (2018) Understanding the enhanced litter decomposition of mixed-species plantations of Eucalyptus and Acacia mangium. Plant Soil 423:141

Sayer J, Sunderland T, Ghazoul J, Pfund JL, Sheil D, Meijaard E, Venter M, Boedhihartono AK, Day M, Garcia C, Oosten CV, Buck LE (2013) Ten principles for a landscape approach to reconciling agriculture, conservation, and other competing land uses. Proc Natl Acad Sci U S A 110:8349–8356

Sedjo R, Sohngen B (2012) Carbon sequestration in forests and soils. Annu Rev Res Econom 4:127–144

Silva JNM, Carvalho JOP, Lopes JCA, Almeida BF, Costa DHM, Oliveira LC, Vanclay JK, Skovsgaard JP (1995) Growth and yield of a tropical rain forest in the Brazilian Amazon 13 years after logging. For Ecol Manag 71:267–274

Silveira JG (2018) Emissões de gases de efeito estufa e estoque de carbono e nitrogênio em área de plantio misto de eucalipto e acácia no norte do Mato Grosso.Universidade Federal de Mato Grosso, UFMT (Mestrado em Ciências Florestais e Ambientais), 94p

Silva CF, Carmo RE, Martins MA, Freitas MSM, Pereira MG, Silva EMR (2015) Deposition and nutritional quality of the litter of pure stands of *Eucalyptus camaldulensis* and *Acacia mangium*. Biosci J 31(4). https://doi.org/10.14393/BJ-v31n4a2015-26297

Singh AN, Zeng DH, Chen FS (2006) Effect of young woody plantations on carbon and nutrient accretion rates in a redeveloping soil on coalmine spoil in a dry tropical environment, India. Land Degrad Dev 17:13–21

Sloan S, Sayer JA (2015) Forest resources assessment of 2015 shows positive global trends but forest loss and degradation persist in poor tropical countries. For Ecol Manag 352:134–145

SNIF (Sistema Nacional de Informações Florestais) (2016) Boletim dos Sistema Nacional Florestal. Ministério do Meio Ambiente, Brazil, 10p. (acesso em 10 de dezembro de 2019). http://www.florestal.gov.br/documentos/publicacoes/2230-boletim-snif-producao-florestal-2016/file

SNIF (Sistema Nacional de Informações Florestais) (2017) Serviço Florestal Brasileiro, Brasil, 32p. (acesso em 10 de dezembro de 2019). http://www.florestal.gov.br/documentos/publicacoes/3230-boletim-snif-2017-ed1-final/file)

Sparovek G, Berndes G, Klug ILF, Barretto AGOP (2010) Brazilian agriculture and environmental legislation: status and future challenges. Env Sci Technol 44:6046–6053

Stallings JR (1990) The importance of understorey on wildlife in a Brazilian eucalypt plantation. Rev Bras Zool 7(3):267–276. https://doi.org/10.1590/S0101-81751990000300008

Stape JL, Binkley D, Ryan MG, Fonseca S, Loos RA, Takahashi EN, Silva CR, Silva SR, Hakamada RE, Ferreira JMA, Lima AMN, Gava JL, Leite FP, Andrade HB, Alves JM, Silva GGC, Azevedo MR (2010) The Brazil Eucalyptus potential productivity project: influence of water, nutrients and stand uniformity on wood production. For Ecol Manag 259:1684–1694

Tarouco CP, Agostinetto D, Panozzo LE, Santos LS, Vignolo GK, Ramos LOO (2009) Períodos de interferência de plantas daninhas na fase inicial de crescimento do eucalipto. Pesqui Agropecu Bras 44(9):1131–1137

Tchichelle SV, Epron D, Mialoundama F, Koutika LS, Harmand JM, Bouillet JP, Mareschal L (2017) Differences in nitrogen cycling and soil mineralization between a eucalypt plantation and a mixed eucalypt and *Acacia mangium* plantation on a sandy tropical soil. South For J For Sci 79(1):1–8

TEEB (2010) The Economics of Ecosystems and Biodiversity: Ecological and Economic Foundations (TEEB). Kumar, P., ed., London (Earthscan)

Tscharntke T, Klein AM, Kruess A, Steffan-Dewenter I, Thies C (2005) Landscape perspectives on agricultural intensification and biodiversity-ecosystem service management. Ecol Lett 8:857–874

Tscharntke T, Clough Y, Bhagwat SA, Buchori D, Faust H, Hertel D, Hölscher D, Juhrbandt J, Kessler M, Perfecto I, Scherber C, Schroth G, Veldkamp E, Wanger TC (2011) Multifunctional shade-tree management in tropical agroforestry landscapes—a review. J Appl Ecol 48:619–629

Turner RK, Paavola J, Cooper P, Farber S, Jessamy V, Georgiou S (2003) Valuing nature: lessons learned and future research directions. Ecol Econom 46(3):493–510

Van der Heijden G, Legout A, Pollier B, Bréchet C, Ranger J, Dambrine E (2013) Tracing and modeling preferential flow in a forest soil—potential impact on nutrient leaching. Geoderma 195:12–22

Vergara W, Scholz SM (eds) (2010) Assessment of the risk of Amazon dieback. World Bank Publications, Washington, DC

Voigtlaender M, Laclau JP, Gonçalves JLM, Piccolo MC, Moreria MZ, Nouvellon Y, Ranger J, Bouillet JP (2012) Introducing *Acacia mangium* trees in *Eucalyptus grandis* plantations: consequences for soil organic matter stocks and nitrogen mineralization. Plant Soil 352:99–111

Voigtlaender M, Brandani CB, Caldeira DRM, Tardy F, Bouillet JP, Gonçalves JLM, Moreira MZ, Leite FP, Brunet D, Paula RR, Laclau JP (2019) Nitrogen cycling in monospecific and mixed-species plantations of *Acacia mangium* and Eucalyptus at 4 sites in Brazil. For Ecol Manag 436:56–67

Westman WE (1977) How much are natures services worth. Science 197(4307):960–964

Wilson MA, Carpenter SR (1999) Economic valuation of freshwater ecosystem services in the United States: 1971–1997. Ecol Appl 9(3):772–783

Zen S (1987) Influência da matocompetição em plantio de *Eucalyptus grandis*. Série Técnica IPEF 4(12):25–35

Chapter 11
The Risk of Invasions When Using *Acacia* spp. in Forestry

Ciro José Ribeiro de Moura, Nina Attias, and Helena de Godoy Bergallo

11.1 Aliens Welcome

Exotic tree species such as *Pinus* and *Eucalyptus* have a great impact on the Brazilian economy, accounting for 6.1% of the Brazilian GDP in 2018 (Industria Brasileira de Árvores, 2019). Nowadays Brazil is the third largest exporter of pulp, accounting for 13.2% of the world market. According to IBGE (2019), planted forests currently occupy ten million hectares, which corresponds to 1% of the country's agricultural lands.

> "*The climate, the soil and the technology we have in Brazil have allowed us to achieve the highest average annual productivity in the world*," explains agronomist João Salomão, general coordinator of Forestry and Livestock Affairs at the Ministry of Agriculture, Livestock and Supply (MAPA).

In 2016, Brazil led the global ranking of forest productivity, with an average of 35.7 m^3 ha^{-1} $year^{-1}$ in eucalypt plantation and 30.5 m^3 ha $year^{-1}$ in planting pine trees, according to data from the Brazilian tree industry (Industria Brasileira de Árvores 2019).

C. J. R. de Moura (✉)
Federal University of Rio de Janeiro, Rio de Janeiro, RJ, Brazil

N. Attias
Department of Animal Biology, Federal University of Mato Grosso do Sul, Campo Grande, MS, Brazil

H. de Godoy Bergallo
Department of Ecology, Institute of Biology Roberto Alcântara Gomes, Rio de Janeiro State University, Rio de Janeiro, RJ, Brazil

E. J. Bran Nogueira Cardoso et al. (eds.), *Mixed Plantations of Eucalyptus and Leguminous Trees*, https://doi.org/10.1007/978-3-030-32365-3_11

However, some of these genera such as *Acacia* and *Pinus* present invasive behavior in some ecosystems, and it is necessary to account for and internalize these impacts economically.

In Brazil, the impacts of all invasive species generate an annual loss of approximately US$ 50 billion per year (Pimentel et al. 2001). However, this estimation is outdated, resulting in an underestimate of the invasion problems caused by several species and its negative environmental and socioeconomic impacts (Rouget et al. 2016). Despite the abundant evidence and warnings of the possible impacts and damages caused by the introduction of exotic species, they continue being introduced to new locations for various purposes (Vitousek et al. 2017).

We estimate that 21% of known plant species in Brazil are exotic (Pimentel et al. 2001), including the main crops such as soybean, coffee, sugarcane, rice, corn, oranges, bananas, coconut, and many others. To control the action of invasive exotic species, management and eradication techniques have to be implemented (van Wilgen et al. 2011) and are surrounded by a high degree of uncertainty, besides elevated costs. Hence, considering the risk of invasion and the costs associated with the control of alien species, the Convention on Biological Diversity (1992) states that, according to the precautionary principle, the prevention of future invasions should be the most efficient form of management.

The perception of the impact of the biological invasion is supplanted by the perception of utility and generation of economic revenues from the use of these species, which is seen as a commodity to feed the people. As a rule, the costs of the invasion impact caused by the invasive alien species are externalized to the whole society. Furthermore, the decision to prioritize caution and control by the society as a whole depends on the perception and type of impact caused by the biological invasion and the willingness to pay for the high control costs.

When an exotic species has been settled for multiple human generations, it becomes part of the landscape memory and can be mistakenly recognized as native species (Diamond 2005). One example of this is the jackfruit *Artocarpus heterophyllus*, originally from Asia and introduced in Brazil in the sixteenth century (Ferrão 1993). The species has been classified as *Artocarpus heterophyllus* by Lam., as published in Encyclopédie Méthodique Botanique in 1789 but, 23 years later, in 1812, the same species was misclassified as *Artocarpus brasiliensis* by Ortega, in Memórias de Mathematica e Phisica da Academia Real das Sciencias de Lisboa 3: 84, portraying how the species had been incorporated into the local landscape.

11.2 Acacia Silviculture

The economic value of eucalypt was recognized from the earliest days of European settlement in Australia, and this stimulated the transfer of seeds of many taxa around the world. However, systematic collection and evaluation of Australian *Acacias* spp. only began in the 1970s. The major agency involved was the Australian Tree Seed Centre that dispatched samples of 322 taxa (1/3 of them *Acacia* spp. native to

Australia) between 1980 and 2010 to 149 countries. Plantations in SE Asia and South Africa supplying the pulp and paper industry also provide logs for solid wood products (Griffin et al. 2011).

This commercially versatile genus can be used in a small scale as a resource of firewood in rural properties or in large scale industrially (Midgley and Turnbull 2003). At local scale, it is used for landscape purposes, such as firebreaks and windbreaks; in the forestation of urban and rural areas; and, in consortium with other legumes, for the *reclamation of degraded soils* (Carvalho et al. 1998, Faria et al. 1998, Souza et al. 2004, Balieiro et al. 2007).

Acacias as N_2-fixing trees have been used in mixed plantations with eucalypt increasing biomass productivity while maintaining the fertility, because of the more efficient use of soils, both physically and chemically (Kleinpaul et al. 2010), highlighting the importance of soil enriched with symbiotically fixed N in substitution of high doses of synthetic N fertilizer (see Chap. 6).

Acacia mangium has become a *widely used species in reforestation programs* in the humid tropical plains of Asia. It also is one of the most common leguminous tree species in plantations in the tropical and subtropical regions of southeastern China (Midgley and Turnbull 2003, Xiong et al. 2008).

In Brazil, *Acacia mangium* is planted with multifunctional characteristics in a consortium with agricultural crops and lends itself to uses in integrated forest and livestock cultures. The first commercial plantation in Brazil was set up in 1930, when 30 kg of seeds were imported from South Africa (Higa et al. 2009).

Around the word, this species is implemented in plantations destined to the production of *firewood* (calorific value of ~4900 kcal/kg) (Sahri et al. 1998), *wood pulp* (Weber et al. 2007), *wood for construction*, and *adhesives* (Hoong et al. 2009).

It began to be used in *reforestation programs as a carbon sequestration plant* (Heriansyah et al. 2007, Tonini et al. 2010). In Brazil, the most common use is in reforestation programs of land reclamation, and this use was incorporated and popularized by Embrapa (the Brazilian Agricultural Research Corporation) in Brazil, especially in the Amazon and in Rio de Janeiro states.

Finally, this tree, in addition to *the various possible uses of its wood*, is approved due to *its melliferous flowers*; that is, its floral nectar is used by exotic bees of the genus *Apis* for the production of honey (Barbosa 2002). Acacia produces a *high amount of wood with low accumulation of nutrients, being a silvicultural option in areas with low soil fertility* (Franco and De Faria 1997, Balieiro et al. 2004).

In 2006 black Acacia (*Acacia mearnsii* De Wild.) was the third most planted forest species in Brazil, surpassed only by Eucalyptus spp. and Pinus spp. (SBS, 2006). Brazil has 9.85 million hectares of planted forests, 75.2% of which in eucalypt and 20.6% in Pinus, according to the survey Plant Production and Forestry 2017 (PEVS in Portuguese) (IBGE, 2017).

The use of *Acacia mearnsii* as a raw material for tannin, cellulose, and charcoal also presents great social importance, mostly in small properties. One strategy of cultivation is at the beginning of the crop and, at 3 years of age, the area is transformed into cattle pasture under the canopies (Mora, 2002).

From 2010 to 2016 the area planted with black Acacia species had been stabilized around 149.000 (±11.85 ha) according to Ibá and Pöyry (2016).

11.3 Success for the Successful

There are approximately 1350 species of the genus *Acacia* (Fabaceae) distributed worldwide (Maslin et al. 2003). We estimate that the area planted with trees of the genus *Acacia* in the world in the year 2000 was close to eight million hectares (FAO 2006). Nevertheless, several factors can influence the successful establishment of an exotic species. In most cases, only a small portion of species introduced are able to thrive in natural environments and indeed become invasive and maintain long-term viable populations and disperse from the area of introduction (Lockwood et al. 2001).

Australian acacias have *great adaptability and rapid growth*, which favors their popularity for introduction in several countries and climatic regimes, with some species introduced in more than 70 countries (Midgley and Turnbull 2003). However, acacias have a consistent history of introduction followed by invasion, which, since 1900, has been recorded throughout Southeast Asia (Richardson et al. 2015), South Africa, Hawaii (Richardson et al. 2011a), and Portugal (Fernandes et al. 2013).

Acacias possess a set of ecological characteristics that can be related to their establishment success such as *short generation time and high seed production* (Richardson et al. 2015). However, it is important to keep in mind that, as for other exotic species, in the journey of success of the acacias, the intentional introduction has aided the species to overcome biological filters (Groom et al. 2006), such as biogeographic isolation and biotic conditions, through human intervention. The probability of a successful establishment can be further increased by the fact that foresters actively select areas where environmental conditions are similar to those in the native ranges of the species (Richardson et al. 2015).

The genus *Acacia* was introduced in Brazil about a 100 years ago (Schneider et al. 1991). Within the genus *Acacia*, the species *A. mangium* Willd. and *A. mearnsii* De Wild. are those that occupy the largest area in Brazil being used in activities such as forestry, land reclamation, and urban and rural afforestation (Attias et al. 2014). In Brazil *A. mangium* has shown a cycle of 8–15 years, and although it has not been invasive in areas already vegetated, either under forest or planted pasture (Franco 2018) under native environments such as the savannas of the state of Roraima, and oligotrophic soils, this species has shown an invasive character (Aguiar Jr et al. 2014).

Other species were also brought to Brazil, as *Acacia auriculiformis* A. Cunn. ex Benth. and *Acacia holosericea* A. Cunn. ex G. Don., and were very successful in their establishment as well.

Coincidently, or not, those are the acacia species most frequently recorded in the regional lists of invasive alien species in Brazil (Attias et al. 2014). The Horus Institute, a Brazilian NGO dedicated to the invasive species study and control, evaluated *A. holosericea* as of high risk of invasiveness in a calculation of the risk of introduction.

According to Balieiro and Tonini (2018), the first experimental mixed plantations with *A. mangium* were carried out in 1979 by the Brazilian Agricultural Research Corporation (Embrapa) in Colombo (Paraná) and, in 1985, plantings were established in Minas Gerais. Around 1993, the Embrapa center at Seropédica (Rio de Janeiro) established the first experimental plantations with the species, which culminated later in the pioneering of that research center in the restoration of degraded areas in Brazil.

In the southeastern states in Brazil, *Acacia* spp. were planted for several purposes as afforestation of roads or highway, landscaping, soil conservation, and land reclamation. Since that time, technicians and specialists noticed the behavior of the species in invading open environments in a very aggressive way, beyond its planting areas (Balieiro, personal communication).

In Brazil, these species have been recorded in areas adjacent to their introduction sites (Mochiutti et al. 2007, Aguiar Jr et al. 2014). Plantations consist of a continuous source of dispersing individuals, increasing the chance of invasion by the species in adjacent areas. Furthermore, the anthropogenic disturbance generated by plantation management activities can modify the environmental conditions and facilitated the invasion process in these areas. However, the organized introduction of acacias in Brazil, by itself, does not explain all the success in the establishment in the new environment.

Some other ecological characteristics can be ascribed to them as their ability to define much of the structure of a new community as a *foundation species* (Ellison et al. 2005), *keystone species* (Mills et al. 1993), and *ecosystem engineer* (Ellison et al. 2005) and probably it is most consistently determined by their *propagule pressure* (Lockwood et al. 2005, Cassey et al. 2018).

Nowadays the evaluation of the risk of invasion of these species is already done by the NGO Horus Institute (2018) providing support for prevention strategies to be adopted before these species cause significant damage in new areas in Brazil, as they already do in other countries (Richardson and van Wilgen 2004).

Moreover, according to the National Database of Invasive Alien Species I3N in Brazil, the genus Acacia is spread across 11 states. The database indicates invasions at different levels by *Acacia auriculiformis*, *Acacia farnesiana*, *Acacia holosericea*, *Acacia longifolia*, *Acacia mangium*, *Acacia mearnsii*, and *Acacia podalyriifolia*.

It is important to consider that in their native environment in Australia, the species of the genus *Acacia* occupy a wide variety of environments, but are particularly prevalent in arid, semiarid, and subtropical dry regions. The seeds are produced in large quantities and have the capacity to remain viable for long periods (ABRS 2001). *Acacia mangium*, however, has very small and not very competitive seeds when competing with already established vegetation.

Acacia species as other plants known for their histories of introductions to new habitats by humans are in the selected group of species that easily become dominant. This can be attributed to their escape from specialist consumers and release from enemies, which is also thought by some researchers to lead to the evolution of increased competitive ability, driven by a decrease in the plant's resource allocation to consumer defense and an increase in allocation to size or fecundity (Callaway and Ridenour 2004).

In Brazil, the deliberate introduction of acacia species for multiple purposes possibly influenced the success of the species. In this sense, the propagule pressure is a determinant for the establishment and success of alien species. The number of individuals introduced to find a new population may have surpassed any obstacle or competitor in favor of the establishment of the acacias (Cassey et al. 2018).

A well-documented case of introduction occurred in 1998, when *A. mangium* was introduced in Roraima as an experimental planting of a thousand seedlings. Due to the apparent success of this planting, Walter Vogel founded Ouro Verde Agrosilvopastoril Ltda. (OVA) and started to invest in commercial acacia forestry in the Boa Vista, RR region, from 1999 (Forest Management Plan—Ouro Verde Project 2007).

During the process of establishing the company, Walter Vogel donated 100 seedlings of *A. mangium* to each public school in Boa Vista. Parts of these seedlings were planted within the school grounds and another part distributed to parents and school employees. Thus, because it provided efficient shading, the species was quickly accepted by the residents and disseminated through the city (Isabela Coutinho, resident of Boa Vista, personal communication), facilitating the invasion process.

In 2007, 80,866 ha of savannah land was occupied by the activities of the OVA company, of which 26,757 ha were planted with acacias, and the production capacity of *A. mangium* sawmills was estimated at 10,000 m^3/year (Forest Management Plan—Ouro Verde Project 2007).

As part of the preparation process of the *A. mangium* planting area, *Cajanus cajan* (L.) Millsp.), another exotic and invasive species in other regions, was planted with a function of wind breaking for the seedlings, facilitating the vertical growth of acacia seedlings.

According to the OVA Management Plan, in recognition of the invasive potential of *A. mangium*, a task force was created to once a year clean up all watercourses of the farms of the company of invasive vegetation of this species (Management Plan Forestry—Ouro Verde Project 2007). However, these measures have not been shown to be totally effective, with *A. mangium* occurring spontaneously outside the plantation area (Aguiar Jr et al. 2014).

11.4 But Why?

Popularly known in Brazil as Australian Acacia, *A. mangium*, or even only Acacia, it has a dense crown and white flowers and can reach up to 30 m in height (Lorenzi et al. 2003, Midgley and Turnbull 2003).

In plantations, flower and seed production starts at 2 years of age and mature pods can be observed 7 months after flowering. Flower pollination occurs by insects, mainly by bees (Midgley and Turnbull 2003). The fruits are spiral-shaped pods, which contain small black seeds that remain in the mature pods fixed by an orange aryl and which are dispersed naturally by wind and birds (Kull and Rangan 2008).

In natural habitat, *A. mangium* individuals are concentrated in lowland coastal areas at altitudes of up to 300 m. They grow on the banks of closed forests, in open forests, woods, and especially areas disturbed by fire (Midgley and Turnbull 2003). The *initial growth rate is directly proportional to the light incidence*, reaching its maximum in open areas. Many physiological characteristics define *A. mangium* as a *pioneer species that is fast growing and with ease of establishment* in a wide variety of environmental conditions, especially in humid tropical areas (Tong and Ng 2008).

Other characteristics of this species favor its dominant establishment in many places as, for example, the *ability to shade competitors* quickly, the *reduced amount of potential pathogens*, and the ability to capture large amounts of rainwater associated with essential nutrients in the runoffs of the trunk. *High tolerance to soils that are compacted* and very acidic (pH 4.2–6.5) and have low nutrient concentration also contributes to this facility (Lorenzi et al. 2003, Midgley and Turnbull 2003, Balieiro and Tonini 2018).

These characteristics guarantee to this species a *great competitive potential in environments under water stress and conditions of low soil fertility* (Faria et al. 1998, Balieiro et al. 2007). It is important to note that these same characteristics (e.g., large seed production, rapid initial growth, and shading capacity of competing species) are common features of invasive alien species (Rejmánek and Richardson 1996, Castro-Diez et al. 2011). However, as commented before, its small seeds are not very successful in establishing at seedlings stage competing in areas densely vegetated.

Despite all characteristics mentioned above, a special one that gives to Acacia species a great competitive advantage is the ability to grow on low-nitrogen soils due to the biological property of forming symbiosis with rhizobial bacteria (Duponnois and Plenchette 2003).

This is especially important in extremely adverse situations found in heavily degraded soils that have lost their upper horizons, where the physical and chemical factors are too restrictive for plant growth. Acacia symbioses with nodulating N_2-fixing bacteria and arbuscular mycorrhizal fungi constitute an efficient strategy to accelerate soil reclamation and initiate natural succession (Chaer et al. 2011, see also Chaps. 6 and 7).

11.5 Acacia and Its Interactions

Complex interactions can be generated as a result of the introduction of exotic plants into a new community. These interactions can cause disturbances in several environmental variables, affecting species, communities, and ecosystems (Richardson et al. 2011a, b, Vilà et al. 2011). In this context, the form and intensity with which an exotic species affects the physical, chemical, and biological environment are described as disturbance (Richardson et al. 2011a, b).

A. mangium plantations are often responsible for changes in various edaphic attributes. Studies conducted in Thailand have indicated that soil moisture within an acacia planting is lower than adjacent open areas, indicating a high rate of water consumption and competition among individuals for this resource (Sakai and Thaingam 1998 cited in Kamo et al. 2009). The concentration of nutrients in these environments can also be altered. Peak plant growth is capable of rapidly absorbing large amounts of nutrients, which can deplete the soil and limit the growth of individuals at advanced ages (Tong and Ng 2008, Nykvist and Sim 2009).

Soil acidification is common under cropping systems depending on N_2 fixation as a source of nitrogen if parts of crop removed contain large quantities of bases. For Nambiar et al. (2014) the effect of Acacia spp. in acidifying soils is questionable because of the confounding problems in the design of the studies. The pH fluctuation in time (or the moment of soil sampling) and the use of a pasture, abandoned land, or forests as reference treatments are some of those factors. In the case of *A. mangium*, if harvest residues and litter are maintained, the only removal of poles should not affect too much the soil pH (Franco, A. A., personal communication).

The amount of litter accumulated in the soil is influenced mainly by the amount of organic matter produced by the plants associated with the litter decomposition rate (Garay et al. 2003). *A. mangium* adult individuals do not have leaves but rather flat stems, called phyllody, with leaflike appearance. When they fall, these phyllodes decay very slowly, accumulating in the litter (Balieiro et al. 2004, Kull and Rangan 2008). The reduction of litter decomposition rate is a common disturbance in invasion processes (Vilà et al. 2011), but very desirable as a soil conditioner in tropical soils for sustainability of the system.

When compared decomposition rates were slower for acacia residues than for Eucalypt residues despite initial higher N and P concentrations in the Acacia residues. The decomposition rates depended on the carbon quality of the litter, primarily in terms of water-soluble compounds and lignin, and on the P availability (Bachega et al. 2016).

In mixed plantations of eucalyptus and Acacia the niche separation of the fine root system's architecture and distribution in the soil horizons, associated with the capacity of biological symbioses with nodulating N_2-fixing bacteria and the different dynamics of phosphorus, can be beneficial for both species (Balieiro, personal communication).

Parrotta and Knowles (1999), in a study of the Brazilian Amazon region, found that a higher litter accumulation and slower humus-layer formation in the natural regeneration and mixed native species plots are due to the relatively slow decomposition of the dominant species in these stands, when compared to other treatments, as *Cecropia* spp., or *Eucalyptus* spp. mixed with *Acacia mangium* in the mixed commercial species treatment. These trends may reflect treatment differences in the development and activity of litter invertebrate communities and other litter decomposers, a topic meriting further research.

The accumulation of litter under individuals of *A. mangium* benefits the recovery of degraded areas by protecting soil from erosion, soil temperature fluctuation, and increasing nutrient reserve (Balieiro et al. 2004). However, this accumulation may also be responsible for increased propensity for burning and inhibition of native seed germination (Parrotta and Knowles 1999, Balieiro et al. 2004, Kull and Rangan 2008).

The modifications observed in the soil can have influence not only on the local vegetation, but also on the edaphic fauna (see Chap. 8). Tsukamoto and Sabang (2005) reported the simplification of the structure of the macroinvertebrate soil community in a 14-year plantation in Malaysia. Compared to the adjacent native forest area, the acacia plantation had a total biomass four times higher, but with a different taxonomic composition and lower diversity. It is interesting to note that one of the dominant species of this community is a species of exotic earthworm that probably was also introduced during planting (Tsukamoto and Sabang 2005).

Another important aspect on soils and soil-fauna is the allelopathic effects attributed to the Acacia species (Lorenzo et al. 2008, Lorenzo et al. 2010). Callaway and Ridenour (2004) proposed that invaders with allelochemical substances have competitive advantages in their new habitats, which they named as "allelopathic advantage against resident species" hypothesis or "AARS."

The same experience as reported in Brazil is also reported in the rest of the world, accordingly with the International Union for Conservation of Nature (IUCN), about several environmental disturbances caused by the introduction of acacia and eucalyptus in Bangladesh since the 1980s. These include competition with native flora, high water consumption, reduction of soil fertility due to the deposition of slowly degradable leaves, inadequacy of their fruits and nectar for the consumption of native fauna, and production of pollen with a potential negative effect on the human respiratory tract (Barua et al. 2001).

Back to Brazil, in Roraima, 3- and 4-year plantations were responsible for the disappearance of native herbaceous vegetation through shading. In this same region, a large increase in the density of exotic bees of the genus *Apis* sp. (Barbosa 2002) was observed. This high density of bees harms the hunting and extraction activities of the indigenous populations that occupy lands near the plantations, generating complaints from them (C. Castilho, personal communication). On the other hand, an experiment of reclamation of mining residue, conducted in Porto Trombetas, Pará state, in a forest predominating matrix, has shown that, after 10 years of planting, several introductions of *A. mangium* have died out completely and have been replaced by a diverse and much superior biomass production than the plots planted with *Eucalyptus* spp. and that no seedling or plant of *A. mangium* was observed in the forest nearby (Campello 1999, Franco 2018).

In savanna areas adjacent to plantations in the state of Roraima, the spontaneous occurrence of adult reproductive *A. mangium* individuals was observed. In this location, the species density was inversely proportional to the distance of the source area, with 900 m being the maximum distance of dispersion observed (Aguiar Jr et al. 2014). According to Richardson et al. (2000), a species may be considered invasive if it is able to generate new reproductive individuals at a distance of more than 100 m from the source individuals in a period of less than 50 years. Thus, the study conducted by Aguiar Jr et al. (2014) shows that this species has invasive behavior in the Cerrado region of Roraima. After only 9 years of planting, *A. man-*

gium was able to disperse over long distances and reach the reproductive stage (Aguiar Jr et al. 2014). The same study found the presence of *Acacia mangium* individuals up to 900 m from the plantation edge 8–9 years after its introduction, independent of life stage or establishment pattern, indicating that this species can naturally disperse over long distances in open areas as natural Amazonian savanna.

11.6 We Are Not Alone: The Pity Comes by Horseback

The black acacia (*A. mearnsii*) is included in the list of the 100 "worst" invasive species in the world according to IUCN (Lowe et al. 2000); the invasion of natural environments by black acacia has not been extensively tested in Brazil, as in other parts of the world, even though it has been planted and used in the South for many years.

In Brazil, this species is found frequently in disturbed environments adjacent to crops, along roadsides and even in protected areas as parks and reserves. In Rio Grande do Sul, it was recorded as invasive in rural environments near cultivated areas. However, because it is a pioneer plant with high light demand, it is not able to establish itself in shady environments, areas of forest, or savannas (capoeiras in Portuguese) (Mochiutti et al. 2007). In this same Brazilian state, Nardelli (2004) recorded occurrences of black acacia in the natural ecosystems adjacent to plantations. The environment of black acacia plantations in Rio Grande do Sul is composed of areas of natural pasture, plains, areas of agricultural cultivation, field areas, regrown natural vegetation, called "capoeiras," and open environments with light availability, suitable for the establishment of the species.

In Paraná state, also located in the south of Brazil with a subtropical climate, *A. mearnsii* was recorded as an invasive species within the Vila Velha State Park (PEVV). At this site, black acacia is found in abundance in areas reforested with *Eucalyptus* spp. and in areas of intensive use (Carpanezzi 2011).

Given the increase in the number of individuals of Acacia and other exotic invasive tree species, as *Pinus elliottii* and *Pinus taeda* in the PEVV (Carpanezzi 2011), the Environmental Institute of Paraná, in partnership with other organizations, managed to promote the withdrawal of 50,000 trees from invasive alien species in 2007 (www.institutohorus.org.br/pr_vilavelha accessed 12/01/2019).

The invasion process of *A. mearnsii* has been responsible for several disturbances in the water balance and edaphic conditions of the invaded sites, affecting local biodiversity and economy (Moyo and Fatunbi 2010). This species has very high levels of water consumption, even when compared to other species of Acacia. This is due to its high rate of evapotranspiration and great capacity to capture rainwater and soil nutrients (Jobbágy and Jackson 2003). In South Africa, researchers estimate that the damage caused by this high water consumption is $ 2.8 million per year (Moyo and Fatunbi 2010). Also in South Africa, Richardson and van Wilgen

(2004) observed that, in riparian environments, commonly colonized by this species, this has a direct consequence in the reduction of the flow of streams.

The canopies of the black acacia, besides shading smaller species, like light-dependent grasses, produce large numbers of leaves. These are deposited in large quantities on the litter and, after their decomposition, can modify the composition of nutrients and minerals of the soil. It was observed that soils in areas dominated by black acacia are drier and more acidic (pH 4.4) when compared to natural areas of grasses (pH 5.3) (Moyo and Fatunbi 2010). The association of these factors, such as the change in chemical composition of the soil, the shading of open areas, and the large layer of litter formed by this species, tends to make it difficult to establish seedlings of native species and, consequently, process of natural succession.

The *biological* characteristics of *A. mearnsii*, such as the *large production of small seeds* and the *short juvenile phase*, allow it to disperse and colonize open areas rapidly. This combination of strategies, among others, makes it an aggressive invader (Rejmánek and Richardson 1996). This species has been recorded as an invasive species in several countries, with the most alarming cases being recorded in South Africa and Hawaii, USA (Daehler and Carino 2000, Richardson and van Wilgen 2004, Henderson 2007, Moyo and Fatunbi 2010).

Currently, *A. mearnsii* can be found in the following 42 countries: Afghanistan, Albania, Angola, Bangladesh, Bosnia and Herzegovina, Botswana, Brazil, Bulgaria, China, Colombia, Croatia, Eritrea, Ethiopia, Indonesia, Iran, Italy, Japan, Kenya, Lesotho, Malawi, Malaysia, Mexico, Mozambique, Myanmar, Namibia, New Zealand, Nicaragua, Pakistan, Panama, Papua New Guinea, Portugal, Romania, South Africa, Sri Lanka, Swaziland, Tanzania, Thailand, Uganda, the United States, Vietnam, Zambia, and Zimbabwe (www.worldagroforestry.org accessed on 07/03/2019).

Black acacia has been officially declared an invasive species in South Africa since 1984. In this country, it is estimated that more than 2.5 million hectares have already been invaded by the species, which is called "green cancer" (Galatowitsch and Richardson 2005).

In African biomes, the species mainly invades areas where there are fires, which stimulate the germination of seeds accumulated in the seed bank (Midgley and Turnbull 2003, Mochiutti et al. 2007, Moyo and Fatunbi 2010). In Hawaii, black acacia has invasive behavior and propagates easily in regions between 600 and 1200 m altitude, with rainfall between 1000 and 1200 mm.

It was only in the late 1930s that 65,000 individuals of this species were introduced into conservation areas in Hawaii. These individuals should be the progenitors of the invading individuals currently found, not only within conservation units but also in other natural environments (Little Jr. and Skolmen 1989, Stone et al. 1992, Frohlich and Lau 2008).

But things can be worse if acacia is planted associated with *Leucaena leucocephala* (Santos et al. 2009), another legume tree species on the list of 100 worst invasive species (Lowe et al. 2000) or other aggressive species.

This chapter presented several case studies in countries that recognized and studied the invasive potential of acacia genera, well documented in scientific literature accumulated over the decades. It is important to consider that we have in Brazil similar environmental conditions to those areas spread across the globe, especially in the tropics and even in Europe. Similarities in the history of invasion by acacia species when the weather and local conditions are favorable for their occurrence may produce the same patterns of invasion, which can happen here in Brazil or, may be, is already happening.

11.7 Conclusions: In Doubt Do Not Overtake

Australian acacias have several characteristic attributes of invasive species. These attributes are considered advantageous in the establishment, dispersal, and population growth phases (Rejmánek 1996, Rejmánek and Richardson 1996, Pysek and Richardson 2007). In addition, these species occupy an extensive geographical area in their native distribution in Oceania, being adapted to a wide range of climatic conditions.

This characteristic favors the establishment of these species in the introduction regions, especially in the tropics and subtropics possibly because these species have a higher capacity to overcome abiotic filters (Castro-Díez et al. 2011). Besides these, several other physiological and biological characteristics of both species are favorable to establishment and invasion.

The invasion process can be influenced equally by factors extrinsic to the species, such as the time elapsed since introduction and the number of forms of use (Pysek and Richardson 2007, Castro-Díez et al. 2011). The most serious problems related to invasion are caused commonly by species widely cultivated for long periods of time (Richardson 1998). The information compiled in this study demonstrates that *A. mangium* was introduced in the Brazilian territory more than 30 years ago (1979) and *A. mearnsii* approximately 100 years ago (1918), both being used for several purposes, and reports of them being invasive under our conditions until now are not very generalized.

Castro-Díez and his collaborators (2011) observed that the number of forms of human use is the most predictive feature of the distribution and abundance of Australian acacias in South Africa, where exotic acacias with the greatest number of uses were proven to be the most abundant and widely distributed. In addition, exotic species introduced through cultivation exert constant pressure of propagules due to the periodic introduction of new individuals. This fact increases the chances of the introduced species to find environments suitable for colonization, besides reducing the influence of environmental heterogeneity as a filter for the establishment of species (Wilson et al. 2009).

In this context, the set of characteristics of the studied species shows its potential invasion in Brazil and in the world. According to Nardelli (2004), Mochiutti et al. (2007), and Aguiar Jr et al. (2014), there are invasion records of *Acacia* spp. in several Brazilian regions as Amazonian savanna (3° north latitude), Brazilian lowland

Atlantic Forest in southeast in Espírito Santo state (19° south latitude), and grasslands (30° south latitude) in the Rio Grande do Sul state.

These are similar to invasion histories of these species in other regions of the world (Daehler and Carino 2000, Barua et al. 2001, Richardson and van Wilgen 2004, Henderson 2007, Moyo and Fatunbi 2010), where they have been used for a longer time, which may serve as an alert. In regions already extensively invaded by these species, such as South Africa, major economic and ecological damages have been reported. The Working for Water program, developed in South Africa, has invested more than US$ 125 million in the mechanical removal of Australian acacia trees in an area of 135,000 ha between 2000 and 2010. However, this program did not obtain success in the eradication or population reduction of the target species in all regions of the action (van Wilgen et al. 2011).

In order to prevent Brazil from having similar losses, preventive and control measures should be adopted, especially in the vicinity of the introduction areas (Richardson and Thuiller 2007). One trend observed in the study of the invasion process is that disturbed environments tend to be more susceptible to invasion (Pysek and Richardson 2007). Areas adjacent to commercial plantations are often deprived of their original vegetation and suffer from various disturbances associated with the management of plantations, such as the movement of machines.

Similarly, areas where species are introduced for the reclamation of degraded soils also tend to be altered environments, especially due to low species diversity. In this sense, these areas would be adequate for the beginning of the process of dispersion and establishment of the exotic species.

The most controversial management issues involve species that have invading behavior, causing serious damage and at the same time providing economic and ecological benefits in specific situations and areas. Australian acacias fit this profile and are managed in different ways in different countries. In South Africa, the planting of Australian acacia for commercial purposes is permitted in demarcated areas. However, all owners are responsible for controlling the dispersal of the species around their lands (van Wilgen et al. 2011).

A strategy to avoid dispersal of species from their places of introduction would be the "encircling" of commercial plantations by native species to the region in question, of fast growth and with great capacity of shadowing. Since both species have high affinity and similar dependence on light, especially in the initial stages of growth, the shading of the perimeter of the plantation could inhibit the germination of the seeds dispersed in the adjacent areas. The maximum dispersion distance of the species should be used as a parameter to define the width of this planting of native species.

In cases where the invasion process has already begun, measures to control invasive populations need to be adopted. In South Africa, young invading individuals are pulled out manually while adult individuals are mechanically cut and treated with herbicides (Garlon 4 or Timbrel) (Dahl et al. 2001). Adult individuals located in areas with steep slopes are ringed to reduce the risk of erosion due to the removal of trees (Department of Water Resources Republic of South Africa—http: //www.dwaf.gov.za/wfw/Control/). Due to the large number of seeds accumulated on the soil and the permanence of plant structures, such as stumps and roots, the removal of individual trees is followed by controlled burning of the region. This process, in addition to inhibiting regrowth, breaks seed dormancy and stimulates their germination. In this way, the seed bank is reduced and the new seedlings can be removed manually. Considering that invasion of riparian environments by these species is common, the use of herbicides should be cautious, since some of its components may persist in the environment for prolonged periods and contaminate water (Dahl et al. 2001).

Correia and Martins (2015) found the presence of *Acacia mangium* in the soil seed bank of a previously native forest area, which followed a reforestation project that contemplates the eradication after the period of planting and after a final clear-cut. This corroborates the proposal that the use of alien species in forestry processes may be, in part of the cases, the risk of contamination for the remaining ecosystems, proving that they can invade native forests.

The good notice that emerges from Correia and Martins (2015) is the low seed density of *Acacia mangium* that germinated under the canopy of this forest with a restoration project. It is plausible and may be inferred that the species is leaving the system for not finding sufficient light levels for germination and establishment. However, the presence of a few seeds on the native forest seed bank illustrates well the problem of the use in forest restoration projects with certain alien invasive species that may contaminate nearby preserved ecosystems.

In this sense, another ecological characteristic of *Acacia mangium* arises, that is, the short life cycle. A personal and empirical observation of a restoration project was conducted in the Piraí municipality, located in SE Brazil over the last two decades, close to the margins of a power plant water reservoir, where the area planted in the 1990s is nowadays almost gone after the trees' dieback. However, the clearings opened by the acacias' death after approximately 25 years after planting are not being colonized by native species, even in the presence of a source of propagules of native species in the forest patches on surroundings.

Considering that planting acacias is a consolidated activity in Brazil, the development of national invasion prevention strategies becomes indispensable. The creation of public policies to reduce the use of these species and regulate their management is essential. Its use for landscaping and the recovery of degraded areas proved to be inefficient and inadequate due to its invading status and should be replaced by native fast-growing species.

Its cultivation for commercial purposes must be regulated, with new policies to delimit the areas of planting and to control the invasion in the surrounding areas, following the example of South Africa. Within this regulation the responsibility of the entrepreneur for the containment and periodic control of the invasion (polluter pays principle) should be highlighted. In addition, detailed records and studies of control of these species in Brazil are required. This type of material is still scarce and not detailed. However, considering that invasion control is much more expensive in comparison to prevention (Thuiller et al. 2005, Richardson and Thuiller 2007, Broennimann and Guisan 2008) and that the invasion process in Brazil is

Table 11.1 Examples of Brazilian native species, which can be used as alternatives to Acacia species for multiple purposes

Acacias	Brazilian native species	
Use or resource	Species equivalent	References
Firewood	*Piptadenia gonoacantha*	Hansted et al. (2016)
Reforestation programs	*Anadenanthera peregrina* *Bombacopsis glabra* *Centrolobium tomentosum* *Citharexylum myrianthum* *Dalbergia nigra* *Peltophorum dubium* *Trema micrantha* *Xylopia brasiliensis*	Suganuma et al. (2014); Rodrigues et al. (2009)
Carbon sequestration plant	Multiple species	Brancalion et al. (2018)
Melliferous	*Mabea fistulifera, Croton sp.*	Oliveira et al. (2004)

already under way (Aguiar Jr et al. 2014), it is necessary that these control measures be systematized with urgency, based on the information available in Brazil and the world.

Although control is of importance, it is highly important to consider the socio-environmental interactions among local people and invader species. Regardless of local people considering the presence and the utility of those species as being positive because of their use as firewood, charcoal, wood, or even a crop, a scientific control will be necessary.

In this sense, Shackleton et al. (2019) recommend that policymakers and managers need to be more reflexive about the ways in which environmental problems are framed and to put those frames more in conversation with local people's experiences in order to productively resolve invasive species management dilemmas.

We propose that it is necessary to investigate the use of equivalent native species as a resource for the substitution of exotic trees, as shown in Table 11.1.

Therefore, the discussion about the use or not of exotic species that can be invasive still is in vogue. In this sense, an important question remains: *Why use exotic species in a mega diverse country*? The book "Ecological Imperialism by Crosby (2004) discusses the biological invasion of the new lands by what they call the "*portable biota*": the set of animals, plants, and diseases that have shipped with the Europeans in the caravels and risk and their analysis in the context of fifteenth century. Hence, after all consequences and all science produced should we still be doing the same thing in the twenty-first century?

In short: Acacias present a great invasive potential, especially in environments with water stress, and where soils are shallow or of low fertility. For being heliophilic, needing high light incidence, they are also competitive with native species in open areas, dominated by ground vegetation. This occurred with *Acacia mangium* in sites of Cerrado (Brazilian savannah) in Roraima and Amapá. The same is true for other acacias and different exotic species, as *Elaeis guineensis*, *Casuarina equisetifolia*, and *Pinus elliottii* among others.

Thus, acacias have a *high invasive potential* in open areas, as the Brazilian Cerrado, and should not be recommended in this biome. Nevertheless, they *show a low invasive potential* in forested areas and pastures, where they cannot compete with established vegetation.

Lastly this chapter is not a monolith engraved with biological xenophobia arguments. We are not proposing to ban or leave out introduced species as coffee, wheat, orange, sugar cane, corn, bananas, eucalyptus, pinus, acacias, most pasture grasses, and many other diverse sources of food but its mandatory to all technicians keep in mind, the intrinsic responsibility on the risks involved in the management of an exotic species, especially those with a large repertoire of invasions across the globe, in order to conserve and preserve local environments as mentioned throughout this chapter.

Acknowledgments CJRM thanks the Coordination of Superior Level Staff Improvement for the doctorate scholarship, NA thanks the National Council for Scientific and Technological Development for the master's scholarship and H.G.B. thanks Carlos Chagas Filho Foundation for Research Support of the State of Rio de Janeiro, National Council for Scientific and Technological Development, and Prociência/UERJ for research grants.

References

ABRS (2001) Flora of Australia, vol 11. CSIRO Publishing, Clayton, p 673

Aguiar A Jr, Barbosa RI, Barbosa JB, Mourão M Jr (2014) Invasion of *Acacia mangium* in Amazonian savannas following planting for forestry. Plant Ecol Divers 7(1–2):359–369

Attias N, Siqueira MF, & de Godoy Bergallo H (2014) Acácias australianas no Brasil: histórico, formas de uso e potencial de invasão. Biodiversidade Brasileira (2):74–96

Bachega LR, Bouillet JP, de Cássia Piccolo M, Saint-André L, Bouvet JM, Nouvellon Y et al (2016) Decomposition of *Eucalyptus grandis* and *Acacia mangium* leaves and fine roots in tropical conditions did not meet the home field advantage hypothesis. For Ecol Manag 359:33–43

Balieiro FDC, Tonini H (2018) Produção científica brasileira (2007–2016) sobre Acacia mangium Willd.: estado da arte e reflexões. Embrapa Solos-Artigo em periódico indexado (ALICE)

Balieiro FDC, Dias LE, Franco AA, Campello EF, de Faria SM (2004) Acúmulo de nutrientes na parte aérea, na serapilheira acumulada sobre o solo e decomposição de filódios de Acacia mangium Willd. Ciência Florestal 14(1):59–65

Balieiro FDC, Franco AA, Fontes RLF, Dias LE, Campello EFC, Faria SMD (2007) Evaluation of the throughfall and stemflow nutrient contents in mixed and pure plantations of Acacia mangium, *Pseudosamenea guachapele* and *Eucalyptus grandis*. Revista Árvore 31(2): 339–346

Barbosa RI (2002) Florestamento dos sistemas de vegetação aberta (Savanas/Cerrados) de Roraima por espécies exóticas (*Acacia mangium* Willd). Conselho Estadual de Meio Ambiente, Ciência e Tecnologia de Roraima, Boa Vista

Barua SP, Khan MMH, Reza AHMA (2001). The status of alien invasive species in Bangladesh and their impact on the ecosystems. In Report of the workshop on alien invasive species, GBF-SSEA, Colombo, Sri Lanka. IUCN Biodiversity Program, pp 1–8

Brancalion PH, Bello C, Chazdon RL, Galetti M, Jordano P, Lima RA et al (2018) Maximizing biodiversity conservation and carbon stocking in restored tropical forests. Conserv Lett 11(4):e12454

Broennimann O, Guisan A (2008) Predicting current and future biological invasions: both native and invaded ranges matter. Biol Lett 4(5):585–589

Callaway RM, Ridenour WM (2004) Novel weapons: invasive success and the evolution of increased competitive ability. Front Ecol Environ 2(8):436–443

Campello EFC (1999) A Influência de Leguminosas Arbóreas Fixadoras de Nitrogênio na Sucessão Vegetal em Áreas Degradadas na Amazônia Viçosa, UFV, 121f. (PhD Thesis) (Forestry Science Institute, University Federal of Viçosa)
Carpanezzi, O.T.B. 2011. Espécies vegetais exóticas no parque estadual de Vila Velha: subsídios para controle e erradicação, p. 67-74. In: Carpanezzi, O.T.B. & Campo, J.B. (orgs.). Coletânea de Pesquisas: Parques Estaduais de Vila Velha, Cerrado e Guartelá. Instituto Ambiental do Paraná: Curitiba, p 374
Carvalho SR, Almeida DL, Aronovich S, Camargo Filho ST, Dias PF, Franco AA (1998) Recuperação de Áreas Degradadas do Estado do Rio de Janeiro. Documentos (76) da Embrapa-CNPAB, p 11
Cassey P, Delean S, Lockwood JL, Sadowski JS, Blackburn TM (2018) Dissecting the null model for biological invasions: A meta-analysis of the propagule pressure effect. PLoS Biol 16(4): e2005987
Castro-Díez P, Langendoen T, Poorter L, López AS (2011) Predicting Acacia invasive success in South Africa on the basis of functional traits, native climatic niche and human use. Biodivers Conserv 20(12):2729–2743
Chaer GM, Resende AS, Campello EFC, de Faria SM, Boddey RM (2011) Nitrogen-fixing legume tree species for the reclamation of severely degraded lands in Brazil. Tree Physiol 31(2):139–149
Correia GG DS, Martins SV (2015) Banco de sementes do solo de floresta restaurada, Reserva Natural Vale, ES. Floresta e Ambiente 22(1):79–87
Crosby AW (2004) Ecological imperialism: the biological expansion of Europe, 900-1900. Cambridge University Press
Daehler CC, Carino DA (2000) Predicting invasive plants: prospects for a general screening system based on current regional models. Biol Invasions 2:93–102
Dahl H, Jakobsen J, Raitzer DA (2001) Wattle eradication via the working for water program, compared with wattle utilization and management for Makomereng, South Africa. SLUSE Report, p 52
de Souza Correia GG, Martins SV (2015) Banco de sementes do solo de floresta restaurada, Reserva Natural Vale, ES. Floresta e Ambiente 22(1):79–87
Diamond J (2005) Collapse: how societies choose to fail or succeed. Divers Distrib 17:788–809
Duponnois R, Plenchette C (2003) A mycorrhiza helper bacterium enhances ectomycorrhizal and endomycorrhizal symbiosis of Australian Acacia species. Mycorrhiza 13(2):85–91
Ellison AM, Bank MS, Clinton BD, Colburn EA, Elliott K, Ford CR, ... Mohan J (2005) Loss of foundation species: consequences for the structure and dynamics of forested ecosystems. Front Ecol Environ 3(9):479–486
FAO (Food and Agriculture Organization) (2006) Global forest resources assessment 2005: progress towards sustainable forest management. Relatório Técnico, p 350
Faria SM, Franco AA, Campello EF, Silva EMR (1998) Recuperação de Solos Degradados com Leguminosas Noduladas e Micorrizadas. Documentos (77) da Embrapa-CNPAB, p 23
Fernandes M, Devy-Vareta N, Rangan H (2013) Plantas exóticas invasoras e instrumentos de gestão territorial. O caso paradigmático do género Acacia em Portugal. Revista de Geografia e ordenamento do território 1(4):83–107
Ferrão JEM (1993) A aventura das plantas e os descobrimentos portugueses, 2nd edn. Instituto de Investigação Científica Tropical, Lisboa
Franco AA, De Faria SM (1997) The contribution of N2-fixing tree legumes to land reclamation and sustainability in the tropics. Soil Biol Biochem 29(5–6):897–903
Franco AA (2018). In: Maria Filipipini Alba J (Editor Técnico.), Recuperação de áreas mineradas. 3ª Edição rev. e ampliada—Brasília, p 456
Frohlich D, Lau A (2008) New plant records from O'ahu for 2007. Bishop Museum Occasional Papers 100:3–12
Galatowitsch S, Richardson DM (2005) Riparian scrub recovery after clearing of invasive alien trees in headwater streams of the Western Cape, South Africa. Biol Conserv 122:509–521
Garay I, Kindel A, Carneiro R, Franco AA, Barros E, Abbadie L (2003) Comparação da matéria orgânica e de outros atributos do solo entre plantações de *Acacia mangium* e *Eucalyptus grandis*. Revista Brasileira de Ciências do Solo 27(1):705–712

Griffin AR, Midgley SJ, Bush D, Cunningham PJ, Rinaudo AT (2011) Global uses of Australian acacias–recent trends and future prospects. Divers Distrib 17(5):837–847

Groom MJ, Meffe GK, Carroll CR (2006) Principles of conservation biology. No. Sirsi i9780878935185. Sinauer Associates, Sunderland

Hansted ALS, Nakashima GT, Martins MP, Yamamoto H, Yamaji FM (2016) Comparative analyses of fast growing species in different moisture content for high quality solid fuel production. Fuel 184:180–184

Henderson L (2007) Invasive, naturalized and casual alien plants in southern Africa: a summary based on the Southern African Plant Invaders Atlas (SAPIA). Bothalia 37(2):215–248

Heriansyah I, Miyakuni K, Kato T, Kiyono Y, Kanazawa Y (2007) Growth characteristics and biomass accumulations of *Acacia mangium* under different management practices in Indonesia. J Trop For Sci 19(4):226–235

Higa RCV, Wrege MS, Mochiutti S, Mora AL, Higa AR, Simon AA (2009) Acácia negra. Embrapa Amapá-Capítulo em livro científico (ALICE)

Hoong YB, Paridah MT, Luqman CA, Koh MP, Loh YF (2009) Fortification of sulfited tannin from the bark of *Acacia mangium* with phenol–formaldehyde for use as plywood adhesive. Ind Crop Prod 30(3):416–421

IBGE (2017) Produção da extração vegetal e da silvicultura. Instituto Brasileiro de Geografia e Estatística

IBGE (2019) Produção da extração vegetal e da silvicultura. Instituto Brasileiro de Geografia e Estatística

Industria Brasileira de Árvores (2016) Anuárioestatístico 2016 ano base 2015

IUCN—ISSG (n.d.) —Base de Dados Global sobre Espécies Exóticas Invasoras. www.issg.org/database. Accessed 10 Jan 2019

Jobbágy EG, Jackson RB (2003) Patterns and mechanisms of soil acidification in the conversion of grasslands to forests. Biogeochemistry 64:205–229

Kamo K, Vacharangkura T, Tiyanon S, Viriyabuncha C, Thaingam R, Sakai M (2009) Response to unmanaged *Acacia mangium* plantations to delayed thinning in north-east Thailand. J Trop For Sci 21(3):223–234

Kleinpaul IS et al (2010) Plantiomisto de Eucalyptus urograndis e *Acacia mearnsii* em sistema agroflorestal: I-Produção de biomassa. Ciência Florestal 20(4):621–627

Kull CA, Rangan H (2008) Acacia exchanges: wattles, thorn trees, and the study of plant movements. Geoforum 39:1258–1272

Little EL Jr, Skolmen RG (1989) Common forest trees of Hawaii (native and introduced), p. 732-733. In: Agriculture Handbook, vol 679. Forest Service of the U.S. Department of Agriculture, Washington, DC, p 321

Lockwood JL, Simberloff D, Mckinney ML, Holle BV (2001) How many, and which, plants will invade natural areas? Biol Invasions 3:1–8

Lockwood JL, Cassey P, Blackburn T (2005) The role of propagule pressure in explaining species invasions. Trends Ecol Evol 20(5):223–228

Lorenzi H, Souza HM, Torres MAV, Bacher L (2003) Árvores Exóticas no Brasil: Madeireiras, ornamentais e aromáticas. Instituto Plantarum, Nova Odessa, p 352

Lorenzo P, Pazos-Malvido E, González L, Reigosa MJ (2008) Allelopathic interference of invasive Acacia dealbata: physiological effects. Allelopath J 22(2):64–76

Lorenzo P, González L, Reigosa MJ (2010) The genus Acacia as invader: the characteristic case of *Acacia dealbata* Link in Europe. Ann For Sci 67(1):101

Lowe S, Browne M, Boudjelas S, De Poorter M (2000) 100 of the world's worst invasive alien species: a selection from the global invasive species database. Hollands Printing Ltd, Auckland, p 11p

Maslin BR, Miller JT, Seigler DS (2003) Overview of the generic status of Acacia (Leguminosae: Mimosoideae). Aust Syst Bot 16(1):1–18

Midgley SJ, Turnbull JW (2003) Domestication and use of Australian acacias: case studies of five important species. Aust Syst Bot 16(1):89–102

Mills LS, Soulé ME, Doak DF (1993) The keystone-species concept in ecology and conservation. BioScience 43(4):219–224

Mochiutti S, Higa AR, Simon AA (2007) Susceptibilidade de ambientes campestres à invasão de acácia-negra (Acacia mearnsii De Wild.) no Rio Grande do Sul. Floresta 37(2):239–253

Mora AL (2002) Aumento da produção de sementes geneticamente melhoradas de Acacia mearnsii De Wild. (Acácia negra) no Rio Grande do Sul. Curitiba. 140 f. Tese (Doutorado em Ciências Florestais)—Setor de Ciências Agrárias. Universidade Federal do Paraná, Curitiba

Moyo HPM, Fatunbi AO (2010) Utilitarian perspective of the invasion of some South African biomes by Acacia mearnsii. Global J Environ Res 4(1):6–17

Nambiar SEK, Hardwood C, Kien ND (2014) Acacia plantations in Vietnam: research and knowledge application to secure a sustainable future. Southern Forests 77:1–10

Nardelli A (2004) Resumo Público do Relatório de Certificação de Manejo Florestal da Tanagro S.A, vol 27. RelatórioTécnico, p 46

Nykvist N, Sim BL (2009) Changes in carbon and inorganic nutrients after clear felling a rainforest in Malaysia and planting with Acacia mangium. J Trop For Sci 21(2):98–112

Oliveira EPF, Castro TMDL, Venâncio SM (2004) Espécies de formigas que interagem com as sementes de Mabea fistulifera Mart.(Euphorbiaceae). Revista Árvore 28(5):733–738

Parrotta JA, Knowles OH (1999) Restoration of tropical moist forests on bauxite-mined lands in the Brazilian Amazon. Restor Ecol 7(2):103–116

Pimentel D, McNair S, Janecka J, Wightman J, Simmonds C, O'Connell C, Wong E, Russel L, Zern J, Aquino T, Tsomondo T (2001) Economic and environmental threats of alien plant, animal, and microbe invasions. Agric Ecosyst Environ 84:1–20

Pysek P, Richardson DM (2007) Traits associated with invasiveness in alien plants: where do we stand. In: Nentwig W (ed) Biological invasions, ecological studies, vol 193. Springer, Berlin, p 441

Rejmánek M, Richardson DM (1996) What attributes make some plant species more invasive? Ecology 77(6):1655–1661

Rejmánek M (1996) A theory of seed plant invasiveness: the first sketch. Biol Conserv 78:171–181

Richardson DM, van Wilgen BW (2004) Invasive alien plants in South Africa: how well do we understand the ecological impacts? S Afr J Sci 100(Feb):45–52

Richardson DM, Thuiller W (2007) Home away from home—objective mapping of high-risk source areas for plant introductions. Divers Distrib 13(3):299–312

Richardson DM (1998) Forestry trees as invasive aliens. Conserv Biol 12(1):18–26

Richardson DM, Carruthers J, Hui C, Impson FAC, Miller JT, Robertson MP, Rouget M, Le Roux JJ, Wilson JRU (2011a) Human mediated introduction of Australian acacia—a global experiment in biogeography. Divers Distrib 17:771–787

Richardson DM, Pysek P, Carlton JT (2011b) A compendium of essential concepts and terminology in invasion ecology, p. 409–420. In: Richardson DM (ed) Fifty years of invasion ecology: the legacy of Charles Elton. Blackwell, Oxford, p 456

Richardson DM, Pysek P, Rejmánek M, Barbour MG, Panetta FD, West CJ (2000) Naturalization and invasion of alien plants: concepts and definitions. Divers Distrib 6(2):93–107

Rodrigues RR, Lima RA, Gandolfi S, Nave AG (2009) On the restoration of high diversity forests: 30 years of experience in the Brazilian Atlantic Forest. Biol Conserv 142(6):1242–1251

Rouget M, Robertson MP, Wilson JR, Hui C, Essl F, Renteria JL, Richardson DM (2016) Invasion debt–Quantifying future biological invasions. Divers Distrib 22(4):445–456

Sahri MH, Ashaari ZA. I. D. O. N, Kader RA, & Mohmod AL (1998) Physical and mechanical properties of Acacia mangium and Acacia auriculiformis from different provenances. Pertanika J Trop Agric Sci, 21(2):73–82

Santos AW, Costa NN, Silva RR, Correa BC, Machado KK, Moura EG (2009) Investigação do potencial sustentável de combinações de leguminosas em aléias para o uso em solos tropicais. Revista Brasileira de Agroecologia 4:2

Schneider PR, Oesten G, Brill A, Mainardi GL (1991) Determinação da produção de casca em acácia- negra, Acacia mearnsii De Wild. Ciência Florestal 1(1):64–75

Shackleton RT, Richardson DM, Shackleton CM, Bennett B, Crowley SL, Dehnen-Schmutz K, Marchante E (2019) Explaining people's perceptions of invasive alien species: A conceptual framework. J Environ Manag 229:10–26

Souza CR, Rossi LMB, Azevedo CP, Lima RMB (2004) Comportamento da *Acacia mangium* e de clones de Eucalyptus grandis x E. urophylla em plantios experimentais na Amazônia Central. Scientia Forestalis 65:95–101

Stone CP, Smith CW, Tunison JT (1992) Alien plant invasions in native ecosystems of Hawaii: management and research. University of Hawaii Press, Honolulu, HI, p 900

Suganuma MS, Assis GB, Durigan G (2014) Changes in plant species composition and functional traits along the successional trajectory of a restored patch of Atlantic Forest. Community Ecol 15(1):27–36

Thuiller W, Richardson DM, Pyšek P, Midgley GF, Hughes G, Rouget M (2005) Niche-based modelling as a tool for predicting the risk of alien plant invasions at a global scale. Glob Chang Biol 11:2234–2250

Tong PS, Ng FSP (2008) Effect of light intensity on growth, leaf production, leaf lifespan and leaf nutrient budgets of *Acacia mangium*, *Cinnamom uminers*, *Dyera costulata*, *Eusideroxylon zwageri* and *Shorea roxburghii*. J Trop For Sci 20(3):218–234

Tsukamoto J, Sabang J (2005) Soil macro-fauna in an *Acacia mangium* plantation in comparison to that in a primary mixed dipterocarp forest in the lowlands of Sarawak, Malaysia. Pedobiologia 49:69–80

van Wilgen BW, Dyer C, Hoffmann JH, Ivey P, Le Maitre DC, Moore JL, Richardson DM, Rouget M, Wannenburgh A, Wilson JRU (2011) National-scale strategic approaches for managing introduced plants: insights from Australian acacias in South Africa. Divers Distrib 17:1060–1075

Vilà M, Espinar JL, Hejda M, Hulme PE, Jorosik V, Maron JL, Pergl J, Schaffner U, Sun Y, Pysek P (2011) Ecological impacts of invasive alien plants: a meta-analysis of their effects on species, communities and ecosystems. Ecol Lett 14:702–708

Vitousek PM, Loope LL, Westbrooks R (2017) Biological invasions as global environmental change. Am Sci 84(5):468

Weber J, Tham FY, Galiana A, Prin Y, Ducousso M, Lee SK (2007) Effects of nitrogen source on the growth and nodulation of *Acacia mangium* in aeroponic culture. J Trop For Sci 19(2): 103–112

Wilson JRU, Dormontt EE, Prentis PJ, Lowe AJ, Richardson DM (2009) Something in the way you move: dispersal pathways affect invasion success. Trends Ecol Evol 24(3):136–144

Xiong Y, Xia H, Li Z, Cai X, Fu S (2008) Impacts of litter and understory removal on soil properties in a subtropical *Acacia mangium* plantation in China. Plant Soil 304:179–188

Chapter 12
Multifunctional Mixed-Forest Plantations: The Use of Brazilian Native Leguminous Tree Species for Sustainable Rural Development

Antonio Carlos Gama-Rodrigues

12.1 Introduction

As a result of global climate change, declining biodiversity, increasing soil degradation, and diminishing water resources, there has been a growing recognition of the relevance of mixed-species plantations for sustainable rural development, integrating actions to ensure water security, energy, and food, in such a way that this forest system can be considered an adequate technique of climate-smart forestry. In this context, there is evidence that mixed-species plantations have high potential to achieving higher productivity than monocultures (Binkley 1992; Wormald 1992; Petit and Montagnini 2006; Kelty 2006; Gama-Rodrigues et al. 2007; Piotto 2008; Piotto et al. 2010; Pretzsch et al. 2017). However, there are limited examples of successful mixed-species plantations, especially mixtures with indigenous tropical tree species (Liu et al. 2018). In Brazil, successful mixed-species plantations have been with fast-growing, exotic, and low-density wood species such as *Eucalyptus* and *Acacia* (Bouillet et al. 2013; Chap. 2), while studies on mixtures of N_2-fixing and non-N_2-fixing native tree species are still quite scarce, despite the high diversity of tree species in all forest types that make up the various Brazilian biomes.

Thus, this chapter focuses on the Brazilian experience, complemented with other experiences in tropical and subtropical countries, on the potential of mixed-species plantations with native tree species and N_2-fixing leguminous species for timber and non-timber products, for the reclamation of degraded lands and for environmental services.

A. C. Gama-Rodrigues (✉)
Darcy Ribeiro State University of Norte Fluminense, Campos dos Goytacazes, RJ, Brazil
e-mail: tonygama@uenf.br

E. J. Bran Nogueira Cardoso et al. (eds.), *Mixed Plantations of Eucalyptus and Leguminous Trees*, https://doi.org/10.1007/978-3-030-32365-3_12

12.2 Planted Forests

The forest plantations in several regions of Brazil are basically with *Eucalyptus* and *Pinus*, but in recent years native and other exotic tree species have been widely used. Currently, the area of forest planted with other species occupies 521.1 thousand ha, corresponding to 7.3% of the total area of existing forest plantations. The species *Acacia mangium* (acácia), *A. mearnsii* (acácia negra), *Hevea brasiliensis* (seringueira), *Tectona grandis* (teca), and *Schizolobium parahyba* var. *amazonicum* (paricá) are established in greater planted area, accounting for 83.4%. Of this total, only *Schizolobium parahyba* occupies a planted area of 87,901 ha, accounting for 16.9% of the total area of plantations with other tree species in Brazil (ABRAF 2013; Cordeiro et al. 2015). *Schizolobium parahyba* has been planted commercially, around 20,000 ha, in the states of Acre, Mato Grosso, Pará, Rondônia, Maranhão, and Tocantins. However, these monospecific plantations with a non-N_2-fixing legume species are usually very heterogeneous and irregular and apparently the results obtained are not satisfactory (Carvalho 2007). As a result, this tree species is used in mixed-species plantations, particularly by association of *Cordia goeldiana* (freijó) and *Swietenia macrophylla* (mogno) (Cordeiro et al. 2015), and also as a shade tree in coffee or cacao plantations in Rondônia and Pará. The planted area of *Acacia mangium* and *A. mearnsii*, both N_2-fixing legume species, is 148.3 thousand ha, concentrated in the states of Amapá, Mato Grosso, Paraná, Roraima, Rio Grande do Sul, and Amazonas (ABRAF 2013). *A. mearnsii* is planted commercially in mixed stands with *Eucalyptus sp.* in Rio Grande do Sul, while *A. mangium* is also used mostly in mixed stands with eucalypt or in ecological restoration areas.

12.3 Mixed Plantations Including Native Species

Brazil has many leguminous tree species which are suitable to produce timber and other products. Knowledge of the functional ecological traits of this tree species group is important to ensure the sustainability of the mixed-species plantations. One of the most important specific attributes is the ability of some species to associate with diazotrophic N_2-fixing bacteria (see Chap. 6). In this case, the Atlantic Forest has 469 species of leguminous trees identified, but only 174 species have positive nodulation registration by N_2-fixing bacteria (Canosa et al. 2012).

The species that compose the Atlantic Forest fragments showed different abilities to absorb nutrients (Leão and Silva 1991; Cunha et al. 2009). When these native tree species were implanted in monocultures, they caused changes in soil physical, chemical, and microbiological properties, differentiating them markedly from the soils under natural forest (Silva 1988; Gama-Rodrigues et al. 1999; Gama-Rodrigues and Barros 2002; Gama-Rodrigues et al. 2008; Gama-Rodrigues et al. 2011a). In these forest plantations, litterfall is season dependent and its decomposition rates are variable (Vinha and Pereira 1983; Vinha et al. 1985; Gama-Rodrigues et al. 2003). Further, the trees grow very slowly in height and diameter for the first

15 years after planting (Vinha and Lobão 1989). An alternative, therefore, to native-species monocultures in tropical regions, both for the reclamation of degraded soils and timber production, is mixed-species plantation systems, which can promote ecophysiological conditions favorable to higher tree growth in addition to providing improvements in soil structure and increasing organic matter content and nutrient availability (Gama-Rodrigues et al. 2007). This type of forest system aims to maximize the ecophysiological attributes of native tree species when in their natural environment.

Based on this, the general hypothesis is that native tree species, when removed from the natural forest and placed in pure plantations, would not present an adequate form (accentuated cymose branching) to commercialize and would always grow below the expectation. There are several reasons for a better development of the native tree species in mixed plantations, as a relatively constant litterfall rate throughout the year, the more varied kinds of litter and of nutrient transfer through litterfall, and the diverse quality of the litter components, leading to a more uniform decomposition rate, besides the complementary role of the different tree species in nutrient uptake, resulting in improved soil fertility (Gama-Rodrigues et al. 2007). A mixed-species plantation reproduces the complex interactions of a natural forest. In such an ecosystem, collective and emergent properties are manifested simultaneously (Salt 1979; Odum 1983). The continuous litterfall rate in the mixed-species plantation is a collective property of the tree species on the site. The litter decomposition rate represents an emergent property, due to the interaction of decomposition processes of its components and not only of the total sum of the individual rates of the tree species (Gama-Rodrigues et al. 2003). Thus, the complementary interactions overlap with those of competition, providing greater stability to the forest ecosystem. Therefore, priority for the establishment of a mixed planting is the combination of species with complementary ecophysiological attributes, such as species with high nutrient cycling rates together with species of high nutrient-use efficiency (Gama-Rodrigues et al. 2007). In this sense, some N_2-fixing leguminous tree species in the stands may be planted solely to improve growing conditions for the target timber tree species. In turn, in monocultures, nutrient cycling is restricted to the litter of the planted tree species. The eventual slow litter decomposition of one particular tree species would diminish the rate of nutrient cycling, which would affect tree growth, independently of any improvement in soil fertility.

12.4 Case Studies

A mixed-species plantation for timber production can be managed in two ways: (1) it may be designed to harvest crop trees all at the same time or (2) the plantation may allow for different species maturation rates, with harvests occurring years to decades apart (Hall and Asthon 2016). In this sense, Gama-Rodrigues et al. (2007) evaluated the biomass and nutrient cycling in 22-year-old pure and mixed stands of six native hardwood species of the Atlantic Forest in southeastern

Table 12.1 Diameter at breast height (DBH), total height (TH), and tree stem volume (VOL) of native forest species, in pure (P) and mixed stands (M) in southeastern Bahia, Brazil

	DBH		TH		VOL	
Species	P (cm)	M (cm)	P (m)	M (m)	P (m^3 tree^{-1})	M (m^3 tree^{-1})
P. angustiflora	16.7	13.1	12.5	15.0	0.114	0.088
C. robustum	14.6	14.6	15.5	19.8	0.125	0.179
A. psilophylla	13.7	14.6	12.7	15.3	0.099	0.155
S. chrysophyllum	18.1	17.5	12.3	18.2	0.127	0.254
C. trichotoma	14.8	16.9	15.5	19.9	0.142	0.222
M. latifolium	16.9	18.0	11.5	16.5	0.166	0.204
Mean	15.8[a]	15.8[b]	13.3	17.5	0.129	0.184

[a]Mean of the pure stands
[b]Mean of the species in the mixed stand (from Gama-Rodrigues et al. 2007)

Table 12.2 Biomass of components of native forest species, in pure (P) and mixed (M) stands

Species	Leaf		Branch		Bark		Bolewood		Total	
	P	M	P	M	P	M	P	M	P	M
	kg tree^{-1}									
P. angustiflora	5.0	7.3	34.0	38.9	8.1	5.7	93	69	140	121
C. robustum	3.0	2.9	11.3	11.5	14.0	16.9	60	81	88	112
A. psilophylla	7.3	20.1	23.0	55.2	6.3	11.1	60	105	95	191
S. chrysophyllum	9.8	9.6	20.9	22.4	8.8	11.8	76	84	116	128
C. trichotoma	1.6	4.2	10.4	15.6	10.0	15.6	69	111	91	147
M. latifolium	6.8	8.8	20.1	22.1	15.0	12.0	84	109	121	152
Mean	5.6[a]	8.8[b]	20.0	27.6	10.4	12.0	74	93	110	142

[a]Mean of the pure stands
[b]Mean of the species in the mixed stand (from Gama-Rodrigues et al. 2007)

Bahia. The following species were studied: *Peltogyne angustiflora* (pau-roxo), *Centrolobium robustum* (putumuju), *Arapatiella psilophylla* (arapati), *Sclerolobium chrysophyllum*(arapaçu), *Macrolobium latifolium* (óleo-comumbá), *Cordia trichotoma* (claraíba). The first five species are leguminous trees. The mixed species outmatched pure stands in height, stem volume, and total biomass (Tables 12.1 and 12.2). Thus, growth and yield in mixed-species stands were higher than in pure stands owing to the combination of species with complementary ecophysiological attributes, consequently improving the efficiency in nutrient use and cycling. In contrast, Silva and Vinha (1991) did not find any differences in tree height of the native species *Arapatiella psilophylla*, *Bombax macrophyllum* (imbiruçu), *Bowdichia virgilioides* (sucupira-preto), and the exotic *Gmelina arborea* (gmelina) between pure and mixed stands under the same environmental conditions. In southern Bahia, Vinha (1992) also reported that in plantations with *Bombax macrophyllum* shaded with *Leucaena leucocephala* (leucena) the survival rate was approximately 90%, whereas in pure plantings it was little more than 10%. Thus, empirical evaluations showed that, in general, the satisfactory further development of native hardwood species in southern Bahia occurred in mixed stands and not in pure stands.

Mixed stands of six hardwood species at the age of 9 years (farinha seca—*Peltophorum dubium*, ipê rosa—*Handroanthus heptaphylla*, ipê amarelo—*Handroanthus chrysotrichus*, mogno africano *Khaya* sp., paineira—*Ceiba speciosa*, and sobrasil—*Colubrina glandulosa*) with 75% density of fast-growing N_2-fixing leguminous trees (acacia auriculada—*Acacia auriculiformis*, acacia—*Acacia mangium*, guachapele—*Pseudosamanea guachapele*, jaracandá da Bahia—*Dalbergia nigra*, orelha de negro—*Enterolobium contortisiliquum*, pau-jacaré—*Piptadenia gonoacantha*, and sabiá—*Mimosa caesalpiniaefolia*) can increase soil P concentrations through cycling and accumulation of organic and inorganic P compounds, and conserving N and organic matter in a red-yellow Latosol, in the state of Rio de Janeiro (Aleixo et al. 2019, *in press*). However, growth and yield of all tree species were not satisfactory (Fig. 12.1).

Fig. 12.1 Mixed stand with six hardwood species and fast-growing N_2-fixing leguminous trees for the reclamation of degraded lands in southern Rio de Janeiro, Brazil (photo: A.C. Gama-Rodrigues)

Fig. 12.2 Consortium of *Acacia mangium* (for energy) and *Schizolobium amazonicum* (for laminate) in Paragominas, state of Pará, Amazon, 10 months after planting (photo: J.L.M. Gonçalves)

In the Amazon region, several consortium combinations have been tested between *Schizolobium parahyba* and other tree species (Fig. 12.2). In southern Pará, mean annual increments in height, diameter at breast height, and volume of *Schizolobium parahyba* were higher in mixed plantations with *Cordia goeldiana* and *Swietenia macrophylla* than in pure plantations (Cordeiro et al. 2015). Souza et al. (2003) reported several consortia tested in the municipality of Igarapé-Açu, Pará: (1) *Schizolobium parahyba* × *Swietenia macrophylla* × *Protium heptaphyllum* (breu sucuruba); (2) *Schizolobium parahyba* × *Tectona grandis* × *Hymenaea courbaril* (jatobá); (3) *Schizolobium parahyba* × *Ceiba pentandra* (sumaúma) × *Ochroma pyramidale* (pau de balsa). The highest growth of the species was in consortium with *Swietenia macrophylla* and *Protium heptaphyllum*. In Amazonas, in a 4-year-old the consortium *Schizolobium parahyba* × *Carapa guianensis* (andiroba) presented the highest growth in height and diameter over the consortia *Schizolobium parahyba* × *Bertholletia excelsa* (castanha-do-pará) and *Schizolobium parahyba* × *Swietenia macrophylla*. Even though *S. parahyba* is a legume tree it is a non-nodulating species and may not contribute with N to the system.

In Costa Rica, 15–16-year-old mixed plantation of *Vochysia guatemalensis*, *Virola koschnyi*, *Jacaranda copaia*, *Terminalia amazonia*, and *Hieronyma alchorneoides* may be the preferred system for reforestation with native species designed for timber production or carbon sequestration because this system is more economically viable and productive than pure plantations (Piotto et al. 2010). Hall and Ashton (2016) describe a project of planting design and growth through time for a mixed-species timber plantation, with four native species in Panama: *Ormosia macrocalyx* (N_2-fixing species), *Dalbergia retusa* (also fixes N_2), *Terminalia amazonia*, and *Hieronyma alchorneoides*. According to the model *T. amazonia* and *H. alchorneoides* can be harvested at around 20 years; while *D. retusa* can be harvested at around 30 years of age.

The use of leguminous tree species associated with N_2-fixing bacteria and arbuscular mycorrhizal fungi is a technique that has shown viability to accelerate land reclamation and initiate natural succession in several regions of Brazil (Gama-Rodrigues et al. 2008; Chaer et al. 2011). However, trees from the genera *Acacia*, *Mimosa*, and *Gliricidia*, among other N_2-fixing species, have been used mainly in pure stands. The focus on commercial use of these species is secondary to improving environmental conditions. Among them, high inputs of organic matter via litterfall (Costa et al. 2014) and increases in the organic C and N contents of the soil and microbial biomass have been reported (Gama-Rodrigues et al. 2008). Advances as higher mineralization rates of C and N of the soil (Nunes et al. 2016), greater contents of soil organic P (Zaia et al. 2008), higher abundance, and richness of the soil fauna (Manhães et al. 2013; Bianchi et al. 2017) were also detected. Such improvements further reduce the risks of erosion, given the permanent cover of the soil by the accumulated litter. In a 7-year-old mixed-species stand with the native Atlantic Forest species *Anadenanthera falcata* (angico—N_2-fixing species), *Myracrodruon urundeuva* (aroeira), *Gochnatia polymorpha* (cambará), and *Tabebuia impetiginosa* (ipê-roxo) uniform litterfall rates enabled a more effective soil cover, even though pure stands of *Gochnatia polymorpha* and *Anadenanthera falcata* had showed the highest litterfall production (Garrido and Poggiani 1982). However, regarding the improvement of soil quality using leguminous trees, whether in pure or mixed systems, the magnitude of edaphic changes is related to the soil buffer capacity. In soils with high contents of organic matter and clay, it is expected that these changes are of low magnitude. In this situation, changes of soil properties will take longer periods of time. Therefore, in addition to soil resilience, the level of degradation and the type and intensity of land use influence the ability of forest species to change the soil attributes, and site productivity.

Rappaport and Montagnini (2014) evaluated the restoration potential of 21 native tree species of the Atlantic Forest 3 years after planting in the understory of an old rubber plantation in southern Bahia. Eight leguminous tree species were tested: *Arapatiella psilophylla* (arapati), *Caesalpinia echinata* (pau-brasil), *Copaifera lucens* (pau-óleo), *Tachigali densiflora* (ingá-açu), *Andira legalis* (angelim), *Swartzia macrostachya* (manga-brava), *Parkia pendula* (faveira), and *Inga heterophylla* (ingá-caixão). Among these eight tree species, *Parkia pendula* and *Tachigali densiflora* grew faster than the other leguminous tree species.

12.5 Use of N_2-Fixing Tree Species in Agroforestry Systems

Multistrata agroforestry systems are considered a good model of mixtures of N_2-fixing and non-N_2-fixing tree species. In this case, agroforestry systems (AFS) based on cacao (*Theobroma cacao*) are the best example of sustainability of shaded

Fig. 12.3 Cacao-cabruca systems in southeastern Bahia, Brazil (photo: A.C. Gama-Rodrigues)

tree-shrub systems because of their high potential for sequestering carbon, recycling nutrients, and providing other environmental services, in order to ensure greater diversity of multiple supply of timber and non-timber products. Brazil is one of the major cacao-producing countries, accounting for 5% of the world's cacao production. The main cacao-producing areas are (1) Amazon region (Pará, Rondônia, Mato Grosso, and Amazonas) in 192,411 ha of planted area and (2) south of Bahia and northern Espírito Santo in 550,000 ha of planted area. The cacao-cabruca and cacao shading leguminous tree systems are the matrices that dominate the landscape in the cocoa-growing region in southern Bahia. In *cabruca* systems, the cacao is planted under thinned natural forest shaded by native tree species (Fig. 12.3), where the cacao stand density is about 600 plants ha^{-1} and the remaining shade trees range from 30 to 70 individuals ha^{-1} with a mixture of N_2-fixing and non-N_2-fixing tree species. This plantation system is established in an area of 385,000 ha^{-1}. In these systems, the leguminous tree *Erythrina* spp. (erythrina) is the most common among such introduced shade trees in areas where all native forest has been removed (Fig. 12.4). In this Atlantic Forest system (AFS) cacao and erythrina are established at densities of 1111 and 32 plants ha^{-1}, respectively, and the canopy of erythrina is not pruned. This plantation system grew out of the 1960s, when CEPLAC (Executive Committee of the Plan of Cacao Farming) initiated a broad program aiming at significant increases in cacao production based on the reduction of cacao shading, eliminating 50–70% of the trees of the Atlantic Forest (Monroe et al. 2016).

Fig. 12.4 Cacao-erythrina agroforestry systems in southeastern Bahia, Brazil (photo: A.C. Gama-Rodrigues)

However, in the last two decades, CEPLAC has recommended that erythrina be replaced by multiple-use tree species such as rubber trees (*Hevea brasiliensis*) to increase the farmers' income. Currently, the planted area of the cocoa-rubber system is 12,000 ha. This AFS is established in double rows of rubber alternated with 4–5 cacao rows and a row of *Gliricidia sepium* (gliricidia), a N_2-fixing legume tree, in the cacao-planting area (Fig. 12.5). These three cacao AFS have high potential for sequestering carbon in the plant-soil system (Gama-Rodrigues et al. 2011b; Monroe et al. 2016), high nutrient cycling rates and soil nutrient stocks (Zaia et al. 2012; Fontes et al. 2014; Aleixo et al. 2017), and high soil biodiversity (Moço et al. 2010; Oliveira et al. 2018). Thus, these ecological processes are excellent evidence of the compatibility and complementarity of different species for the sustainability of multistrata production systems.

There are several models of agroforestry systems based on the shaded cacao with a mixture of N_2-fixing and non-N_2-fixing tree species in the Amazon region (Müller and Gama-Rodrigues 2012). Figure 12.6 shows the combination of cocoa with the tree species *Schizolobium amazonicum* (paricá), *Cordia alliodora* (freijó-louro), *Bagassa guianensis* (garrote), *Swietenia macrophylla* (mogno), *Bertholletia excelsa* (castanha-do-pará), and *Tabebuia heptaphylla* (ipê-roxo), all of them non N_2-fixing species. In Rondônia, this model has been used since 1973 and totals 30,650 hectares. It can be considered as an artificial cabruca system. Another plantation system design is the combination of cacao with coconut (*Cocos nucifera*) and gliricidia plus banana. In this system, the planting densities are cocoa 740 plants ha^{-1}, coconut 123

Fig. 12.5 Cacao-rubber agroforestry systems with gliricidia in southeastern Bahia, Brazil (photo: A.C. Gama-Rodrigues)

plants ha^{-1}, banana 740 plants ha^{-1}, and gliricidia 247 plants ha^{-1} (Fig. 12.7). Currently, there are about 500 ha of this mixed system. On the other hand, Almeida et al. (2009) evaluated the richness of tree species and timber potential, present in some cacao plantations established in Ouro Preto do Oeste, Rondônia, since the 1980s. The number of tree species found during the survey is described in Table 12.3. The leguminous tree species accounted for 22% of all tree species surveyed. The largest number of leguminous tree species was Caesalpinioideae (six species), followed by Mimosoideae (five species) and Papilionoideae (three species). Most shade tree species came from natural regeneration and only 2–20% from seedling planting, with *Schizolobium amazonicum* being the most common tree species, with 45% of individuals inventoried and timber average volume of 51 m^3 ha^{-1}. Additionally, these cacao AFS with this profile of floristic diversity could be considered in the Amazon as areas of forest restoration in the properties with demand of environmental liabilities.

Forest-based fallow systems enriched with fast-growing N_2-fixing leguminous tree species appear to be a viable option for managing N despite the large amount of N removed from the system as timber and fuelwood (Gama-Rodrigues 2011). This technique has been used as an alternative to the use of fire and reduction of a fallow period in slash-and-burn systems adopted by family farming in northeast Pará (Kato et al. 2006). The selected leguminous tree species to accelerate the accumulation of biomass and nutrients are *Acacia angustissima*, *A. auriculiformis*, *Acacia mangium*,

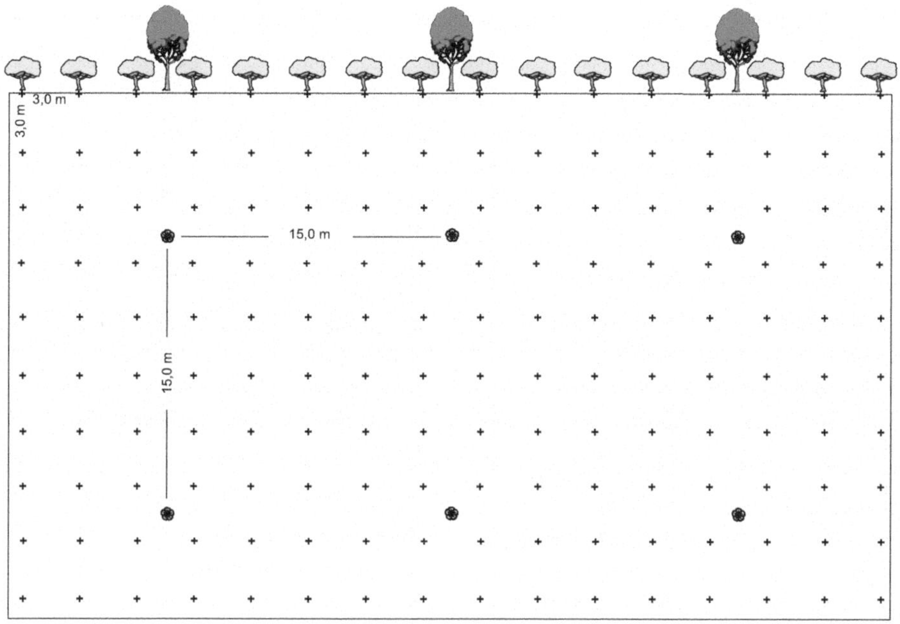

Fig. 12.6 Schematic diagram showing the spatial arrangement of cacao-based (+) association with native tree species (❀): *Schizolobium amazonicum*, *Cordia alliodora*, *Bagassa guianensis*, *Swietenia macrophylla*, *Bertholletia excelsa*, and *Tabebuia heptaphylla* (from Müller and Gama-Rodrigues 2012)

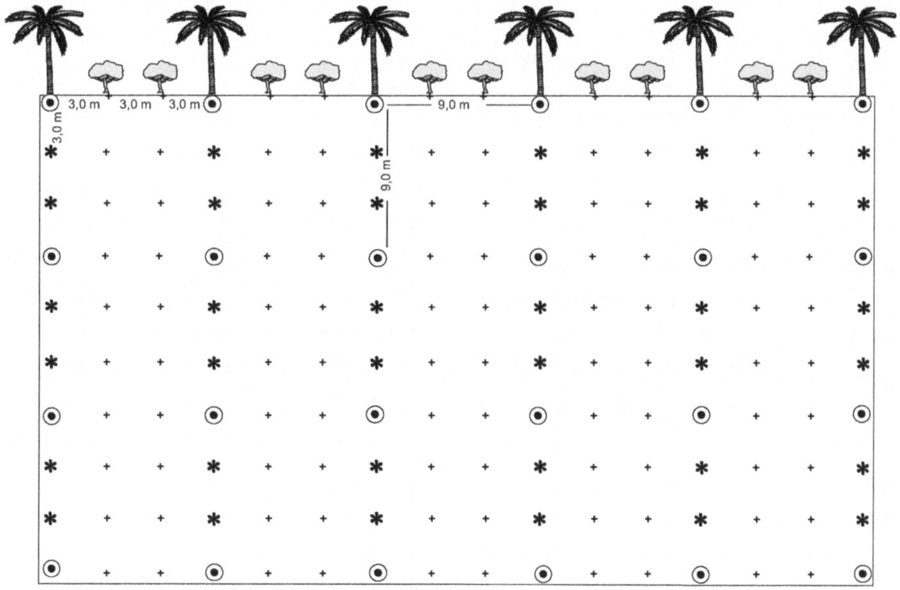

Fig. 12.7 Schematic diagram showing the spatial arrangement of cacao-based (+) association with coconut (●) and gliricidia (∗) (from Müller and Gama-Rodrigues 2012)

Table 12.3 Shading species present in four cocoa plantations in Ouro Preto do Oeste, Rondônia, Brazil

Common name (Portuguese)	Family	Genus or species
Abacateiro	Lauraceae	*Persea americana* Mill.
Algodoeiro ou Imbiruçu	Bombacaceae	*Eriotheca* sp.
Amoreira	Moraceae	*Maclura tinctoria* (L.) D. Don ex Steud
Angelim-saia	Fabaceae-Mimosoideae	*Parkia pendula* (Willd.) Benth. ex Walp.
Angico-branco	Fabaceae-Mimosoideae	*Piptadenia foliolosa* Benth
Angico-rosa	Fabaceae-Mimosoideae	*Parapiptadenia rigida* (Benth.) Brenan
Babaçu	Arecaceae	*Attalea speciosa* Mart. ex Spreng
Babão ou Coco Babão	Arecaceae	*Syagrus comosa* (Mart.) Mart.
Bacurizeiro	Arecaceae	*Attalea phalerata* Mart. ex Spreng
Bajinha	Fabaceae-Caesalpinioideae	*Pterogyne* sp.
Bandarra ou Paricá	Fabaceae-Caesalpinioideae	*Schizolobium parahyba* var. amazonicum (Huber ex Ducke) Barneby
Bolão	Sapotaceae	*Pouteria pachycarpa* Pires
Cajazinho ou Taperebá	Anacardiaceae	*Spondias mombin* L.
Canela	Lauraceae	*Nectandra* sp.
Castanheira-do-brasil	Lecythidaceae	*Bertholletia excelsa* Humb. & Bonpl.
Cebolão	Unknown	Unknown
Cedro-rosa	Meliaceae	*Cedrela odorata* L.
Cerejeira	Fabaceae-Papilionoideae	*Amburana acreana* Ducke (A.C.Sm.)
Coração-de-negro	Fabaceae-Papilionoideae	*Swartzia panacoco* Cowan
Camaruzeiro	Fabaceae-Papilionoideae	*Dipteryx* sp.
Embireira	Timeleaceae	*Daphnopsis* sp.
Farinha-seca	Chrysobalanaceae	*Parinari coriaceum* Benth
Feijão-cru	Fabaceae-Mimosoideae	*Pithecellobium saman* var. acutifolium Benth
Figueira	Moraceae	*Ficus* sp.
Freijó-cinza	Boraginaceae	*Cordia goeldiana* Huber
Garapa	Fabaceae-Caesalpinioideae	*Apuleia leiocarpa* (Vogel) J.F.Macbr
Garrote	Moraceae	*Bagassa guianensis* Aubl.
Gmelina ou Melina	Verbenaceae	*Gmelina arborea* Roxb.
Goiabeira	Myrtaceae	*Psidium guajyava* L.
Imbaúba	Cecropiaceae	*Cecropia* sp.
Ingazeira	Fabaceae-Mimosoideae	*Inga* sp.
Ipê-amarelo	Bignoniaceae	*Tabebuia incana* A. Gentry
Ipê-champagne	Bignoniaceae	*Tabebuia* sp.
Ipê-roxo	Bignoniaceae	*Tabebuia* sp.
Ipê-tabaco	Bignoniaceae	*Tabebuia serratifolia* (Vahi) Nichols.

(continued)

Table 12.3 (continued)

Jangada	Rubiaceae	*Guettarda viburnoides* Cham. &Schltdl.
Jaqueira	Moraceae	*Artocarpus integra* L.
Jatobá	Fabaceae-Papilionoideae	*Hymenaea oblongifolia* Huber
Jenipapeiro	Rubiaceae	*Genipacaruto* H.B.K.
Laranjeira	Rutaceae	*Citrus sinensis* (L.) Osbeck
Leiteira	Moraceae	*Brosimum* sp.
Limeira	Rutaceae	*Citrus bergamia* Risso
Limoeiro	Rutaceae	*Citrus limon* (L) Burm
Louro	Lauraceae	*Ocotea* sp.
Mamica-de-porca	Tutaceae	*Zanthoxylum acreanum* (Krause) J.F. Macbr.
Mangueira	Anacardiaceae	*Mangifera indica* L.
Maparaíba	Unknown	Unknown
Mogno	Meliaceae	*Swietenia macrophylla* KING
Mutamba	Sterculiaceae	*Guazuma* sp.
Paineira	Bombacaceae	*Chorisia* sp.
Pau d'alho	Phytolaccaceae	*Gallesia integrifolia* (Spreng.) Harms
Pau-sangue	Guttiferae	*Vismia brasiliensis* Choisy
Pintadinho	Fabaceae-Caesalpinioideae	*Poeppigia procera* Presl
Ponkan	Rutaceae	*Citrus reticulata* Blanco
Pupunheira	Arecaceae	*Bactris gasipaes* H.B.K.
Seringueira	Euphorbiaceae	*Hevea brasiliensis* (Willd. ex. Adr. De Juss.) Muell. Arg.
Sete Camadas	Unknown	Unknown
Sumaumeira	Bombacaceae	*Ceiba pentandra* (L) Gaertn.
Tarumã	Berbenaceae	*Vitex* sp.
Tauari	Lecytidaceae	*Couratari* sp.
Tucumazeiro	Arecaceae	*Astrocaryum* sp.
Unha-de-vaca	Fabaceae-Caesalpinioideae	*Bauhinia* sp.
Urtigão	Unknown	Unknown

From Almeida et al. 2009

Clitoria racemosa, *Inga edulis*, and *Sclerolobium paniculatum*. The fallow vegetation enriched with *A. mangium* is the one that presented the greatest potential for sequestering carbon 2 years after fallow.

12.6 Outlook and Conclusions

Mixed-forest plantations have high potential to be an alternative to conventional monoculture systems to address environmental, social, and economic issues in Brazil. But their integration in agricultural landscapes requires a qualitative SWOT (strengths-weaknesses-opportunities-threats) analysis of local stakeholders' perceptions. A priori, lack of systematic research and technology inputs to improve the system would be the major weaknesses, while sustainability, multi-functionality, and high sociocultural values would be common strengths. Potential government support, climate-smart forestry, and climate-change mitigation would constitute good opportunities, but lack of knowledge on economic viability and strong pressure towards high productivity would be threat factors. Thus, SWOT analyses should take from climate-smart forest to climate-smart landscapes. To achieve climate-smart landscapes, future research needs should prioritize forest breeding and biotechnology for multifunctional mixed-forest plantations integrated into the biological management of soil fertility and modeling of plant-soil interactions. Therefore, the technology produced based on ecological processes can support farmers as managers of complex social-ecological systems better than those based on technological packages. Thus, integrated management of natural resources can increase rural prosperity through better communication of results between different stakeholders.

References

ABRAF (2013) Anuário Estatístico ABRAF 2013 ano base 2012. ABRAF, Brasília, p 148

Aleixo S, Gama-Rodrigues AC, Costa MG, Sales MVS, Gama-Rodrigues EF, Marques JRB (2017) P transformations in cacao agroforests soils in the Atlantic forest region of Bahia, Brazil. Agrofor Syst 91:423–437

Aleixo S, Gama-Rodrigues AC, Gama-Rodrigues EF, Campello EFC, Silva EC, Furlan DA, Schripsema J (2019) Can tree legumes increase soil phosphorus availability? A link between the P and N cycles in the Brazilian Atlantic Forest. (in press)

Almeida CMVC, Locatelli M, Lima AA, Xavier IP, Cidin ACM (2009) Diversidade de espécies arbóreas e potencial madeireiro em sistemas agrossilviculturais com cacaueiro em Ouro Preto do Oeste, vol 21. Agrotrópica, Rondônia, Brasil, pp 73–82

Bianchi MO, Scoriza RN, Resende AS, Campello EFC, Correia MEF, Silva EMR (2017) Macrofauna edáfica como indicadora em revegetação com leguminosas arbóreas. Floresta e Ambiente 24:e00085714. https://doi.org/10.1590/2179-8087.085714

Binkley D (1992) Mixture of nitrogen-fixing and non-nitrogen-fixing tree species. In: Cannell MGR, Malcolm DC, Robertson PA (eds) The ecology of mixed-species stands of trees. Blackwell Scientific Publications, Oxford, pp 99–123

Bouillet JP, Laclau JP, Gonçalves JLM, Voigtlaender M, Gava JL, Leite FP, Hakamada R, Mareschal L, Mabiala A, Tardy F, Levillain J, Deleporte P, Epron D, Nouvellon Y (2013) *Eucalyptus* and *Acacia* tree growth over entire rotation in single-and mixed-species plantations across five sites in Brazil and Congo. Forest Ecol Manag 301:89–101

Canosa GA, Faria SM, Moraes LFD (2012) Leguminosas florestais da mata Atlântica brasileira fixadoras de nitrogênio atmosférico. Embrapa Agrobiologia. Comunicado Técnico 144:12

Carvalho PER (2007) Paricá (*Schizolobium amazonicum*). Embrapa Florestas. Circular Técnica 142:8

Chaer GM, Resende AS, Campello EFC, Faria SM, Boddey M (2011) Nitrogen-fixing legume tree species for the reclamation of severely degraded lands in Brazil. Tree Physiol 31:139–149

Cordeiro IMCC, Barros PLC, Lameira OA, Gazel Filho AB (2015) Avaliação de plantios de paricá *Schizolobium parahyba* var. *amazonicum* (Huber ex Ducke) Barneby de diferentes idades e sistemas de cultivo no município de aurora do Pará—PA (Brasil). Ciência Florestal 25:679–687

Costa MG, Gama-Rodrigues AC, Zaia FC, Gama-Rodrigues EF (2014) Leguminosas arbóreas para recuperação de áreas degradadas com pastagem em Conceição de Macabu, Rio de Janeiro, Brasil. Scientia Forestalis 42:101–112

Cunha GM, Gama-Rodrigues AC, Gama-Rodrigues AC, Velloso ACX (2009) Biomassa e estoque de carbono e nutrientes em florestas montanas da Mata Atlântica na região norte do Estado do Rio de Janeiro. R Bras Ci Solo 33:1175–1185

Fontes AG, Gama-Rodrigues AC, Gama-Rodrigues EF, Sales MVS, Costa MG, Machado RCR (2014) Nutrient stocks in litterfall and litter in cocoa agroforests in Brazil. Plant Soil 383:313–335

Gama-Rodrigues AC (2011) Soil organic matter, nutrient cycling and biological dinitrogen-fixation in agroforestry systems. Agrofor Syst 81:191–193

Gama-Rodrigues AC, Barros NF (2002) Ciclagem de nutrientes em floresta natural e em plantios de eucalipto e de dandá no sudeste da Bahia, Brasil. R Árvore 26:193–207

Gama-Rodrigues AC, Barros NF, Comerford NB (2007) Biomass and nutrient cycling in pure and mixed stands of native tree species in southeastern Bahia, Brazil. R Bras Ci Solo 31:287–298

Gama-Rodrigues AC, Barros NF, Mendonça ES (1999) R. Bras. Alterações edáficas sob plantios puros e misto de espécies florestais nativas do sudeste da Bahia, Brasil. R Bras Ci Solo 23:581–592

Gama-Rodrigues AC, Barros NF, Santos ML (2003) Decomposição e liberação de nutrientes do folhedo de espécies florestais nativas em plantios puros e mistos no sudeste da Bahia. R Bras Ci Solo 27:1021–1031

Gama-Rodrigues EF, Gama-Rodrigues AC, Barros NF, Moço MKS (2011a) The relationships between microbiological attributes and soil and litter quality in pure and mixed stands of native tree species in southeastern Bahia, Brazil. Can J. Microbiol 57:887–895

Gama-Rodrigues EF, Gama-Rodrigues AC, Nair PKR (2011b) Soil carbon sequestration in cacao agroforestry systems: a case study from Bahia, Brazil. In: Kumar BM, Nair PKR (eds) Carbon sequestration potential of agroforestry systems: opportunities and challenges, advances in agroforestry, vol vol. 8. Springer/Dordrecht Heidelberg, London/New York, pp 85–99

Gama-Rodrigues EF, Gama-Rodrigues AC, Paulino GM, Franco AA (2008) Atributos químicos e microbianos de solos sob diferentes coberturas vegetais no norte do Estado do Rio de Janeiro. R Bras Ci Solo 32:1521–1530

Garrido MAO, Poggiani F (1982) Avaliação da quantidade e do conteúdo de nutrientes do folhedo de alguns povoamentos puros e misto de espécies indígenas, vol vol. 15/16. Silvicultura São Paulo, Piracicaba, pp 1–22

Hall JS, Ashton MS (2016) Guide to early growth and survival on plantations of 64 tree species native to Panamá and the Neotropics. Smithsonian Tropical Research Institute, Republic of Panama, p 173

Kato OR, Kato MSA, Carvalho CJR, Figueiredo RO, Camarão AP, Sá TDA, Denich M, Vielhauer K (2006) Uso de agroflorestas no manejo de florestas secundárias. In: Gama-Rodrigues AC, Barros NF, Gama-Rodrigues EF et al (eds) Sistemas agroflorestais: bases científicas para o desenvolvimento sustentável. Universidade Estadual do Norte Fluminense Darcy Ribeiro, Campos dos Goytacazes, pp 119–158

Kelty MJ (2006) The role of species mixtures in plantation forestry. Forest Ecol Manage 233:195–204

Leão AC, Silva LAM (1991) Bioelementos na cobertura vegetal e no solo do ecossistema dos tabuleiros costeiro do sudeste da Bahia, Brasil. Agrotrópica 3:87–92

Liu CLC, Kuchma O, Krutovsky KV (2018) Mixed-species versus monocultures in plantation forestry: development, benefits, ecosystem services and perspectives for the future. Glob Ecol Conserv 15:e00419. https://doi.org/10.1016/j.gecco.2018.e00419

Manhães CMC, Gama-Rodrigues EF, Moço MKS, Gama-Rodrigues AC (2013) Meso- and macrofauna in the soil and litter of leguminous trees in a degraded pasture in Brazil. Agrofor Syst 87:993–1004

Moço MKS, Gama-Rodrigues EF, Gama-Rodrigues AC, Machado RCR, Baligar VC (2010) Relationships between invertebrate communities, litter quality and soil attributes under different cacao agroforestry systems in the south of Bahia, Brazil. Appl Soil Ecol 46:347–354

Monroe PHM, Gama-Rodrigues EF, Gama-Rodrigues AC, Marques JRB (2016) Soil carbon stocks and origin under different cacao agroforestry systems in Southern Bahia, Brazil. Agr Ecosyst Environ 221:99–108

Müller MW, Gama-Rodrigues AC (2012) Sistemas agroflorestais com cacaueiro. In: Valle RR (ed) Ciência, tecnologia e manejo do cacaueiro. CEPLAC/CEPEC/SEFIS, Brasília, pp 407–435

Nunes DAD, Gama-Rodrigues EF, Barreto PAB, Gama-Rodrigues AC, Monroe PHM (2016) Carbon and nitrogen mineralization in soil of leguminous trees in a degraded pasture in northern Rio de Janeiro, Brazil. J For Res 27:91–99

Odum EP (1983) Ecologia, 1.ed. edn. Guanabara Koogan S.A, Rio de Janeiro, p 434

Oliveira PHG, Gama-Rodrigues AC, Gama-Rodrigues EF, Sales MVS (2018) Litter and soil-related variation in functional group abundances in cacao agroforests using structural equation modeling. Ecol Indic 84:254–262

Petit B, Montagnini F (2006) Growth in pure and mixed plantations of tree species used in reforesting rural areas of the humid region of Costa Rica, Central America. For Ecol Manage 233:338–343

Piotto D (2008) A meta-analysis comparing tree growth in monocultures and mixed plantations. For Ecol Manage 255:781–786

Piotto D, Craven D, Montagnini F, Alice F (2010) Silvicultural and economic aspects of pure and mixed native tree species plantations on degraded pasturelands in humid Costa Rica. New For 39:369–385

Pretzsch H, Forrester DI, Bauhus J (2017) Mixed-species forests: ecology and management. Springer-Verlag, Germany, p 653

Rappaport D, Montagnini F (2014) Tree species growth under a rubber (*Hevea brasiliensis*) plantation: native restoration via enrichment planting in southern Bahia, Brazil. New For 45:715–732

Salt GW (1979) A comment on use of the use of the term "emergent properties". Am Nat 113:145–148

Silva LF (1988) Alterações edáficas provocadas por essências florestais implantadas em solos de tabuleiro no Sul da Bahia. Rev Theobroma 18:259–267

Silva LF, Vinha SG (1991) Influência da matéria orgânica no comportamento de espécies florestais, em plantio puro e misto, em solos de tabuleiro do sudeste baiano. Agrotrópica 3:93–99

Souza CR, Rossi LMB, Azevedo CP, Vieira AH (2003) Paricá *Schizolobium parahyba* var. *amazonicum* (Huber ex Ducke) Barneby. Embrapa Amazônia Ocidental. Circular Técnica 18:12

Vinha SG (1992) Espécies nativas da Mata Atlântica sul baiana uma opção para reflorestamento. In: Novaes AB, José ARS, Barbosa AA, Souza IVB (eds) Reflorestamento no Brasil. UESB, Vitória da Conquista-BA, pp 56–73

Vinha SG, Lobão DEVP (1989) Estação ecológica do Pau-Brasil, Porto Seguro, Bahia. CEPLAC/CEPEC. 40p, Ilhéus

Vinha SG, Pereira RC (1983) Produção de folhedo e sua sazonalidade em 10 espécies arbóreas nativas no Sul da Bahia. Rev Theobroma 13:327–341

Vinha SG, Carvalho AM, Silva LAM (1985) Taxa de decomposição do folhedo de dez espécies de árvores nativas no Sul da Bahia, Brasil. R Theobroma 15:207–212

Wormald TJ 1992 Mixed and pure forest plantations in the tropics and subtropics. FAO Forestry Paper 103. Rome, FAO Technical Papers. Food and Agriculture Organization of the United Nations, p 152

Zaia FC, Gama-Rodrigues AC, Gama-Rodrigues EF (2008) Formas de fósforo no solo sob leguminosas florestais, floresta secundária e pastagem no Norte Fluminense. R Bras Ci Solo 32:1191–1197

Zaia FC, Gama-Rodrigues AC, Gama-Rodrigues EF, Moço MKS, Fontes AG, Machado RCR, Baligar VC (2012) Carbon, nitrogen, organic phosphorus, microbial biomass and N mineralization in soils under cacao agroforestry systems in Bahia, Brazil. Agrofor Syst 86:197–212

Chapter 13
The Brazilian Legal Framework on Mixed-Planted Forests

Luiz Fernando Duarte de Moraes, Renata Evangelista de Oliveira, Maria Jose Brito Zakia, and Helena Carrascosa Von Glehn

13.1 Introduction

Forests have been playing a very important role in the establishment of land-use policies worldwide. Forest cover changes directly affect biodiversity, global carbon budget, and ecosystem functions. In many countries in Latin America, historically there has been a contrasting dynamics between rates of deforestation and reforestation. From 2001 to 2010, Brazil lost hundreds of thousands of hectares of forests, and simultaneously the country witnessed the greatest expansion of woody vegetation gain (Aide et al. 2013). A significant part of this gain was likely due to planting forests with exotic and native species all over the country.

Planting forests in Brazil has two main motivations throughout history: as an economic activity, supplying raw materials for construction and furniture making, and as actions for the fulfillment of legal obligations. Landowners plant forests as an option to restore permanent preservation areas and legal reserves and so meet the legal requirements placed by the current legal framework on land use (Oakleaf et al. 2017). Recently, planted forests have also become relevant to mitigate the effects of climate change.

Natural and planted forests have multiple benefits, contributing to production, protection and conservation, and environmental and social services (SFB 2013;

L. F. D. de Moraes (✉)
EMBRAPA Agrobiology, Brazilian Agricultural Research Corporation, Seropédica, RJ, Brazil
e-mail: luiz.moraes@embrapa.br

R. E. de Oliveira
Department of Rural Development, Federal University of São Carlos, Araras, SP, Brazil

M. J. B. Zakia
Department of Forest Sciences, São Paulo State University, Botucatu, SP, Brazil

H. C. Von Glehn
São Paulo Environmental System, São Paulo, SP, Brazil

E. J. Bran Nogueira Cardoso et al. (eds.), *Mixed Plantations of Eucalyptus and Leguminous Trees*, https://doi.org/10.1007/978-3-030-32365-3_13

Yao et al. 2016). Additionally, different forms of reforestation may be used for many objectives (Baral et al. 2016), for instance, depending on the purpose (industrial use, environmental, agroforestry, and farm forestry) or species composition (monoculture or mixed species, hardwood or softwood, native or exotic species). Forest plantations can be used to restore biodiversity and to provide goods and services, but different goals require different strategies, planting, and management models as well, including the use of more or less species. In Brazil, most planted forests are tree monospecific plantations of exotic species, mainly *Pinus* and *Eucalyptus*.

According to Lamb et al. (2005), if we intend to supply goods and ecological services, tree plantation monocultures are efficient for timber or food production, but in most circumstances they are less successful in supplying services.[1] Mixed-species plantations with native species can potentially supply a wider range of goods and services than monocultures, since biodiversity gains are expected to be greater.

Whether for economic reasons or for requirement of legal compliance, there is a high demand and the enormous challenge of identifying opportunities offered by laws and public policies for the planting of mixed forests.

In this chapter, we discuss the legal framework of planted forests, focusing on federal Brazilian rules for use and management of forest plantations, with either exotic or native species. We intend to clarify when, where, and how those forests can be established according to legal rules, to a more effective provision of timber and non-timber products, and services as well.

We focus on the legal framework of the so-called multifunctional mixed-planted forests; neither native forest remnants nor monocultures will be the object of discussion.

13.2 Concepts

New definitions on "forest" have been necessary to have policies promoting effective forest conservation; minor changes in traditional definitions may distinguish native forests from plantations, ensuring that planted forests will protect biodiversity and contribute to sustainable development (Sasaki and Putz 2009).

For the Brazilian legal framework, the concept of forest has not been established yet. According to FAO, forest is as a "land spanning more than 0.5 hectares with trees higher than 5 meters and a canopy cover of more than 10 %, or trees able to reach these thresholds in situ. It does not include land that is predominantly under agricultural or urban land use"; planted forests are defined as being predominantly composed of trees established through planting and/or deliberate seeding (FAO, 2012).

[1] Services supplying means conserving ecological and hydrological processes.

Since FAO's definition is not suggested to meet the minimum requirements to attend public policies, a new definition was adopted under the Kyoto Protocol: "Forest is a minimum area of land of 0.05–1.0 hectares, with tree crown cover (or equivalent stocking level) of more than 10–30%, with trees with the potential to reach a minimum height of 2–5 m at maturity in situ." A forest may consist of either closed forest formations where trees of various storeys and undergrowth cover a high proportion of the ground or open forest. Young natural stands and all plantations which have yet to reach either a crown density of 10–30% or tree height of 2–5 m are included under forest. Those are considered areas normally forming part of the forest area which are temporarily unstocked, as a result of human intervention such as harvesting or natural causes but which are expected to revert to forest (Kant 2006).

During the World Congress of Forests in 1990, new concepts and management practices were proposed to distinguish the so-called forest plantations (simple trees cultivation/silviculture) from "planted forests" (which brings the comprehension of these forests as ecosystems):

> *The success of the forest plantations depends on the species fitness, its origin, and on their objectives. Further than the dogmatic controversial issues concerned to the introduction of exotic species, management priorities should aim the maintenance of soil productivity potential, as well of some biodiversity and sustainable income. The management of forest plantations must aim changing plantations into forests.*

In fact, this simple change of words brings distinct concepts: while "forest plantations" focus on timber production, "planted forests" are planned and managed for both forest production and ecosystem services, like water regulation and soil and biodiversity conservation.

FAO's recommendation in the 1990s may be seen as a way of integrating forestry production (through plantations) to the maintenance of ecological and hydrological processes, i.e., to social and environmental values (ensured by planted forests, in this case). As we can see, there is not a clear relationship between "forest plantations" (single-tree cultivation) and "planted forests," since the concept of planted forests is suggested to aggregate social and environmental issues into planning, decision-making, and forest management (Lima and Zakia 2006).

Anyway, a new concept has been proposed: *multifunctional forests* are suggested to embrace issues further than structure, concerning species composition and ecological functioning of forests, closely related to the provision of ecosystem services, as described in Fig. 13.1. As suggested in Fig. 13.1, a mixed plantation consisting of *Eucalyptus* and *Acacia* (only two species and both exotic) is not expected to provide multiple functions (or services) to the environment.

The Brazilian legal framework (Federal Decree 8375/2014) defines planted forests as those "consisting predominantly of trees, established by planting or seedling, cultivated with economic purposes and for trades, out of permanent preservation, restrict use and legal reserve areas."

Planted forests in Brazil are placed by legislation according to two aspects: the portion of either the landscape or the rural property they are located on, and the

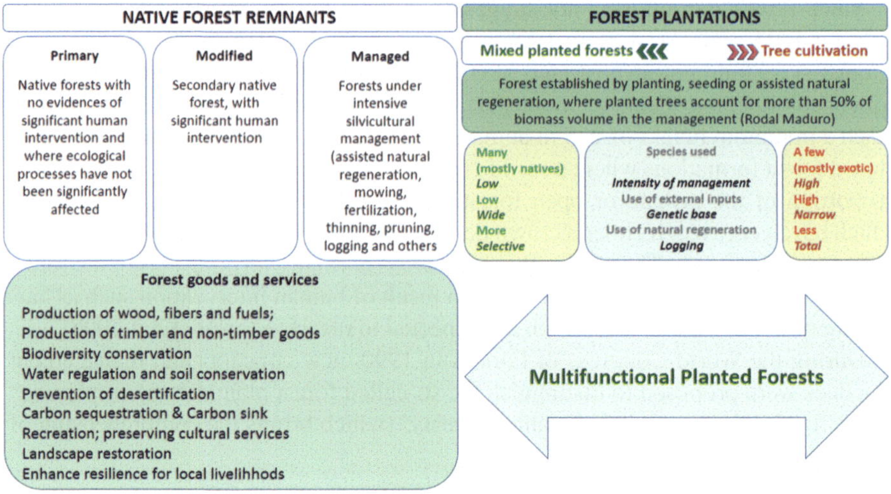

Fig. 13.1 Framework to introduce the concept of multifunctional mixed forests (adapted from Thiel, 2017). Hans Thiel, Close to Nature Planted Forests (CTNPF). World Bank/FAO Collaborative Program (CP) Initiative. Report, 2017 (not published)

objectives of planting forests. Forests can be planted on permanent preservation areas and legal reserve, which are mandatory protected natural areas (Silva and Ranieri 2014); on areas of restricted use; and on areas available for agriculture and forestry with no restrictions. Legal definitions of each type of area described below, as well as their closeness to different kinds of planted forests, are given below:

- *Permanent Preservation Area (APP)*: A protected area, covered or not by native vegetation, with its environmental functions of preserving water resources, landscape, geological stability, and biodiversity; improving the fauna and flora gene flow; protecting the soil; and ensuring the well-being of human populations.
- *Legal Reserve*: An area located within a property or a rural possession, under the terms of article no. 12, with the purpose of ensuring the economic and sustainable use of the rural property's natural resources, keeping and restoring ecological processes, and promoting biodiversity conservation, which includes shelters, and protection of wild fauna and native flora.
- *Consolidated Rural Area*: A rural property area occupied by human beings, existing prior to July 22, 2008, with its buildings, improvements, or agricultural and forestry activities, which may include a fallow system.
- *Areas of Restricted Use*: Areas with specific characteristics, which are:
  - Pantanal areas and plains, and areas where ecologically sustainable exploitation is allowed, and where technical recommendations from official research authorities must be observed. New removals of native vegetation for alternative use of the soil will depend on the approval of the state's environmental authority.

  - Slope areas between 25° and 45° where sustainable forest management and farming, cattle breeding, and forestry activities are allowed, as well as the maintenance of the physical infrastructure connected to the development of these activities, provided that good agricultural practices are observed. The conversion of new areas is forbidden, except when declared of public interest.
- *Area for Common Use*: Non-protected area, available for agriculture, cattle raising, and forestry (alternative soil uses).

An overview of the evolution of legal framework on forests may allow understanding how the regulatory environment affects the perception on the alternatives for the use and management of forests in rural properties. To meet the main objective of this book, it is necessary to state that mixed *Eucalyptus* and *Acacia* planted forests are not allowed to fulfill legal environmental requirements (to recover permanent preservation areas or legal reserves, for instance). The legal framework related to native and planted forests in Brazil, in a historical perspective, is presented below.

13.3 Brazilian Federal Legal Framework

The historical development of legal system regulating the use and management of forests in Brazil may be divided into historical periods.

The first regulatory mechanisms on forests in Brazil were based on the Portuguese legal system. The Colonial period, from 1500 to the beginning of the nineteenth century, was characterized by an unsustainable exploitation of timber products. A good example was the intense withdrawal of the valuable "pau-brasil" (Brazil wood), exclusively traded with the Portuguese Crown that adopted the first legal acts (as the "Pau-Brasil Act, in 1605) to protect natural resources, basically with economic purposes (Medeiros 2006; Bacha 2004). By the end of the eighteenth century, a Royal Charter stated the need to conserve native forests and prohibited the unauthorized cutting of valuable hardwood tree species (Medeiros 2006). From that period to the end of the nineteenth century the plantation of forests had basically ornamental and scientific purposes (Hora 2015).

During the Imperial period (nineteenth century), Brazilian Atlantic Forest faced an intense deforestation for coffee farming expansion, which led to the creation of command and control policies. In the first decades of the Brazilian Republic, established in the end of the nineteenth century, the government promoted the forest sector with economic purposes, by introducing *Eucalyptus* (the potential of native tree species was poorly known) plantations, which resulted in great deforestation.

Following this period, government initiatives showed some concerns on environmental aspects, and created the Brazilian Forest Service, in the 1920s, and the first Forest Code, in 1934, which proposed four categories of forests: protective, remnants, model, and production (Thomas and Foleto 2013; Bacha 2004). The 1934 forest code implied the "obligation of large consumers of forest products (such as steel companies and transportation) to keep the cultivation of forests for firewood or charcoal supply (spare)" (Bacha 2004). Imprisonment, detention, and fines were some of the penalties imposed by the 1934 forest code to those responsible for deforestation, burning forests, and invading public lands (Moretto et al. 2010).

Environmental concern increased during the 1960s, when an intensification of command-and-control regulation to stop deforestation was observed. No tools to stimulate the conservation of native forests were created though. A second version of the Brazilian Forest Code was passed in 1965, replacing the 1934 version. Some modifications of the first version included the requirement of authorization of the public authorities to explore all native forests, the requirement for forest replacement for all consumers of forest products, and the requirement for management plans to explore the forests in some regions of the country (Bacha 2004). From the beginning of the twentieth century to the 1960s, although forests were planted for economic purposes, planted forest expansion had not promoted the development of forestry (Hora 2015).

Despite requiring forest replacement in deforested areas, legal framework had not defined which species could be planted (Moretto et al. 2010), which resulted in the intense introduction of exotic and invasive species. By that time (1960s), there was a conflict between the legal framework related to forests and other laws, like the Land Statute, which stated that the landowner could benefit from deforestation (Bacha 2004). Consequently, rural landowners were authorized, even when funded, to replace native forests by homogeneous stands (Moretto et al. 2010; Bacha 2004). It resulted in the expansion of *Eucalyptus* and *Pinus* plantations over native forests (Moretto et al. 2010).

The area of planted forest faced a great expansion from 1965 to 1986, due to public financial incentives and strengthening of the legal framework (Hora 2015); in 1970, planted forests covered 1.66 million hectares, reaching up to almost six million hectares in 1985 (Bacha 2004). There were no clear concerns on the costs of planting forests in that period (Hora 2015).

The expansion of forest-planted area in Brazil from 1975 to 2000, however, was about 5% of the deforested area in Amazonia, in the same period (Bacha 2004). In summary, forest plantations were used in that period as a tool for economic development, putting aside other benefits forests can provide, especially those related to biodiversity conservation and ecosystem service delivery.

The economic development approach in that period was evidenced by the creation, in 1967, of the Brazilian Institute of Forest Development (IBDF), to regulate afforestation and reforestation activities, evaluating projects applied to access public funds. Most of the reforestation projects supported by the IBDF had economic purposes and used mostly exotic species. Only by the end of the 1980s the legal framework brought the concern on prioritizing native species in reforestation made by legal compliance (Moretto et al. 2010).

In 1988, the new Brazilian Constitution allowed states to also create specific laws and legislate on the management and use of forest resources. On the other hand, public policies for the economic development remained a great threat to forest conservation, and deforestation indices remained high (Bacha 2004). As the destruction of forest resources could affect the economic development of the country, new policies based on command and control have been established to reduce deforestation (Bacha 2004). During the Earth Summit in 1992, in Rio de Janeiro, intergovernmental and international agreements were made to use sustainable development to protect forests (Nazo and Mukai 2001).

The lack of financial incentives, although forestry was considered a profitable activity, was followed by a reduction in the planted area, to almost five million ha in 2000. As it was difficult to expand planted forests, the sector invested, from 1990 to 2000, in the development of tools and techniques to improve efficiency in forestry (Hora 2015).

Up to 2006, Brazilian legal frames targeting forest remained excessively protectionist, but the Law 11284/2006, which aimed to protect the Atlantic Forest, promoted sustainable use and conservation in public forests. That law provided the basis of a forest-based development model, by taking into account several issues, like ecosystem and biodiversity protection, rights of traditional communities, an efficient and rational forest use, and conditions to stimulate long-term investment (Bustamante et al. 2018). Importance of economic goals (e.g., profitability, productivity, efficiency) became most visible when the responsibility for and coordination of planted forests were transferred from the Ministry of the Environment to the Ministry of Agriculture (Bustamante et al. 2018).

There is a clear expansion of planted forests over the last decades in Brazil (Hora 2015), and a concern that forestry expansion may negatively affect biodiversity. The current Brazilian legal framework provides a system for the protection and forestry regulation with laws mostly focusing on native vegetation (Brazil holds six biomes[3]), water resources, and climate. Four legal tools,[2] as below, rule planted forests:

(a) The National Policy on Climate Change (Federal Law 12187/2009), established to consolidate and expand protected areas, fostering reforestation and revegetation of degraded areas.
(b) Conflicts between biodiversity conservation and social and economic interests pushed the establishment of the most recent legal framework in Brazil: the Law for the Protection of Native Vegetation (12651/2012). This law establishes rules for the protection, restoration, and compensation of native vegetation, and defines rules for forest exploitation and controls the origin of forest products. It regulates extractive activities and management of wood and non-wood products in native and planted forests, in conformity to a previously approved sustainable forest management plan (Oliveira and Sais 2017; Zakia and Guedes Pinto 2013; Kuntschik 2012). Forest plantation in non-protected areas is considered as agriculture.
(c) Most recently, Brazil has established an Agriculture Policy for Planted Forests (Decree 8375/2014), and the National Policy for the Recovery of Native Vegetation (Decree 8972/2017). Both policies offer interesting opportunities for the development and expansion of planted forests in Brazil.

[2] Federal laws can be consulted in http://www4.planalto.gov.br/legislacao/

13.4 The National Plan on Climate Change, the Federal Law and the National Policy for the Protection of Native Vegetation, and National Policy on Planted Forests: Opportunities and Challenges

The goals placed by legal framework and forest-based Brazilian policies offer significant opportunities for the expansion of multifunctional mixed-planted forests, but some bottlenecks need to be addressed.

One policy that has a great potential in fostering the expansion of mixed-planted forest is the Brazilian National Climate Change Plan (NCCP), which aims "to make the economic and social development compatible with the protection of the climate system and to promote the reduction of greenhouse gas emissions by encouraging the use of clean energy" (Brasil 2008). One of the goals of the NCCP was to eliminate the net loss of forest coverage in Brazil by 2015, which meant avoiding deforestation, and upscaling of forest plantations to 11 million ha in 2020; two million ha have been expected to be plantations of native species to replace degraded pastures (Brasil 2008).

The most important Brazilian law on forestry (Federal Law 12651/2012) establishes that no previous authorization is necessary for reforestation, using either exotic or native species (Brazil 2012). In 2017, the decree 8.972/2017 created the National Policy for Native Vegetation Recovery (Proveg), to articulate, integrate, and promote policies, programs, and actions that encourage forest recovery and other native vegetation forms. In order to implement that policy, the National Plan for Recovery of Native Vegetation (Planaveg) aims to have at least 12 million ha of forests and other forms of native vegetation restored in Brazil, up to 2030. Among the guidelines of Planaveg are the following: to foster society's awareness of the benefits of recovering native vegetation, and to improve the regulatory environment and increased legal certainty for the recovery of native vegetation with economic exploitation (Brazil, 2017).

The principles of the National Plan on Climate Change are to stimulate the production of forest goods and services for the social and economic development of the country, and mitigate the effects of climate change (Brazil, 2014). The Brazilian Ministry of Agriculture is in charge of creating the "National Plan for the Development of Planted Forests," which will establish forest production goals in Brazil, and the actions to be taken to achieve them.

Principles, objectives, and goals above are assumed as voluntary commitments from Brazil for the restoration of degraded areas, which may result in the expansion of planted forests. Multifunctional mixed forests have the conditions to fulfill legal requirements, provide adequate environmental services, and deliver economic benefits. An overview on the regulation of multifunctional mixed forests in Brazil is summarized in Table 13.1.

According to the Brazilian legal framework, multifunctional mixed forests consisting exclusively of native tree species can be planted to recover all these sites in rural properties: permanent preservation areas (within small, medium, and large properties), legal reserves, restricted-use areas, and areas that do not need any specific regulation (Table 13.1).

Table 13.1 Summary of permissions and possibilities placed by Brazilian legal framework for the implementation of multifunctional planted forests in rural properties

	Recovery of APP in small properties (see Box 13.1)	Recovery of APP in medium and great properties (see Box 13.1)	Recovery of legal reserve	Recovery of restricted-use area	Non-protected area (suitable for agriculture)[a]
Plantations of exotic trees only	Not allowed	Not allowed	Not allowed	Not allowed	Allowed
Plantations combining native and exotic trees	Allowed	Not allowed	Allowed	Allowed	Allowed
Plantations of native trees only	Allowed	Allowed	Allowed	Allowed	Allowed
Agroforestry systems with exotic trees only	Not allowed	Not allowed	Not allowed	Not allowed	Allowed
Agroforestry systems with native trees only	Allowed	Allowed	Allowed	Allowed	Allowed
Agroforestry with native and exotic species	Allowed	Not allowed	Allowed	Allowed	Allowed
Plantations for economic purposes	Allowed	Not allowed	Allowed	Allowed	Allowed
Environmental function established by legal framework	Preserve water resources, landscape, geological stability and biodiversity, facilitate gene flow of fauna and flora, protect soil	Preserve water resources, landscape, geological stability and biodiversity, facilitate gene flow of fauna and flora, protect soil	Assist the conservation and rehabilitation of ecological processes and promote the conservation of biodiversity, as well as shelter and protection of wildlife and native flora	Assist the conservation and rehabilitation of ecological processes and promote the conservation of biodiversity, as well as shelter and protection of wildlife and native flora	Water and soil conservation
Eligible for the payment of ecosystem services	Yes	Yes	Yes	Yes	No legal prediction

(continued)

Table 13.1 (continued)

	Recovery of APP in small properties (see Box 13.1)	Recovery of APP in medium and great properties (see Box 13.1)	Recovery of legal reserve	Recovery of restricted-use area	Non-protected area (suitable for agriculture)[a]
Legal rules	Federal Law 12.651/2012 Federal Decree 7850/2012	Federal Law 12.651/2012 Federal Decree 7850/2012	Federal Law 12.651/2012 Federal Decree 7850/2012	Federal Law 12.651/2012	Federal Law 12.651/2012; Federal Decree 8.375/2014
Allowed use and management for economic purposes	Needs regulation	Not applicable	Needs regulation	Needs regulation	–

APP permanent preservation areas

[a]Here is a bill under discussion in the Brazilian Congress (6411 PL/2016) that proposes that silvicultural activities should not require environmental licensing, in case of planting and management of either native or exotic trees, for logging and forest resource extraction, in consolidated rural areas, located in APP, or in degraded lands due to human activities, since those lands are not located on the APP or the legal reserve area

Brazilian rural properties may be classified into small, medium, or large, according to the corresponding fiscal modules (in units of area—see Box 13.1). Small properties and familiar-based farmers have some extra benefits related to the use and management of their lands in Brazil. There is no legal restriction for native planted forests to meet ecological requirements, but no exploitation is allowed in recovered permanent preservation areas (APP) located in medium and large properties. As we can see, Brazilian current legal framework allows small farmers (Box 13.1, below) to manage planted forests in their APP.

The use of noninvasive exotic species in planted forests to meet legal requirements is allowed for the recovery of APP only in small farms and to recover legal reserves, always combined with native species. No systems consisting exclusively of exotic trees—like the so-called *Eucalyptus-Acacia* system—are allowed to recover legally protected areas. This indicates a great concern on the potential invasiveness of some exotic species, as *Acacia* species.

Plantations for economic purposes are also allowed for the recovery of restricted-use areas and legal reserves, which enhances the potential multifunctionality of the models designed for those areas (Fig. 13.2). These plantations are allowed in restricted-use areas only if the use of area is consolidated (according to Federal Law 12651/2012), and under the adoption of water and soil conservation practices that include the *Eucalyptus-Acacia* system.

Since it is mandatory, the recovery of legal reserve areas offers the best opportunity for the expansion of multifunctional mixed-planted forests, since environmental functions and economic benefits can be obtained simultaneously. It is important to notice that mixed-planted forests in the legal reserve may contain noninvasive

Box 13.1 Zakia and Guedes-Pinto (2013)

Rural Module?

A rural module is calculated for each rural property, and its area reflects the kind of exploitation or utilization prevailing in the rural property.

Fiscal Module

A *fiscal module* is a land measuring unit in Brazil, established by Law no. 6746 of December 10, 1979. It is expressed in hectares, is variable, and set for each municipality according to:

- Kind of exploitation prevailing in the municipality
- The income brought in by the main exploitation business
- Other existing exploitation businesses in the municipality which, even if not prevailing, are significant for the income they bring, or the area they use
- Concept of family property

A *fiscal module* should not be confused with a *rural module*.

A fiscal module equals the minimum area required for a rural property to run a viable exploitation business. Depending on the municipality, a fiscal module may vary from 5 to 110 ha. In metropolitan regions, a rural module is usually quite smaller than in rural areas farther away from major urban centers.

A fiscal module also serves as a standard to define beneficiaries at PRONAF (small family farm producers, owners, sharecroppers, legal holders, partners, or tenants of up to four (4) fiscal modules).

PROVISION OF FOREST GOODS AND SERVICES	PLANTED FORESTS			
	Mixed planted forests ⟪⟪			⟫⟫ Tree cultivation
	Recovered Permanent Preservation Areas	Recovered Legal Reserve Areas	Mixed plantations (Forestry, Agroforestry)	Silviculture (monocultures)
	Economical use allowed only for small landowners (under specific rules)	Economical use allowed under specific rules (Forest Management Plan is required)	Economical use allowed. In the case of using native trees, the plantation needs to be registered	
Production of wood, fibers and fuels	MEDIUM	HIGH	HIGH	HIGH
Production of timber /non-timber goods	MEDIUM	MEDIUM	HIGH	LOW
Biodiversity conservation	MEDIUM	MEDIUM	MEDIUM	LOW
Water regulation and soil conservation	HIGH	HIGH	MEDIUM	LOW
Prevention of desertification	HIGH	HIGH	HIGH	MEDIUM
Carbon sequestration	MEDIUM	MEDIUM	MEDIUM	HIGH
Carbon sink	LOW	MEDIUM	MEDIUM	MEDIUM
Recreation; preserving cultural services	MEDIUM	LOW	HIGH	LOW
Landscape restoration	HIGH	MEDIUM	MEDIUM	LOW
Enhance resilience for local livelihoods	MEDIUM	HIGH	HIGH	LOW

Fig. 13.2 Opportunities of economic uses in planted forests and potential of providing forest services and goods according to the legal framework (adapted from Thiel, 2017). Hans Thiel, Close to Nature Planted Forests (CTNPF). World Bank/FAO Collaborative Program (CP) Initiative. Report, 2017 (not published)

exotic species, which still needs legal regulation, especially for the management and use of native trees.

The presence of native forest remnants has been costly and sometimes punitive to rural landowners. Brazilian laws historically have restricted the use and management of native forests, resulting in an understanding that "an overprotective approach may fail in effective protection." The management practices and uses recently allowed in legal reserve are a chance to change that paradigm.

In the definition of legal reserve given by the Federal Law 12651/2012, the economic purposes are not allowed just to mitigate costs of the reforestation (restoration), but also to offer new incomes for the landowner. In this new approach, the ecological function of the legal reserve concerns rather to ecosystem services than the biodiversity conservation or community structure. The model that better fits to the new objectives (as to restoration purposes) of the legal reserve is that related to the concept of multifunctional planted forest (see Table 13.1).

The law states that economic-based alternatives to be proposed for the use and management of legal reserve must target the maintenance of ecological functions. As other legal tools also require the legal reserve to attend ecological functions, there is no reason to avoid the sustainable use of forest resources in the legal reserve. It is a matter of searching a balance between biodiversity, conservation, and economic sustainability, and enabling the dual functions of the legal reserve.

In São Paulo state, this is especially challenging. Most of the rural properties do not have enough native vegetation and need reforestation actions to fulfill the minimum requirements of a legal reserve. Landowners may either protect, restore, or offset the legal reserve to meet the legal requirements (Oakleaf et al. 2017). The restoration of legal reserve to comply with the legal framework may be done by either assisting natural regeneration or planting forests in the rural properties. The combination of restoration and offset may offer a great opportunity to allow the expansion of forest cover in strategic areas for water conservation and to reconnect isolated forest remnants (implementation of ecological corridors). Lands in São Paulo state have the greatest opportunity costs of land use in Brazil, both in the Atlantic Forest and in Cerrado, resulting in high costs of restoration of native forests as well. A proper regulation may provide economic sustainability to multifunctional planted forests in the legal reserve, independent of legal obligations.

13.5 Final Comments

Brazilian legal framework places a good range of opportunities for the expansion of planted forests. However, environmental threats, such as global warming and biological invasions, recommend extreme care on the use of exotic and invasive tree species, as is the case of the *Eucalyptus-Acacia* consortium; their use must be planned considering rather a multifunctional approach. For the establishment of public policies, however, mixed-planted forests should be multifunctional and prioritize the use of multiple native species in mixed combinations (diversity is always welcome!). This is an opportunity for the development of the silviculture of native species.

Planting mixed forests for the recovery of permanent preservation areas in small properties and legal reserves needs regulation to enable the delivery of economic benefits to landowners, as well as environmental services to the landscape.

Governance is also a key issue. The opportunities discussed in this chapter, considering all the possible uses and services from multifunctional forests, will only be achieved if governments maintain and support the legal framework and public policies listed here.

Government actions must support the consolidation of the legal environment, and new policies should avoid conflicts with formerly published policies. Conflicts can affect the effectiveness of regulation, hindering the achievement of policy-related goals. Further, since most of Brazilian forest-related policies are associated to international agreements concerning global issues, governance should contribute to a collective international effort, and try not to favor specific sectors.

It should be noted that government actions must *always* strengthen sustainable development policies and strategies and that all care must be taken to avoid changes in government and/or political directions negatively affecting them (unfortunately, recent decisions of the Brazilian Government apparently have pointed in another direction). It is also a duty of Brazilian civil society to watch out for any misconduct in this regard.

Finally, it is important to remember that the forest cover, hopefully fulfilled with the so-called multifunctional forests, is located in multiple and heterogeneous rural landscapes, where the socioeconomic and cultural aspects are essential for the permanence (or not) of these forests, and their participation in the effectiveness of rural development in Brazil.

References

Aide TM, Clark ML, Grau HR, López-Carr D, Levy MA, Redo D, Bonilla-Moheno M, Riner G, Andrade-Núñez MJ, Muñiz M (2013) Deforestation and reforestation of Latin America and the Caribbean (2001–2010). Biotropica 45(2):262–271

Bacha CJC (2004) O Uso de Recursos Florestais e as Políticas Econômicas Brasileiras - Uma Visão Histórica e Parcial de um Processo de Desenvolvimento. Estudos Econômicos 34(2):393–426

Baral H, Guariguata MR, Keenan RJ (2016) A proposed framework for assessing ecosystem goods and services from planted forests. Ecosyst Serv 22:260–268

Brasil (2008) Inter-ministerial committee on climate change. National plan on climate change. Executive summary. http://www.mma.gov.br/estruturas/208/_arquivos/national_plan_208.pdf. Accessed 4 Dec 2018

Bustamante JM, Stevanovc M, Krotta M, Carvalho EF (2018) Brazilian state forest institutions: implementation of forestry goals evaluated by the 3L model. Land Use Policy 79:531–546

FAO (2012) Forest Resources Assessment (FRA) 2015: Terms and definitions. Forest resources assessment working paper 180. 36 p.

Hora AB (2015) Análise da formação da base florestal plantada para fins industriais no Brasil sob uma perspectiva histórica. BNDES Setorial 42:383–426

Kant P (2006) Definition of forests under the Kyoto Protocol: choosing appropriate values for crown cover, area and tree height for India. http://www.amity.edu/igwes/3.pdf Accessed 20 Jan 2019

Kuntschik DP (2012) Propostas para subsidiar um plano de implantação de florestas nativas com viabilidade econômica e ecológica. Instituto de Pesquisas e Estudos Florestais. Coordenadoria de Biodiversidade e Recursos Naturais da Secretaria de Estado de Meio Ambiente da SMA/SP. http://www.ipef.br/pcsn/documentos/relatorio_sintese_workshoppreliminar.pdf. Accessed 19 Dec 2018

Lamb D, Erskine PD, Parrotta JA (2005) Restoration of degraded tropical forest landscapes. Science 310:1628–1632

Lima WP, Zakia MJB (2006) As florestas plantadas e a água. Implementando o conceito da microbacia hidrográfica como unidade de planejamento. RiMa, São Carlos, p 218

Medeiros R (2006) Evolução das tipologias e categorias de Áreas Protegidas no Brasil. http://www.scielo.br/pdf/asoc/v9n1/a03v9n1.pdf. Accessed 17 Dec 2018

Moretto SP, Carvalho MMX, Nodari ES (2010) A Legislação Ambiental e as Práticas de Reflorestamento em Santa Catarina. V Encontro Nacional da Anppas, Florianópolis

Nazo GN, Mukai T (2001) O direito ambiental no Brasil: evolução histórica e a relevância do direito internacional do meio ambiente. http://bibliotecadigital.fgv.br/ojs/index.php/rda/article/view/48313. Accessed 27 Dec 2018

Oakleaf JR, Matsumoto M, Kennedy CM, Baumgarten L, Miteva D, Sochi K, Kiesecker J (2017) Legal GEO: Conservation tool to guide the siting of legal reserves under the Brazilian Forest Code. Appl Geogr 86:53–65

Oliveira RE, Sais AC (2017) Análise de instrumentos jurídicos para áreas de reserva legal em São Paulo: 2001 a 2016. Espacios 38(41):31–44

Sasaki N, Putz FE (2009) Critical need for new definitions of "forest" and "forest degradation" in global climate change agreements. Conserv Lett 2:226–232

SFB—Brazilian Forest Service (2013) Brazilian forests at a glance—2013: data from 2007 to 2012. Brazilian Forest Service, Brasília, p 188

Silva JS, Ranieri VE (2014) The legal reserve areas compensation mechanism and its economic and environmental implications. Ambiente Sociedade 17(1):115–132

Thomas BL, Foleto EM (2013) A evolução da legislação ambiental no âmbito das Áreas Protegidas brasileiras. Rev Eletr Curso de Direito da UFSM 8:734–735

Yao RT, Harrison DR, Velarde SJ, Barry LE (2016) Validation and enhancement of a spatial economic tool for assessing ecosystem services provided by planted forests. Forest Policy Econ 72:122–131

Zakia MJB, Guedes Pinto LF (2013) Guia para aplicação da nova lei florestal em propriedades rurais. IMAFLORA, São Paulo, p 32

Index

E. J. Bran Nogueira Cardoso et al. (eds.), *Mixed Plantations of Eucalyptus and Leguminous Trees*, https://doi.org/10.1007/978-3-030-32365-3

N

The manufacturer's authorised representative in the EU is Springer Nature Customer Service Centre GmbH, Europaplatz 3, 69115 Heidelberg, Germany. If you have any concerns regarding our products, please contact ProductSafety@springernature.com

Printed and bound by CPI Group (UK) Ltd, Croydon, CR0 4YY
30/09/2026
02225462-0009